Advances in GaN, GaAs, SiC and Related Alloys on Silicon Substrates

Materials Research Society
Symposium Proceedings Volume 1068

Advances in GaN, GaAs, SiC and Related Alloys on Silicon Substrates

Symposium held March 24–28, 2008, San Francisco, California, U.S.A.

EDITORS:

Tingkai Li
Sharp Laboratories of America, Inc.
Camas, Washington, U.S.A.

Joan M. Redwing
The Pennsylvania State University
University Park, Pennsylvania, U.S.A.

Michael Mastro
U.S. Naval Research Laboratory
Washington, D.C., U.S.A.

Edwin L. Piner
Nitronex Corporation
Durham, North Carolina, U.S.A.

Armin Dadgar
Otto-von-Guericke-Universitaet Magdeburg
Magdeburg, Germany

Materials Research Society
Warrendale, Pennsylvania

CAMBRIDGE
UNIVERSITY PRESS

University Printing House, Cambridge CB2 8BS, United Kingdom

One Liberty Plaza, 20th Floor, New York, NY 10006, USA

477 Williamstown Road, Port Melbourne, VIC 3207, Australia

314-321, 3rd Floor, Plot 3, Splendor Forum, Jasola District Centre, New Delhi - 110025, India

79 Anson Road, #06-04/06, Singapore 079906

Cambridge University Press is part of the University of Cambridge.

It furthers the University's mission by disseminating knowledge in the pursuit of education, learning and research at the highest international levels of excellence.

www.cambridge.org
Information on this title: www.cambridge.org/9781605110387

Materials Research Society
506 Keystone Drive, Warrendale, PA 15086
http://www.mrs.org

First published 2008
First paperback edition 2012

Single article reprints from this publication are available through University Microfilms Inc., 300 North Zeeb Road, Ann Arbor, MI 48106

CODEN: MRSPDH

A catalogue record for this publication is available from the British Library

ISBN 978-1-605-11038-7 Hardback
ISBN 978-1-107-40856-2 Paperback

CONTENTS

*Invited Paper

GaN BASED OPTICAL DEVICES ON SILICON

*Invited Paper

CONVENTIONAL III-V MATERIALS AND DEVICES ON SILICON

*Invited Paper

SILICON AND OTHER MATERIALS ON SILICON

*Invited Paper

PREFACE

Symposium C, "Advances in GaN, GaAs, SiC and Related Alloys on Silicon Substrates," was held March 24–28 at the 2008 MRS Spring Meeting in San Francisco, California. To meet increasingly challenging and complex system requirements as well as staying cost effective, it is not enough to use one single semiconductor materials system. Therefore, major efforts have been expended in recent years to combine the low cost and well established Si-based CMOS processing attributes with the superior performance attributes of Compound Semiconductors (CS). Such a combination—marrying the best of both worlds—will enable performance superior to that achievable with CS and CMOS alone with CMOS affordability. With an approach that directly integrates the CS into the CMOS wafer, only one wafer is processed to achieve a finished chip. Therefore, great efforts have been made to achieve direct integration of III-V materials systems such as GaN, GaAs, SiC and related alloys with Si. Many different approaches have been developed to overcome major lattice and thermal expansion mismatches between these systems with great success as illustrated by the multitude of excellent work presented, and the fact that excellent device performance has been achieved and are starting to become commercially available.

The *GaN Based Electronic Device and Sensors on Silicon* chapter gives an overview of the exciting new high power GaN-on-Si high electron mobility transistor (HEMT) devices, field effect transistor (FET) devices and sensors, and also reported on the new results on crack-free AlGaN/GaN HEMT devices grown on 6" Si substrates with low sheet resistance. The high efficiency GaN base light emission diode (LED), rare earth co-doping LEDs, and other optical devices on silicon are discussed in the *GaN Based Optical Devices on Silicon* chapter. The *GaN and Related Alloys on Silicon Growth and Integration Techniques* chapter focuses on materials challenges one faces when trying to grow GaN and Related Alloys respectively onto Si substrates due to lattice mismatch and the relative large thermal expansion mismatches. Discussed are various approaches taken to overcome those challenges and how they affect electrical and optical properties, followed by device performance. The chapter also presents very promising results on H-MOVPE growth of thick crack-free GaN on nanopatterned Si substrates, and InGaN thin film growth on Si(111) by the newly developed ENABLE technique (Energetic Neutral Atomic-Beam Lithography and Epitaxy) for electronic and optoelectronic devices, including solar cells. The *Conventional III-V Materials and Devices on Silicon* chapter describes a wide variety of III-V materials integrated on silicon, as well as high speed III-V devices introducing HEMTs, modulation doping and quantum well structures. The devices evolved from GaAs-based to InP-based HEMTs and HBTs, leading to next generation Sb based devices. In turn, various concepts were introduced, and examples were discussed, including fabrication processes and the challenges of producing good ohmic contacts due to the doping limitations in GaAs. The chapter also presents the growth of semiconductor nanowires as a route to combining heavily mismatched materials with focus on the growth of III-V nanowires on group IV substrates. The *Silicon and Other Materials on Silicon* chapter includes the current states of SiC and other materials integrated on silicon for electronics and optical devices applications.

The strong and increasing interest in GaN, GaAs, SiC and related alloys on silicon substrates indicates the worldwide importance of these materials and devices. This symposium proceedings represents the latest technical advancements and information on III-V materials and devices on silicon substrates from universities, national laboratories and industries. It also provides insight into emerging trends in these exciting technologies.

Tingkai Li
Joan M. Redwing
Michael Mastro
Edwin L. Piner
Armin Dadgar

July 2008

ACKNOWLEDGMENTS

The symposium chairs are pleased to acknowledge the following contributed and invited authors for the outstanding quality of their presentations and proceedings manuscripts, including:

Erik Bakkers
Ken Elliott
Ulrich Heinle
Fabrice Letertre
Edward Y. Chang
Augusto Gutierrez
Wayne Johnson
Armand Rosenberg

Authors who gave invited presentations included:

Takashi Egawa
Olga Kryliouk
Katherine Herrick
Fabian Schulze

The tutorial instructors are: Hongxing Jiang and Edward Y. Chang

The symposium chairs are also deeply indebted to the session chairs for their efforts in overseeing the sessions, guiding subsequent discussions, and reviewing the proceedings manuscripts:

Katherine Herrick
Ken Elliott
John Roberts
Ulrich Heinle
Edward Y. Chang
Olga Kryliouk
Augusto Gutierrez
Fabian Schulze
Fatemeh Shahedipour-Sandvik
Takashi Egawa
Mark Fanton

The symposium chairs also wish to express their gratitude to Aixtron AG who provided financial support, enabling us to present the "Advances in GaN, GaAs, SiC and Related Alloys on Silicon Substrates" symposium.

A special thanks is extended to the Materials Research Society staff, as well as the 2008 MRS Spring Meeting Chairs, for the development of the outstanding conference.

MATERIALS RESEARCH SOCIETY SYMPOSIUM PROCEEDINGS

Volume 1066 — Amorphous and Polycrystalline Thin-Film Silicon Science and Technology—2008, A. Nathan, J. Yang, S. Miyazaki, H. Hou, A. Flewitt, 2008, ISBN 978-1-60511-036-3
Volume 1067E —Materials and Devices for "Beyond CMOS" Scaling, S. Ramanathan, 2008, ISBN 978-1-60511-037-0
Volume 1068 — Advances in GaN, GaAs, SiC and Related Alloys on Silicon Substrates, T. Li, J. Redwing, M. Mastro, E.L. Piner, A. Dadgar, 2008, ISBN 978-1-60511-038-7
Volume 1069 — Silicon Carbide 2008—Materials, Processing and Devices, A. Powell, M. Dudley, C.M. Johnson, S-H. Ryu, 2008, ISBN 978-1-60511-039-4
Volume 1070 — Doping Engineering for Front-End Processing, B.J. Pawlak, M. Law, K. Suguro, M.L. Pelaz, 2008, ISBN 978-1-60511-040-0
Volume 1071 — Materials Science and Technology for Nonvolatile Memories, O. Auciello, D. Wouters, S. Soss, S. Hong, 2008, ISBN 978-1-60511-041-7
Volume 1072E —Phase-Change Materials for Reconfigurable Electronics and Memory Applications, S. Raoux, A.H. Edwards, M. Wuttig, P.J. Fons, P.C. Taylor, 2008, ISBN 978-1-60511-042-4
Volume 1073E —Materials Science of High-k Dielectric Stacks—From Fundamentals to Technology, L. Pantisano, E. Gusev, M. Green, M. Niwa, 2008, ISBN 978-1-60511-043-1
Volume 1074E —Synthesis and Metrology of Nanoscale Oxides and Thin Films, V. Craciun, D. Kumar, S.J. Pennycook, K.K. Singh, 2008, ISBN 978-1-60511-044-8
Volume 1075E —Passive and Electromechanical Materials and Integration, Y.S. Cho, H.A.C. Tilmans, T. Tsurumi, G.K. Fedder, 2008, ISBN 978-1-60511-045-5
Volume 1076 — Materials and Devices for Laser Remote Sensing and Optical Communication, A. Aksnes, F. Amzajerdian, N. Peyghambarian, 2008, ISBN 978-1-60511-046-2
Volume 1077E —Functional Plasmonics and Nanophotonics, S. Maier, 2008, ISBN 978-1-60511-047-9
Volume 1078E —Materials and Technology for Flexible, Conformable and Stretchable Sensors and Transistors, 2008, ISBN 978-1-60511-048-6
Volume 1079E —Materials and Processes for Advanced Interconnects for Microelectronics, J. Gambino, S. Ogawa, C.L. Gan, Z. Tokei, 2008, ISBN 978-1-60511-049-3
Volume 1080E —Semiconductor Nanowires—Growth, Physics, Devices and Applications, H. Riel, T. Kamins, H. Fan, S. Fischer, C. Thelander, 2008, ISBN 978-1-60511-050-9
Volume 1081E —Carbon Nanotubes and Related Low-Dimensional Materials, L-C. Chen, J. Robertson, Z.L. Wang, D.B. Geohegan, 2008, ISBN 978-1-60511-051-6
Volume 1082E —Ionic Liquids in Materials Synthesis and Application, H. Yang, G.A. Baker, J.S Wilkes, 2008, ISBN 978-1-60511-052-3
Volume 1083E —Coupled Mechanical, Electrical and Thermal Behaviors of Nanomaterials, L. Shi, M. Zhou, M-F. Yu, V. Tomar, 2008, ISBN 978-1-60511-053-0
Volume 1084E —Weak Interaction Phenomena—Modeling and Simulation from First Principles, E. Schwegler, 2008, ISBN 978-1-60511-054-7
Volume 1085E —Nanoscale Tribology—Impact for Materials and Devices, Y. Ando, R.W. Carpick, R. Bennewitz, W.G. Sawyer, 2008, ISBN 978-1-60511-055-4
Volume 1086E —Mechanics of Nanoscale Materials, C. Friesen, R.C. Cammarata, A. Hodge, O.L. Warren, 2008, ISBN 978-1-60511-056-1

MATERIALS RESEARCH SOCIETY SYMPOSIUM PROCEEDINGS

Volume 1087E —Crystal-Shape Control and Shape-Dependent Properties—Methods, Mechanism, Theory and Simulation, K-S. Choi, A.S. Barnard, D.J. Srolovitz, H. Xu, 2008, ISBN 978-1-60511-057-8
Volume 1088E —Advances and Applications of Surface Electron Microscopy, D.L. Adler, E. Bauer, G.L. Kellogg, A. Scholl, 2008, ISBN 978-1-60511-058-5
Volume 1089E —Focused Ion Beams for Materials Characterization and Micromachining, L. Holzer, M.D. Uchic, C. Volkert, A. Minor, 2008, ISBN 978-1-60511-059-2
Volume 1090E —Materials Structures—The Nabarro Legacy, P. Müllner, S. Sant, 2008, ISBN 978-1-60511-060-8
Volume 1091E —Conjugated Organic Materials—Synthesis, Structure, Device and Applications, Z. Bao, J. Locklin, W. You, J. Li, 2008, ISBN 978-1-60511-061-5
Volume 1092E —Signal Transduction Across the Biology-Technology Interface, K. Plaxco, T. Tarasow, M. Berggren, A. Dodabalapur, 2008, ISBN 978-1-60511-062-2
Volume 1093E —Designer Biointerfaces, E. Chaikof, A. Chilkoti, J. Elisseeff, J. Lahann, 2008, ISBN 978-1-60511-063-9
Volume 1094E —From Biological Materials to Biomimetic Material Synthesis, N. Kröger, R. Qiu, R. Naik, D. Kaplan, 2008, ISBN 978-1-60511-064-6
Volume 1095E —Responsive Biomaterials for Biomedical Applications, J. Cheng, A. Khademhosseini, H-Q. Mao, M. Stevens, C. Wang, 2008, ISBN 978-1-60511-065-3
Volume 1096E —Molecular Motors, Nanomachines and Active Nanostructures, H. Hess, A. Flood, H. Linke, A.J. Turberfield, 2008, ISBN 978-1-60511-066-0
Volume 1097E —Mechanical Behavior of Biological Materials and Biomaterials, J. Zhou, A.G. Checa, O.O. Popoola, E.D. Rekow, 2008, ISBN 978-1-60511-067-7
Volume 1098E —The Hydrogen Economy, A. Dillon, C. Moen, B. Choudhury, J. Keller, 2008, ISBN 978-1-60511-068-4
Volume 1099E —Heterostructures, Functionalization and Nanoscale Optimization in Superconductivity, T. Aytug, V. Maroni, B. Holzapfel, T. Kiss, X. Li, 2008, ISBN 978-1-60511-069-1
Volume 1100E —Materials Research for Electrical Energy Storage, J.B. Goodenough, H.D. Abruña, M.V. Buchanan, 2008, ISBN 978-1-60511-070-7
Volume 1101E —Light Management in Photovoltaic Devices—Theory and Practice, C. Ballif, R. Ellingson, M. Topic, M. Zeman, 2008, ISBN 978-1-60511-071-4
Volume 1102E —Energy Harvesting—From Fundamentals to Devices, H. Radousky, J. Holbery, B. O'Handley, N. Kioussis, 2008, ISBN 978-1-60511-072-1
Volume 1103E —Health and Environmental Impacts of Nanoscale Materials—Safety by Design, S. Tinkle, 2008, ISBN 978-1-60511-073-8
Volume 1104 — Actinides 2008—Basic Science, Applications and Technology, B. Chung, J. Thompson, D. Shuh, T. Albrecht-Schmitt, T. Gouder, 2008, ISBN 978-1-60511-074-5
Volume 1105E —The Role of Lifelong Education in Nanoscience and Engineering, D. Palma, L. Bell, R. Chang, R. Tomellini, 2008, ISBN 978-1-60511-075-2
Volume 1106E —The Business of Nanotechnology, L. Merhari, A. Gandhi, S. Giordani, L. Tsakalakos, C. Tsamis, 2008, ISBN 978-1-60511-076-9
Volume 1107 — Scientific Basis for Nuclear Waste Management XXXI, W.E. Lee, J.W. Roberts, N.C. Hyatt, R.W. Grimes, 2008, ISBN 978-1-60511-079-0

Prior Materials Research Society Symposium Proceedings available by contacting Materials Research Society

GaN Based Electronic Device and Sensors on Silicon

Mater. Res. Soc. Symp. Proc. Vol. 1068 © 2008 Materials Research Society 1068-C04-01

GaN-on-Si HEMTs: From Device Technology to Product Insertion

Wayne Johnson, Sameer Singhal, Allen Hanson, Robert Therrien, Apurva Chaudhari, Walter Nagy, Pradeep Rajagopal, Quinn Martin, Todd Nichols, Andrew Edwards, John Roberts, Edwin Piner, Isik Kizilyalli, and Kevin Linthicum
Nitronex Corporation, 2305 Presidential Drive, Durham, NC, 27703

ABSTRACT

In the last decade, GaN-on-Si has progressed from fundamental crystal growth studies to use as a platform technology for a line of reliable, commercially availability RF power transistors. This paper will briefly review progression of the GaN-on-Si material system and device processing, then present performance of the technology for commercial and military RF wireless applications.

INTRODUCTION

Future modulation schemes and air interfaces for mobile wireless communications require base stations with several watts of linear output power and instantaneous bandwidth up to 15% at frequency bands extending to ~6 GHz. Translated to the transistor level, this implies excellent linearity and thermal stability with simultaneous high power and high frequency capability. Additionally, advanced switch-mode amplifier architectures require output stage power transistors with cutoff frequencies >20 GHz and high voltage capability to support drain modulation designs. Military communications (e.g., JTRS - Joint Tactical Radio System) and electronic warfare (e.g., broadband jammers) systems also present significant challenges due to requirements of high output power and bandwidth up to or exceeding one decade. In such units, highly efficient broadband power transistors can reduce component count and decrease weight and/or footprint.

Appropriately designed GaN-on-Si HEMTs are excellent candidates to meet the performance targets in commercial and military insertion opportunities described above. The GaN material system has long been lauded for its high frequency and power handing capability. The HEMT structure provides excellent transconductance and linearity. GaN-on-Si offers GaN performance attributes in a cost-competitive platform. While the high-quality, low-cost, and large-diameter of Si substrates are well understood, other inherent advantages of GaN-on-Si include the ability to leverage established Si processes for wafer grinding & polishing, via-hole formation, and AuSi eutectic die attach.

Through thermal design and advanced packaging, thermal management of GaN-on-Si – sometimes cited as a limiting feature of the technology – can reach levels similar to GaN-on-SiC. High frequency operation has been demonstrated in both discrete and MMIC implementations. A family of packaged GaN-on-Si power transistors has been qualified and commercially released. These products operate at frequencies from DC to 6GHz and voltages up to 28V. Qualification of a product family operating to 48V is in progress and will extend the power levels and bandwidth of the existing 28V products, enabling high power, extremely portable systems. Such devices will facilitate improved communications transmit distance and extend the umbrella size of electronic protection units.

MATERIAL GROWTH AND DEVICE PROCESSING

All Nitronex GaN-on-Si products are grown by MOCVD on 100 mm float-zone Si (111) substrates. Proprietary, strain-compensating (Al,Ga)N transition layers and an amorphous $Si_xAl_{1-x}N_y$ nucleation layer are employed to accommodate lattice and thermal expansion mismatch between the substrate and the epilayers. Proper AlN nucleation conditions and reactor flow dynamics during GaN growth have been shown to play a critical role in overall device performance. Optimization of these has significantly reduced microwave loss to the substrate [1] and has improved RF power and efficiency of HEMTs [2]. Growth conditions have been optimized to result in a high level of thickness uniformity across the 100 mm substrate. Typical thickness uniformity of the stack is ~5% with 5 mm edge exclusion. 2DEG sheet resistance uniformity is <5%. Surface roughness as measured by 5 µm x 5 µm AFM scan is typically in the 5 - 10 Å. Threading dislocation density as measured by both XSTEM and AFM is $\sim 2 \times 10^9$ cm^{-2}.

Various $Al_xGa_{1-x}N$ barrier layers have been explored, ranging in composition from $x = 0.2$ to 0.3 with barrier thicknesses from 160 - 300 Å. The epi structure used for all products consists of a 180 Å $Al_{0.26}Ga_{0.74}N$ barrier and thin GaN capping layer. All layers of the material stack are nominally undoped. Transport characteristics of the channel have been evaluated by several methods. Room temperature Hall measurements result in 2-dimensional electron gas (2DEG) mobility of 1500 cm^2/V-s range and sheet charge density of 8.5×10^{12} cm^{-2}. Typical 2DEG sheet resistance of a passivated van der Pauw cross-bridge structure is 490 Ω/square.

HEMT fabrication begins with Ti/Al/Ni/Au (250Å / 1000Å / 400Å / 1500Å) ohmic metallization and RTA in flowing N_2 at ~850°C. Devices are immediately passivated in by SiN_x PECVD deposition at 300°C baseplate temperature. The passivation step is followed by a nitrogen ion implantation step to define active areas of the device. The implant was simulated to produce $>10^{20}$ cm^{-3} vacancy concentration from the surface to a depth of ~0.5 µm. Energy / dose conditions are 30 keV / 6×10^{12} cm^{-2}, 160 keV / 1.8×10^{13} cm^{-2}, and (doubly-charged) 400 keV / 2.5×10^{13} cm^{-2}. Typical sheet resistance of implanted material is $\sim 10^{11}$ Ω/square. Ion implantation has previously been shown to produce highly isolated material regions with resistivity unaffected by subsequent device processing steps. Additionally, implantation produces a planar surface more conducive to MMIC realization.

Openings for gate metallization are formed by a low-damage ICP plasma etch of the SiN_x passivant. The etch accurately transfers a 0.5 µm gate CD to the underlying (Al,Ga)N surface. Etch conditions include 3 W RF power and 60 W ICP power with 20 sccm flowrate of SF_6. This process typically achieves ~3% etchrate uniformity. Schottky gates of Ni/Au (200Å / 5kÅ) are deposited by a separate lithography step to allow overlap of the gate metallization onto the surface of the passivant, forming a dielectrically-defined T-gate that can be optimized to reduce peak electric field. A PECVD SiN_x encapsulation layer is then deposited to protect the device and serve as the capacitor dielectric for source field plate (SFP) deposition. The source field plate is deposited by evaporation and extends from the source contact – over the gate electrode – and terminates in the midpoint of the drift region. The SFP was optimized to reduce RF dispersion as measured by pulsed-DC and load pull measurements. The SFP reduces feedback capacitance, and thus increases microwave gain, relative to devices with field plates formed by overlap of the gate electrode on the SiN_x passivant. Devices are interconnected by electroplated Au airbridges to complete frontside processing. A cross-sectional micrograph of the active regions of a GaN-on-Si HEMT is shown in Figure 1.

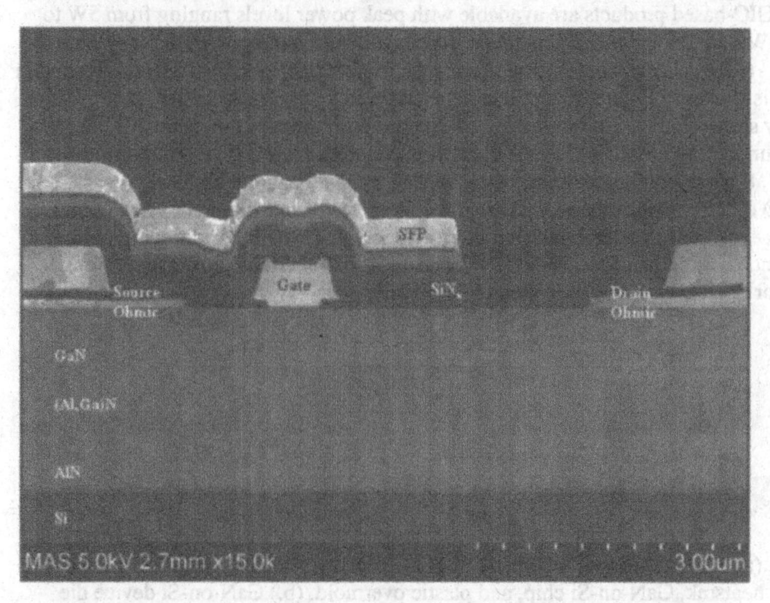

Figure 1. Cross-sectional STEM of GaN-on-Si HEMT, showing epilayers and device construction.

Backside processing consists of grinding and CMP using standard Si techniques. Final wafer thickness is product-specific, and products with chip thickness as low as 2 mils are under development. Through-wafer vias are formed Bosch etch and exhibit near-vertical sidewalls. Frontside source pads are grounded to the backside of the wafer by sputtered seed metal deposition and thick Au backside electroplating. All wafer are probed and screened for both DC and RF performance.

Packaging solutions include low-cost plastic SOIC and thermally-enhanced LDMOS-style air cavity outlines, providing a broad cross-section of performance and pricing options.

GaN-on-Si IN LOW-COST PLASTIC PACKAGING

The economics of GaN-on-Si combined with low-cost plastic packages enable GaN performance to reach price points required for many driver and pre-driver stage applications, areas not often considered for GaN devices. The plastic packaging process utilizes a copper heatsink, thermally-conductive and electrically-conductive epoxy die attach, and RoHS-compliant plastic overmolding. Figure 2 gives an overview of the plastic package and assembly.

SOIC-based products are available with peak power levels ranging from 5W to nearly 50W. NPTB00004 is a general purpose broadband transistor rated for operation from DC - 6GHz. Large-signal RF performance from 900MHz to 3500MHz measured in load pull is shown in Figure 3. Optimizing source and load impedance for each frequency shown in Fig. 3 produces 5 - 7 W output power across this range. When operated under 2500 MHz WiMAX modulation (single carrier OFDM waveform 64-QAM 3/4, 8 burst, continuous frame data, 10MHz channel bandwidth, peak/avg. = 10.3dB @ 0.01% probability on CCDF) at 2% error vector magnitude (EVM), output power exceeds 28dBm with associated drain efficiency of >20%. Additionally, noise performance is excellent for this device. From 1400 - 2000 MHz, the noise figure is <1.5dB for operation at V_{DS} = 28V and I_D = 50 mA.

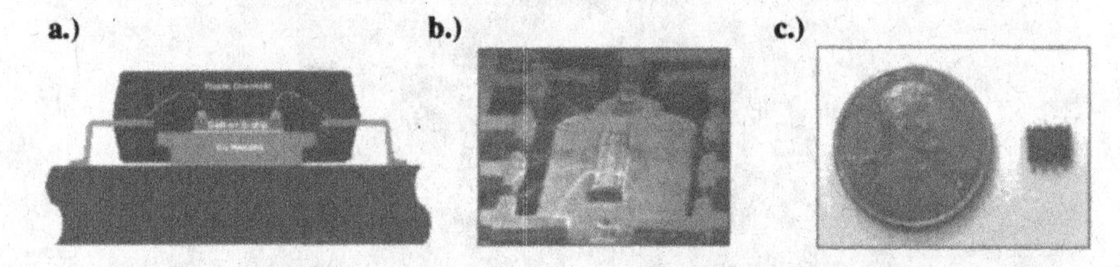

Figure 2. (a.) Cross-sectional schematic of SOIC package showing input and output leads, Cu heatsink, GaN-on-Si chip, and plastic overmold. (b.) GaN-on-Si device die bonded and wirebonded in SOIC package immediately prior to overmolding. The chip dimension is 1.65 mm × 0.55 mm. (c.) GaN-on-Si product in 8-pin SOIC package with size reference.

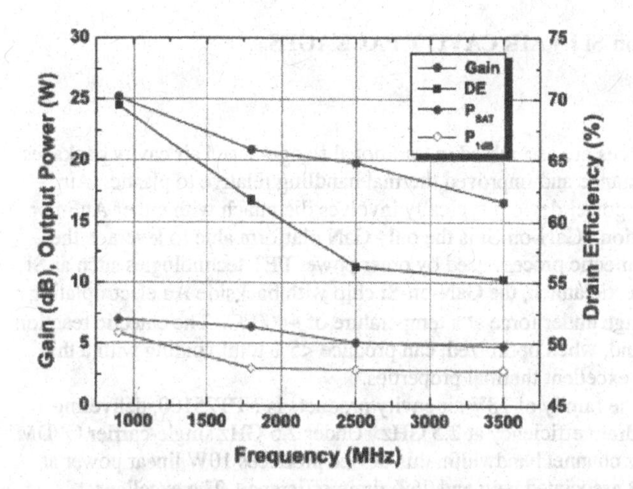

Figure 3. NPTB00004 CW performance summary across frequency from 900MHz to 3500MHz. Optimized source and load impedances were presented to the device at each frequency. $V_{DS} = 28V$.

Combining the high power density of GaN with the small footprint of an SOIC package, the recently-released NPT1004 delivers 45W of peak envelope power (PEP) at 2500 MHz. Producing 5W linear power, >13dB gain, and 27% drain efficiency at 2% EVM across the 2500 - 2700 MHz WiMAX band, this device is also an excellent general purpose 28V driver to a high power GaN final stage. Products with even higher output power in plastic packages are believed to be possible through thermal optimization including non-epoxy die attach.

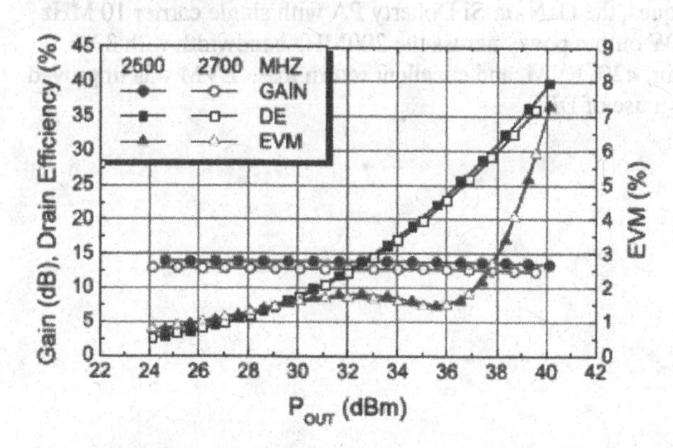

Figure 4. NPT1004 WiMAX performance across 200MHz bandwidth from 2500 - 2700 MHz. $V_{DS} = 28V$.

HIGH POWER GaN-on-Si IN AIR CAVITY PACKAGES

28V Class

High power devices are assembled in traditional flange-based air cavity packages for excellent RF performance and improved thermal handling relative to plastic. Air cavity assembly of high power devices typically involves die attach with either AuSn or AuSi by eutectic formation. GaN-on-Si is the only GaN platform able to leverage the well-established AuSi eutectic process used by other power FET technologies such as Si LDMOS. For AuSi eutectic attach, the GaN-on-Si chip with backside Au electroplating is scrubbed onto the flange under force at a temperature of ~400°C. The eutectic reaction proceeds very quickly and, when optimized, can produce <5% total voiding with a thin (few μm) bondline and excellent thermal properties.

The flagship in the family of 28V air cavity products is NPT25100, delivering 125W PEP with >60% drain efficiency at 2.5 GHz. Under 2.5 GHz single-carrier OFDM modulation and 10 MHz channel bandwidth, this device produces 10W linear power at 2.0% EVM with 16.5dB associated gain and 26% drain efficiency. The excellent bandwidth of the technology enables the same product to operate at UMTS band from 2.11 - 2.17 GHz, producing >20W average power at an adjacent channel power ratio (ACPR) of -35 dBc. This performance across both UMTS and WiMAX bands is attributable to the low inherent output capacitance and high output impedance (Z_{OUT}) of GaN HEMTs relative to comparably-sized LDMOS FETs. These properties obviate the need for add ional L-C matching components such as MOS capacitor chips in the output network formed inside the package. Such components tend to reduce native bandwidth available from the FET chip in order to obtain impedances that can be easily matched by external circuitry. Representative W-CDMA sweeps of NPT25100 from 2110 - 2170 MHz are given in Figure 5. A 2500 – 2700 MHz Doherty power amplifier has also been designed and fabricated using two NPT25100 devices. Utilizing digital pre-distortion (DPD) linearization techniques, the GaN-on-Si Doherty PA with single carrier 10 MHz OFDM signal produced 20W output power across the 200MHz bandwidth with 33% drain efficiency, >11dB gain, <2% EVM, and excellent return loss. EVM was improved by ~1.5% (absolute) through use of DPD.

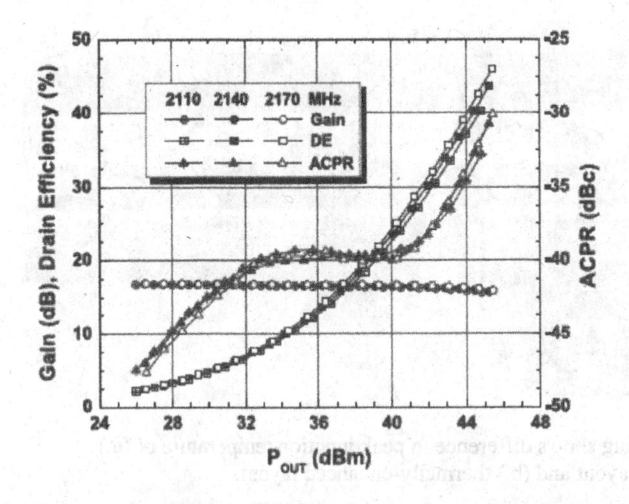

Figure 5. Typical NPT25100 W-CDMA performance in load pull system. V_{DS} = 28V.

48V Class

Thermal design must be carefully considered when optimizing GaN devices for high voltage and high power. Without appropriate thermal management at both the chip and package level, the high power density of GaN can cause performance degradation and/or fundamental limitations in device safe operating area. Thermal optimization at the chip level can be attained through device layout and through aggressive thinning. Improvement to the thermal properties of the package are also critical in order to maximize the power and efficiency available from the GaN-on-Si HEMT. All of these techniques have been employed to develop a family of 48V GaN-on-Si broadband transistors delivering power levels from 40W to 200W in a compact package. In the highest power case, package-level "power density" (defined as peak output power divided by package volume) reaches ~650 W/cm^3.

Chip-level optimization began with modification of the FET layout. In traditional, bar-style power FET geometries, gate contacts act as line sources of heat. Thermal coupling between adjacent gates can elevate peak junction temperature. A FET layout was designed and constructed to minimize this coupling by separating unit cells of 2mm gate width into a staggered or 'tiled' configuration. This significantly reduced thermal coupling and resulted in thermal improvements summarized in Figure 6. In this figure, a constant DC dissipated power was applied and peak temperature was measured using infrared imaging. Peak temperature of the optimized layout shown in 6(b.) is 25°C lower than for 6(a.) at the same dissipated power, leading to a 25% reduction in thermal impedance (R_{TH}).

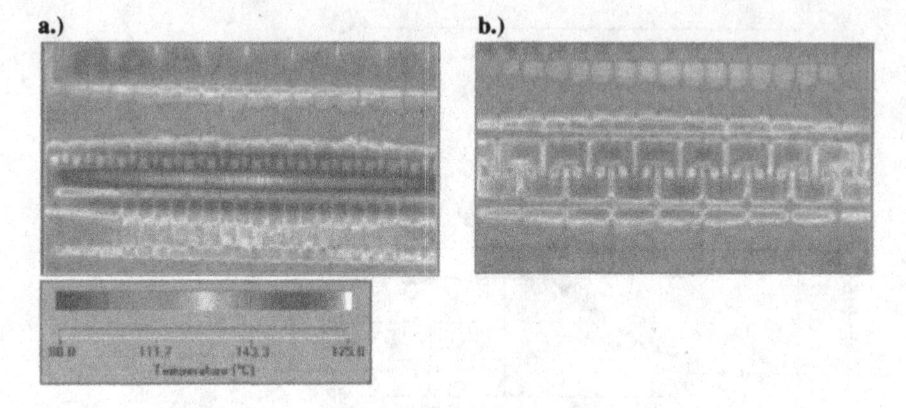

Figure 6. Infrared imaging shows difference in peak junction temperature of (a.) standard, bar-style FET layout and (b.) thermally-enhanced layout.

In addition to the layout modifications, more aggressive wafer thinning was employed to further reduce R_{TH}. Initial products launched at 28V employed thinning to 6 mils. Thermally-enhanced 28V products such as NPT25100 decreased substrate thickness to 4 mils. At this thickness, 48V devices with optimized layout produce a normalized thermal impedance of ~35°C-mm/W (R_{TH} multiplied by FET total gate width). Thinning to 2 mils produces an additional 15% reduction in R_{TH} to near 30°C-mm/W. These values compare favorably to GaN-on-SiC results, clearly demonstrating that thermal properties of Si (as compared to SiC) will not limit or preclude realization of high power GaN-on-Si FETs.

Optimization of packaging materials is also critical for thermal design. Thermal impedance from substrate to package can be comparable to R_{TH} from junction to substrate, underscoring the importance of the package for any high power device technology. For 48V GaN-on-Si, novel Cu-flanged air cavity packages were developed and implemented. Integration of Cu as a heat sink presents 2 main challenges: (1) the CTE of Cu precludes the use of conventional Al_2O_3 ceramic materials as package dielectric; and (2) the CTE mismatch between Cu and Si – combined with the ductility of Cu – can cause package bowing during assembly. Item #1 was addressed by package supplier(s) through development of organic materials with suitable dielectric properties and thermal expansion closely matched to Cu. Most of these organic materials cannot withstand AuSi eutectic die attach temperatures, so AuSn eutectic attach is currently used for all devices packaged in Cu. Item #2 is mitigated by optimizing thickness and/or pre-bow (intentionally bowing the package prior to assembly so it becomes flat post-assembly). This is a multivariate problem involving chip size and thickness, flange size and thickness, # of components, die attach method, and even lid attach conditions. These materials challenges were successfully addressed and enabled packaging and product development of high-power 48V GaN-on-Si devices.

Devices ranging in output power from 40W to 200W are currently in product qualification. These devices employ the epi and device fabrications details described earlier in conjunction with thermal improvements in layout, thinning, and packaging. A

2-tone sweep at 900 MHz (1MHz tone spacing) is shown in Figure 7. At 48V, PEP of ~200W with 22dB associated gain and >60% drain efficiency are achieved. For this device, R_{TH} ~ 1°C/W. This power level – in an outline suitable for highly portable systems – enables improved military communications transmit distance and extends the protective umbrella produced by broadband jammers. Such devices are expected to have widespread impact in future military systems.

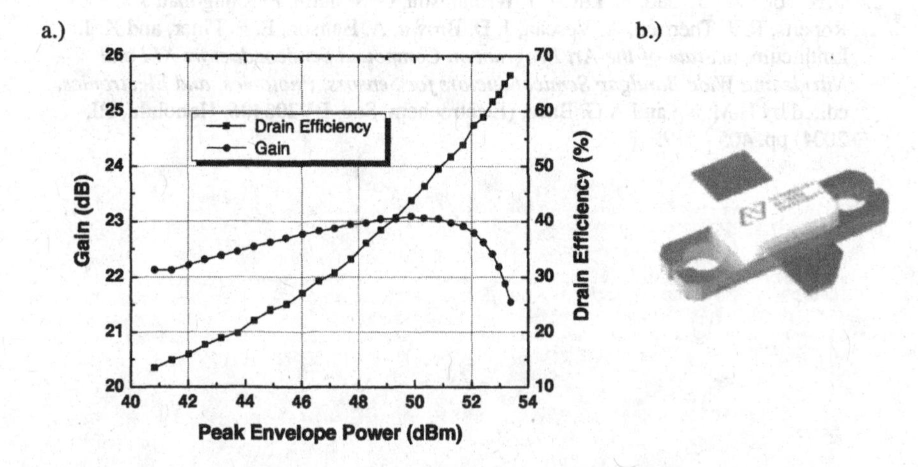

Figure 7. (a.) 2-tone performance of GaN-on-Si HEMT at 900MHz and V_{DS} = 48V and (b.) photograph of packaged 48V device. Flange dimensions are 0.80" × 0.23".

CONCLUSIONS

GaN-on-Si has been clearly shown to be a successful heteroepitaxial material system for RF wireless power transistors. Materials challenges at the wafer and package level have been addressed and state-of-the-art HEMTs for both commercial and broadband military applications are now commercially available. Future directions for GaN-on-Si technology include qualification of 48V and higher frequency (X-band and beyond) discrete devices. In addition, monolithic integration of passive components such as spiral inductors, thin film and epi resistors, MIM capacitors, and transmission lines is expected to play a key role in future GaN-on-Si products. In this area, the advantages of the low cost substrate will be even more pronounced as chip size increases to accommodate passive circuitry.

REFERENCES

1. P. Rajagopal, J.C. Roberts, J.W. Cook Jr., J.D. Brown, E.L. Piner, and K.J. Linthicum in *GaN and Related Alloys*, edited by H.M. Ng, M. Wraback, K. Hiramatsu, and N. Grandjean (Mat. Res. Soc. Proc. **798**, Boston, MA, 2003) Y7.2.

2. J.W. Johnson, J. Gao, K. Lucht, J. Williamson, C. Strautin, P. Rajagopal, J. C. Roberts, R. J. Therrien, A. Vescan, J. D. Brown, A. Hanson, E. L. Piner, and K. J. Linthicum, in *State-of-the-Art Program on Compound Semiconductors XLI and Nitride and Wide Bandgap Semiconductors for Sensors, Photonics, and Electronics*, edited by H.M. Ng and A.G. Baca, (Electrochem. Soc. **PV200406**, Honolulu, HI, 2004) pp. 405 - 419.

Mater. Res. Soc. Symp. Proc. Vol. 1068 © 2008 Materials Research Society 1068-C04-08

GaN Electrochemical Probes and MEMS on Si

Ulrich Heinle[1], Peter Benkart[1], Ingo Daumiller[2], Mike Kunze[2], and Ertugrul Sönmez[2]
[1]MicroGaN GmbH, Albert-Einstein-Allee 45, Ulm, 89081, Germany
[2]MicroGaN GmbH, Lise-Meitner-Strasse 13, Ulm, 89081, Germany

ABSTRACT

AlGaN /GaN HEMT structures were used for liquid and gas sensors. The sensitivity of pH sensors was determined after clean-in-place tests. A multigas measurement at 330°C is presented. A piezo-driven mechanical actor is demonstrated and used as a variable capacitor. GaN on silicon logic circuitry is demonstrated at 250°C.

INTRODUCTION

Gallium nitride based materials are suitable for use at high temperatures and in harsh environments due to their physical and chemical properties. These attributes, combined with piezoelectric and mechanical characteristics, are the base of a considerable interest in sophisticated GaN sensors and actors. Furthermore, the market need for electrochemical sensors, mechanical sensors/actors, as well as electronic circuits, has led to a rapid progress of GaN based devices. Basic construction elements for electrochemical and mechanical sensors/actors, as well as the corresponding GaN-based circuitry, will be presented. The exemplary devices presented are robust and highly sensitive fluid and gas detectors and for the first time a mechanical GaN-based tunable capacitor. AlGaN/GaN heterostructures on silicon are the bases for all presented devices.

EXPERIMENTAL DETAILS

pH sensors

The AlGaN/GaN material system is suitable for the fabrication of chemical sensors due to its chemical inertness. The principle operation is based on the modulation of the HEMT's 2DEG by surface charges caused by adsorption of ions on the device surface. GaN-based ion-sensitive field effect transistors (ISFETs) have been demonstrated using Ga_xO_z as ionsensitive layers [1].

These devices are very sensitive to small variations of pH values, but the native or thermal gallium oxides need to be replaced by other dielectrics because their chemical stability is not sufficient for cleaning procedures in industrial applications, like food processing. It is not practical to remove the sensors from the processing equipment before cleaning and mounting them afterwards, because this will increase cost of ownership and that is not acceptable. Therefore, the development of chemically very stable dielectrics is indispensable.

The cleaning processes are simulated with clean-in-place (CIP) tests. During these tests the sensors are exposed to aggressive acids and bases in a cyclic manner. In our case the sensors are exposed to nitric acid (pH 1) and caustic soda (pH 13) for 30 cycles. The acid cycle lasts 5 minutes at room temperature and the base cycle takes 15 minutes at 60°C.

Figure 1 shows the basic layout of an AlGaN/GaN ISFET with an deposited metal oxide layer. In principle we have a HEMT structure where the area between the ohmic contacts is covered with an ionsensitive dielectric. High demands are also made to the packaging material. It has not only to withstand chemically aggressive liquids but seal the device perfectly in order to avoid corrosion of the electrical contacts.

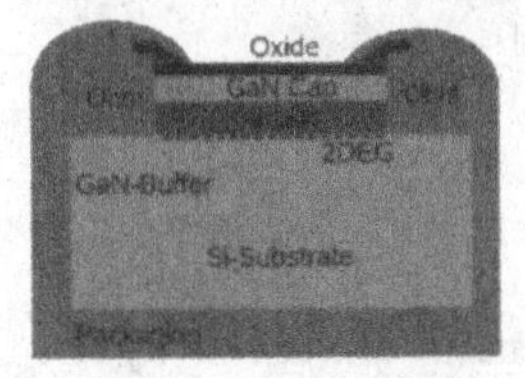

Figure 1. Basic layout of the AlGaN/GaN ISFET used in this study.

The measurement setup consists of the AlGaN/GaN ISFET, a Pt electrode, a reference electrode (Saturated Calomel Electrode), and a potentiostat (see figure 2). The potential between ISFET and SCE is varied periodically and the current between ISFET and Pt electrode is measured. Thus, no current flows through the SCE.

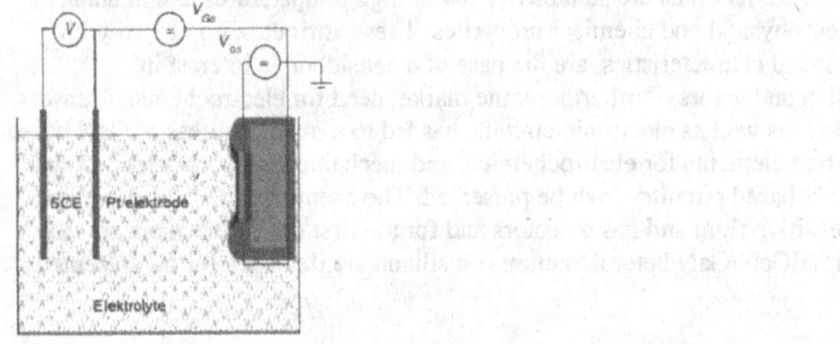

Figure 2. Measurement setup consisting of the ISFET, Pt electrode, reference electrode (SCE), and potentiostat.

The sensitivity and the stability of the devices were tested by cyclic IV-measurements. The graph of figure 3 shows the response curves before and after after a CIP test with 30 cycles. The absence of drift and the linearity does clearly show the stability and sensitivity of the sensors. The extracted sensitivity corresponds to 59 mV/pH.

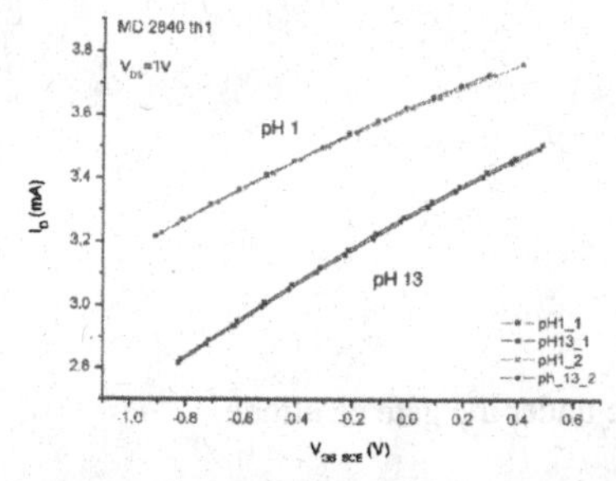

Figure 3. IV-characteristic of the pH sensor before and after CIP tests.

Figure 4 shows the IV-characteristics of the sensors after a soak in NaOH and HNO_3, respectively. After the soak we do still not see a drift behaviour, but an offset between the curves before the soak and afterwards. The offset is more pronounced for nitric acid exposure than for NaOH. The cause of the offset is not clear yet. It could be that the interface between the dielectric and the semiconductor is altered by diffusion processes, which modify the channel region. In any case the sensitivity remains very high.

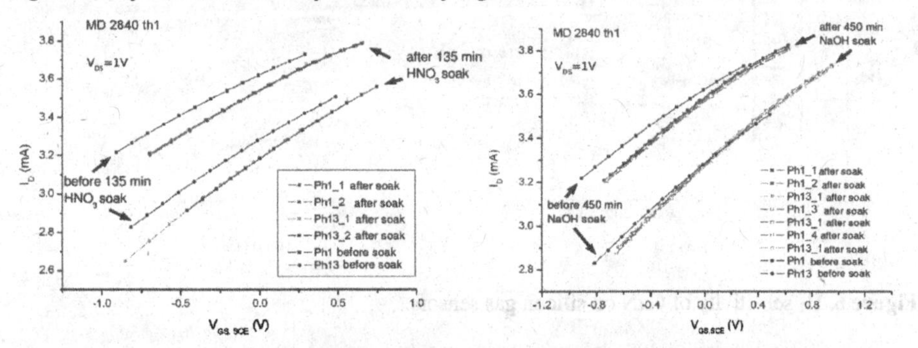

Figure 4. IV-characteristics of the sensors after a soak in HNO_3 and NaOH respectively. The observed offset is more pronounced for HNO_3.

Gas sensors

The effect of the ion-sensitivity of GaN based HEMT structures is not limited to the described pH-detection in liquids. Thus, also gas sensors can be fabricated using GaN based devices. They should be capable of operation in harsh environments, like corrosive ambients, and at elevated temperatures.

The principle operation of gas sensitive semiconductor devices has already been demonstrated in the 1970s [2]. Hydrogen sensitivity has been achieved by silicon MOS devices with platinum electrodes. Schalwig et al. have reported gas sensitive GaN based Schottky diodes on sapphire substrates [3].

In contrast to these devices we are using MicroGaN's core unit cell, a GaN based HEMT structure on silicon substrate, and deposit dielectrics before the formation of the gate electrode by platinum evaporation. Figure 5 shows the schematic layout of our devices.

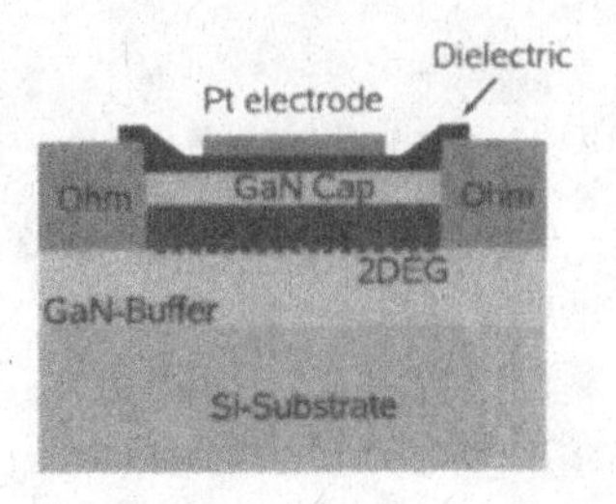

Figure 5. Schematic layout of gas sensors with a dielectric under the gate electrode.

The sensors are very sensitive to variations of H_2, as can be seen in figure 6. They reveal a wide thermal operation range from 150°C to 300°C and a detectable hydrogen concentration range from 5 ppm to 2600 ppm, thus spanning over 3 orders of magnitude. Thus, the sensors show a very high sensitivity to variation in hydrogen concentration.

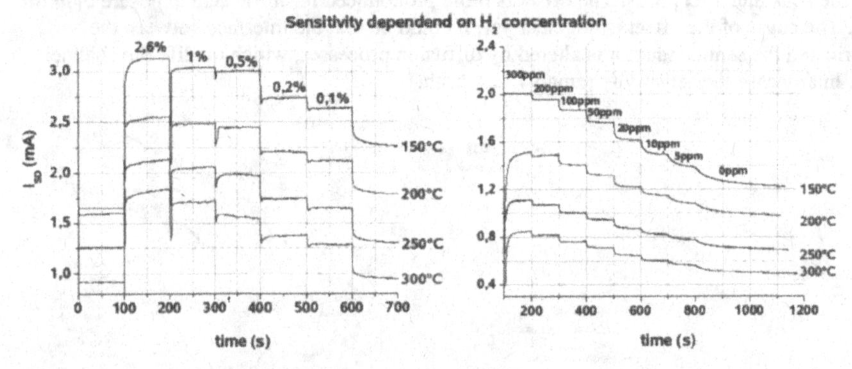

Figure 6. H_2 sensitivity of GaN on silicon gas sensors.

In addition to the H_2 sensitivity, our sensors are sensitive to different gases, like ammonia, propylene and a variety of oxygen based gases. Figure 7 shows a measurement at 330°C. It can be seen that the devices are very sensitive to oxygen based gases with a very high change of the gate voltage.

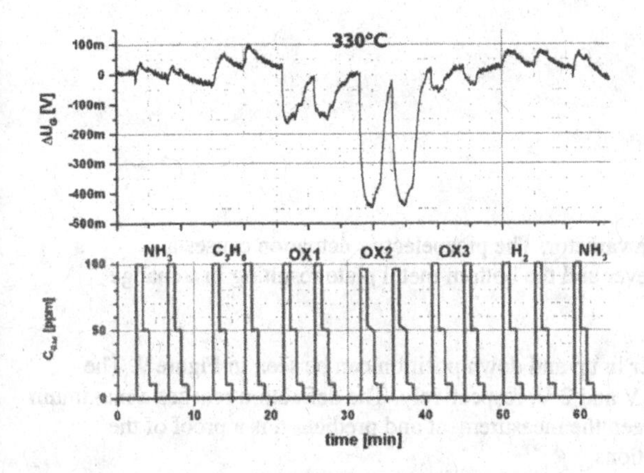

Figure 7. Multigas measurement at 330°C. The sensors are very sensitive towards oxygen based gases.

Piezo-driven mechanical actors

Aditionally to the described modulation of the HEMT's 2DEG by surface charges for gas or fluid detection, mechanical stress leads to modulations of the 2DEG. This effect can be used to fabricate robust mechanical sensor devices. Again, these devices are based on MicroGaN's core unit cell, the GaN based HFET on Si substrate. Cantilever based devices for sensor and actor applications are fabricated to demonstrate the capability of such mechanical sensor systems. For this, the availability of high quality GaN on silicon material [4] is a prerequisite for the successful fabrication of freestanding gallium nitride structures. Only this growth technology makes it possible to remove the substrate fast and cost effective. Capacitance pressure sensors [5] and freestanding cantilever structures [6] have been reported in the literature.

We will, to our knowledge for the first time, present a piezo-driven mechanical actor with the gallium nitride material system. These a tunable capacitors (varactors) and RF-MEMS switches are necessary devices for software defined radio architectures, impedance matching and filter switching.

Extended FEM simulations have been performed to determine the layer composition and the layout of the HEMT and cantilever, respectively. The simulations predict a deflection of a 100 x 800 μm cantilever of more than 5 μm with an applied voltage of -8V, which is at least a factor of 6 lower compared to electrostatic actors. This applied voltage would deflect the cantilever with a force of 20 μN.

The sensor structure itself is fabricated with a modified HEMT process where an additional deep GaN etch step is included in order to define the cantilever. At the end of the process the silicon is etched from the backside with an anisotropic silicon etch. The actor consists of a HEMT structure and a cantilever. A variation of the capacitance is realized by changing the distance between the cantilever and a metal plate. A schematic drawing is shown in figure 8.

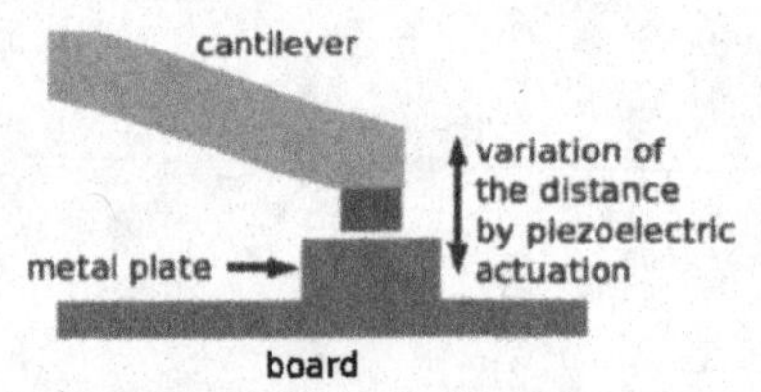

Figure 8. Prinziple configuration of a varactor. The piezoelectric actuation causes a distance variation between the cantilever and the bottom metal plate resulting in a change of the capacitance.

REM pictures of the cantilever in up and down position can be seen in Figure 9. The corresponding applied voltage are -8 V and 2 V, respectively. The deflection reaches a maximum of 5.6 µm. The good agreement between the measurement and prediction is a proof of the accuracy of the finite element simulations.

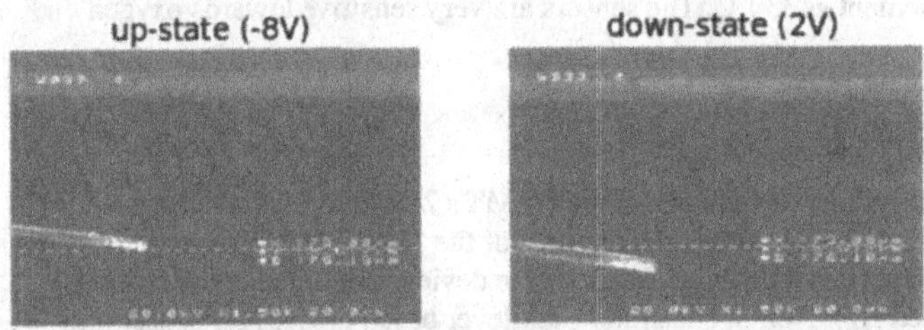

Figure 9. REM pictures of a cantilever in the up- and down-state. The measured maximum deflection is 5.6 µm.

If the cantilevers are placed above a metal plate a capacitance change can be measured by deflecting them. Figure 10 shows two measurements of two different varactors. It can be seen that the measured capacitance follows the input signal in a linear manner. The varactors do not show any mechanical instabilities, like pull-in or snap-down. The total capacitance swing is 40 fF.

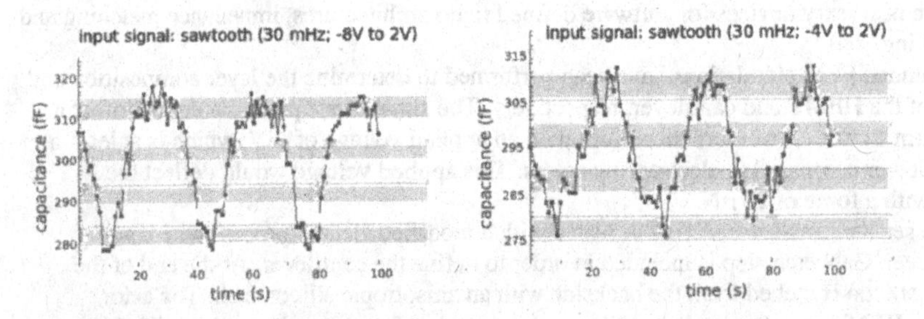

Figure 10. Measured variation of the capacitance in dependence of the input signal for two different devices.

Logic circuits

In order to combine sensors for harsh environments and add functionality to sensor systems, digital circuitry with equal capability in terms of robustness needs to be implemented in gallium nitride technology. The wide band gap and other GaN material properties are the bases for signal processing and control applications in harsh environments. In the following we will demonstrate that GaN on silicon is extremely well suited for the integration of sensor systems and logic circuitry.

Logic circuitry is usually realized with CMOS technology where the maximum operating temperature is limited to 150°C. This limitation has been overcome with Silicon-on-insulator based products. But the increase of the intrinsic carrier concentration of silicon at high temperatures restricts the use of these circuits below 250°C.

AlGaN/GaN heterostructures are suitable for digital control applications at elevated temperatures due to their wide bandgap. The drawback of this technology is the lack of enhancement FETs, which makes it impossible to use CMOS logic, but require the use of buffered FET logic (BFL). This technology was first demonstrated by van Tuyl and Liechti in GaAs technology [7] and led to a mature 4 GHz logic family [8].

Figure 11 shows exemplary for the variety of circuits we have demonstrated, an 2-to-1 multiplexer consisting of a NOT gate, an OR gate and two AND gates. From the truth table we can see that the output a equals input e_0 for S=0 and input e_1 for S=1. Our circuits have been designed that a logic 1 equals 1V and a logic zero -3V. The corresponding measurement shows a fully functional device at 250°C and an input frequency of 1 kHz. Minor layout adjustments of the circuits are necessary to further increase operation temperatures of our circuits. Thus, operation at temperatures even up to 400°C was achieved.

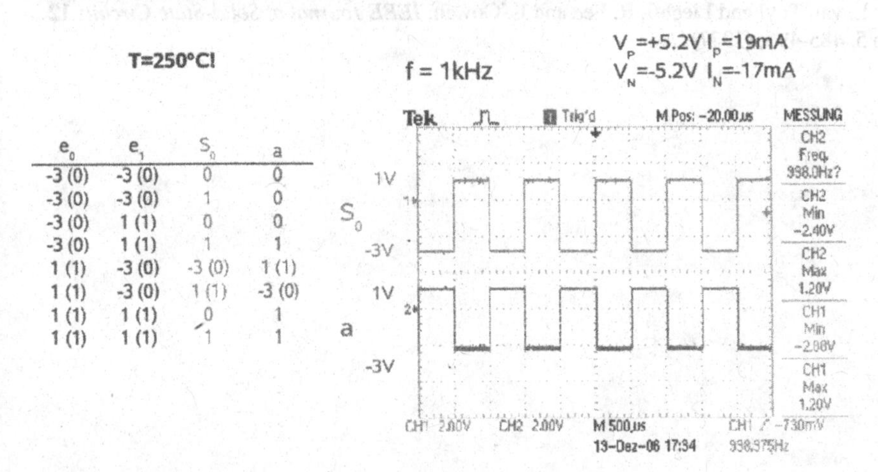

e_0	e_1	S_0	a
-3 (0)	-3 (0)	0	0
-3 (0)	-3 (0)	1	0
-3 (0)	1 (1)	0	0
-3 (0)	1 (1)	1	1
1 (1)	-3 (0)	-3 (0)	1 (1)
1 (1)	-3 (0)	1 (1)	-3 (0)
1 (1)	1 (1)	0	1
1 (1)	1 (1)	1	1

Figure 11. Schematic layout of the multiplexer and the corresponding truth table. The measurement has been performed at 250°C and a frequency of 1 kHz.

RESULTS

A full set of single devices, such as sensor or actor devices, was developed using MicroGaN's core unit cell, the GaN based HEMT on silicon substrate. For their possible use in a variety of applications, which require the outstanding material properties of the wide bandgap system GaN, an associated logic circuitry was demonstrated to meet complex customer requirements.

REFERENCES

1. G. Steinhoff, M. Herrmann, W. J. Schaff, L. F. Eastman, M. Stutzmann, and M. Eickhoff, *Appl. Phys. Lett.* **83**, 177 (2003).
2. I. Lundström, M. S. Shivaraman, C. Svensson, I. Lundkvist, Appl. Phys. Lett. **26** 55–57 (1975).
3. J. Schalwig, G. Müller, O. Ambacher, M. Stutzmann, *Phys. Stat. Sol. (Part A)* **1851** 39–45 (2001).
4. A. Krost and A. Dadgar, *Phys. Stat. Sol. A* **194**, no. 2, 361–375, (2002).
5. B. S. Kang, J. Kim, S. Jang, F. Ren, J. W. Johnson, R. J. Therrien, P. Rajagopal, J. C. Roberts, E. L. Piner, K. J. Linthicum, S. N. G. Chu, K. Baik, B. P. Gila, C. R. Abernathy, and S. J. Pearton, *Appl. Phys. Lett.* **86**, no. 25, (2005).
6. T. Zimmermann, M. Neuburger, P. Benkart, F. J. Hernández-Guillén, C. Pietzka, M. Kunze, Daumiller, A. Dadgar, A. Krost, and E. Kohn, *IEEE Electron Device Letters* **27**, no. 5, (2006).
7. R. L. van Tuyl and C. A. Liechti, *IEEE Journal of Solid-State Circuits* **9**, no. 5, 269-276, (1974).
8. R. L. van Tuyl and Liechti, R. Lee and E. Gowen, *IEEE Journal of Solid-State Circuits* **12**, no.5, 485-495, (1977).

Mater. Res. Soc. Symp. Proc. Vol. 1068 © 2008 Materials Research Society 1068-C03-17

An Analytical Compact Direct-Current and Capacitance Model for AlGaN/GaN High Electron Mobility Transistors

Miao Li, Xiaoxu Cheng, and Yan Wang
Institute of Microelectronics, Tsinghua Univ., Beijing, 100084, China, People's Republic of

ABSTRACT

We develop an analytical model for the Direct-Current (DC) and capacitance-voltage (CV) characteristics of AlGaN/GaN High Electron Mobility Transistors (HEMTs), providing accurate predictions of the transconductance and the gate capacitance in both linear and velocity saturation regions. The models provide an accurate and smooth connection in the area near the knee point for the DC characteristics, which is attributed to precise descriptions of the channel charge. An accurate model for the low-field mobility of the two dimensional electron gases (2DEG) has been developed, considering the nonmonotonic dependence of the carrier velocity on the electric field perpendicular to the channel for the first time. The calculated transconductance and output conductance are proved accurate and to have high order continuity, especially in large voltage biases, which is hard for some other analytical or numeral models. The gate capacitance has been obtained analytically and verified by experimental data. The slight decrease in the measured gate-to-source capacitance over the velocity saturation region which indicates the results of neutralization of donors and the contribution of the free electron in the AlGaN layer has been modeled accurately and smoothly for the first time. The predicted cut-off frequency is in excellent agreements with measured data over the full range of applied biases. The models are implemented into the HSPICE simulator for the DC, AC and transient simulations, with good speed and convergence characteristics.

INTRODUCTION

The high electron mobility transistor (HEMT) fabricated in AlGaN/GaN materials has shown promising performance for high power and high temperature microwave applications since its attractive electronic material properties [1]. High performance GaN-based MMICs, such as voltage controlled oscillators (VCO), mixers, and low noise amplifiers (LNA) have been developed recently [2-4]. Along with the fast development of the GaN-based devices and circuits, reliable and predictive compact models are needed, which are scalable, physics based, and applicable to parameters optimization and extraction. However there is still no mature device model for GaN-based HEMTs in commercial circuit simulators, and BSIM3 for Si MOSFET or some empirical models have been used as substitutes [2]. Gangwani *et al* have developed a physical based accurate model for the *CV* characteristics of AlGaN/GaN HEMTs in which the analysis was carried out in strong inversion region, based on the approximation of full depletion in the AlGaN layer [5]. Manju *et al* have presented a gate capacitance model considering temperature effect and strain relaxation, as well as the donor neutralization and free carrier generation in the AlGaN layer [6]. However the model does not provide the distribution of the capacitance between source and drain.

In this paper, an analytical model for the DC and CV characteristics of AlGaN/ GaN HEMTs has been presented. Analytical descriptions for the drain current as well as the transconductance and the output conductance in linear and velocity saturation regions are provided. We also develop a new gate charge model which incorporates effects of the donor neutralization and free carrier generation in the AlGaN layer under a wide range of gate and drain biases. Then the gate-to-source capacitance (C_{gs}) and gate-to-drain capacitance (C_{gd}) have been obtained analytically under various applied biases and the cut-off frequency can be predicted. The model has been implemented into the HSPICE simulator and the results have been verified by experimental data.

THEORY

We have developed a threshold-voltage based two-dimensional analytical model for the DC characteristics of AlGaN/GaN HEMTs, which provided accurate descriptions for the current-voltage characteristics as well as transconductance [7]. Proper modifications to the low-field mobility had been made to expand the bias scope of application [8]. Now we appropriately simplify the original model to make it more applicable for the calculation of gate charge and capacitance. The onset of saturation is marked by the channel electric field approaching the critical field E_T. We suppose $\lambda = w\varepsilon\mu_0 / d$, where μ_0 is the low field mobility, w is the gate width, ε and d are the permittivity and the total thickness of the AlGaN layer respectively. In the low-field region the drain current I can be obtained as:

$$I = \frac{\lambda E_T (2V_{gt} - V_{ds})V_{ds}}{2(aV_{ds} + E_T L)} \quad (1)$$

where $V_{gt}=V_{gs}-V_{th}$, V_{gs} is the gate-to-source bias, V_{th} is the threshold voltage, V_{ds} is the drain-to-source bias, and L is the gate length. In the velocity saturation region, we have to solve (2) to obtain normalized electric field (by E_T) e_0 at the source. Then the drain current can be obtained by (3). The meanings of undetermined parameters a and p can be found in [7].

$$V_{gt}(1 - \frac{(1+a)e_0}{1+ae_0}) + pE_T \sinh((L - \frac{V_{gt}(1-e_0^2)}{2e_0(1+ae_0)E_T})/p) = V_{ds} \quad (2)$$

$$I = \frac{V_{gt}\lambda E_T e_0}{1+ae_0} \quad (3)$$

Differentiating the drain current with respect to the gate-to-source and drain-to-source biases, we can acquire the transconductance and output conductance analytically.

We can obtain the charge balance equation of the AlGaN/GaN system when ignoring the substrate charge:

$$qn_g + q\int_{-d}^{0} (N_D^+ - n_1)dy + q(\sigma - n_s) = 0 \quad (4)$$

where n_g is the gate charge sheet density, N_D^+ and n_1 are the ionized donor and free electron concentration (per unit volume), σ is the polarization charge sheet density, and n_s is the sheet charge density of the channel. When the applied gate voltage or the doping concentration in the AlGaN layers is high, the full depletion approximation conventionally used [5] may cause discrepancies when calculating the gate charge and capacitance [6]. George *et al* have presented an effective method which includes an approximation to the Fermi-Dirac integral to deal with the one-dimensional situation for AlGaAs/GaAs HEMTs [9]. However there are still some problems. Firstly, since the threshold voltage is usually determined under the assumption of full depletion approximation, the whole system including the gate, AlGaN layer, and the 2DEG in the channel, may not be self-consistent. Particularly, the potential at the edge between the doped AlGaN layer and the spacer layer can only be obtained using full depletion approximation (see $\psi(-d_{sp})$ in [6] or $\psi(0)$ in [9]). The error of gate charge caused by the potential discrepancy grows exponentially along with the increase of gate bias. Secondly, the method provided the square of the electric field at the gate rather than directly determined the field. So the sign of the field cannot be directly determined. When an AlGaN/GaN HEMT switches from depletion to positively biased condition, n_g increases dramatically, while the direction of the electric field at the gate changes. So the original method limited the gate-bias scope application of the model. Last but not the least the calculation process is complex and the method can hardly be used to obtain the gate-to-source and gate-to-drain capacitances directly and analytically.

We develop a concise compact gate charge model and obtain the CV characteristics analytically. First of all, the quasi-one-dimensional Poisson's equation in the low field region of AlGaN layer is solved numerically and the self-consistent potential is acquired. Then the electric field as well as charge density at the gate is calculated. At last we fit the gate charge density from the numerical result:

$$n_g = n_s - \sigma - N_D d_d + \alpha \frac{a_0 + a_1 V_{gt}}{1 + \exp(a_2 V_{gt} + a_3)} \tag{5}$$

where N_D is the doping concentration in AlGaN layer, d_d is the thickness of the doped AlGaN layer, a_0~a_3 are undetermined parameters, and α is a factor evaluating the affection of traps and deep donors. The total gate charge Q_g can be obtained by integrating n_g along the channel analytically. Differentiating Q_g with respect to the gate-to-source and gate-to-drain voltage, we can obtain the C_{gs} and C_{gd} respectively. The cut-off frequency can be obtained as [5], where g_m is the transconductance.

$$f_T = \frac{g_m}{2\pi(C_{gd} + C_{gs})} \tag{6}$$

RESULT AND DISCUSSION

To assess the validity of our model, we compare the output characteristics, transconductance, and output conductance under diverse gate biases and drain biases with experimental data for an $Al_{0.15}Ga_{0.85}N$/GaN HEMT [10], which are shown in Figure 1, Figure 2, and Figure 3, respectively. In the velocity saturation region and the area near the knee point, our model provides smooth and accurate descriptions for the drain current, which can be attributed to the good continuity of the charge control models between the low-field and the saturation region. It can be noticed that our model provides a good fit, which has high order continuity, for the transconductance in both low field region and velocity saturation region. Even under high gate-to-source biases the model agrees with measured data well, which is hard for some other analytical or numeral models [11] and [12]. The output conductance from our model is proved accurate under a wide range of drain-to-source biases. The small discrepancy at the knee point may come from the approximation in the simplification.

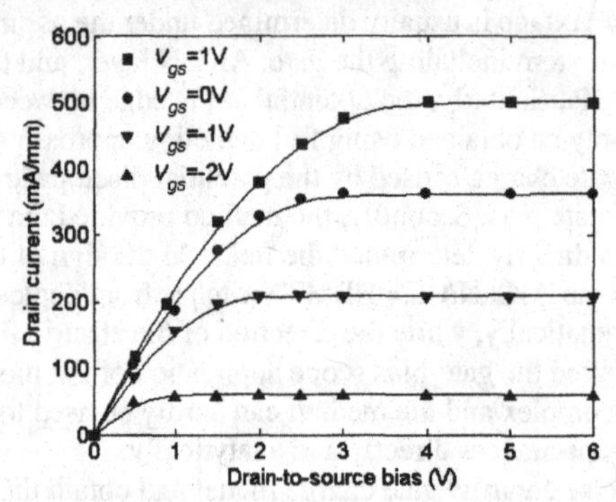

Figure 1. Simulated (solid line) and experimental (symbol) drain current for an $Al_{0.15}Ga_{0.85}N$/GaN HEMT. The gate-to-source bias is swept from -2V to 1V.

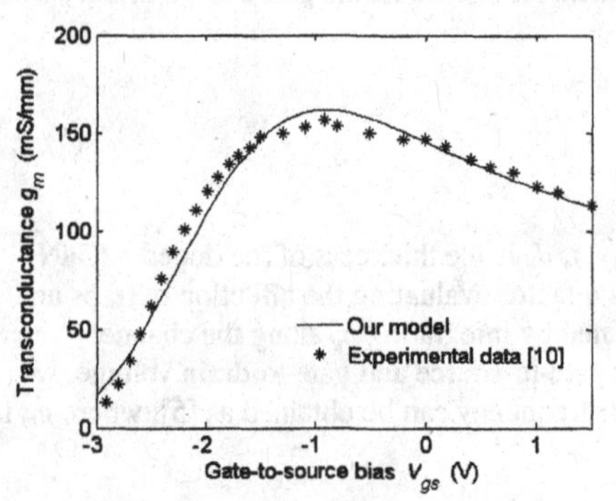

Figure 2. Simulated and experimental transconductance for an $Al_{0.15}Ga_{0.85}N$/GaN HEMT. The drain-to-source bias is 5V.

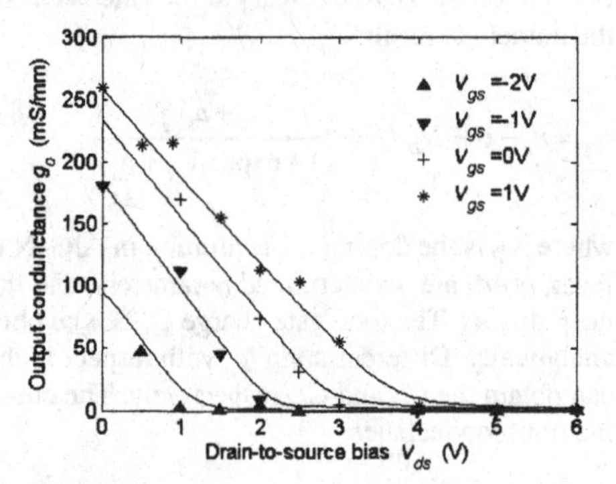

Figure 3. Simulated (solid line) and experimental (symbol) output conductance for an $Al_{0.15}Ga_{0.85}N$/GaN HEMT.

The gate charge obtained by our model is compared with numerical result as well as calculation under full depletion approximation, shown in Figure 4. When the gate-to-source bias is low, gate charge is dominated by the carriers in the 2DEG. When gate voltage increasing, gate charge will rapidly increase due to the generation of free carriers and neutralization of donors in the AlGaN layer [13]. We can observe that the model agree with the numerical result, which means we can accurately simulate the influence of the ionized donor and free electron in the AlGaN layer on the gate charge. The simulated and measured [14] *CV* characteristics have been presented in Figure 5 and Figure 6. In Figure 5 the slight decrease of C_{gs} over the velocity saturation region, which indicates the contribution of the donor neutralization and free electron in the AlGaN layer, has been modeled accurately and smoothly for the first time. In Figure 6 our model provides a good agreement for the C_{gd} with measured data in a wide span of biases except the deep velocity saturation region, where the parasitic capacitance can no longer be ignored. The cut-off frequency respect to the drain-to-source biases has been calculated and compared with measured data, shown in Figure 7. The prediction of f_T from our model is accurate and almost remains constant in the saturation region.

The model has been implemented into the HSPICE simulator by VerilogA and is applicable to DC, AC and transient simulations. The model has a 2nd-order continuity of the drain current and channel charge and functions as fast as the BSIM4 models under similar conditions and provides good convergence characteristics.

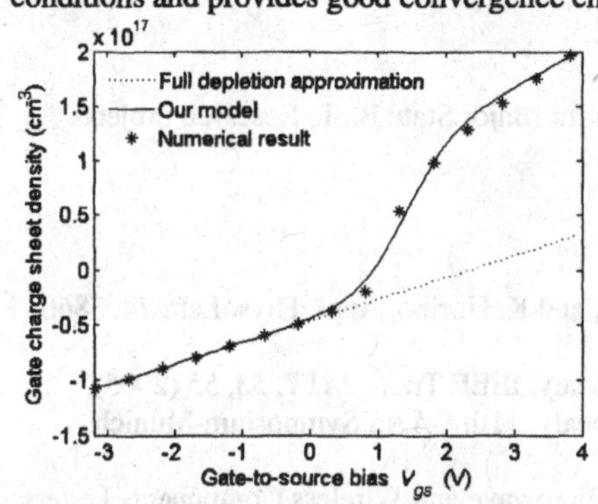

Figure 4. Gate charge density from numerical simulation and our model for an $Al_{0.15}Ga_{0.85}N/GaN$ HEMT, compared with full depletion approximation result.

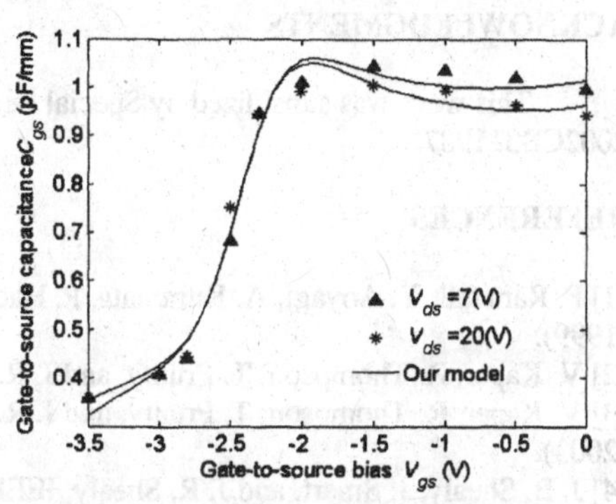

Figure 5. Simulated (solid line) and experimental (symbol) [14] gate-to-source capacitance for an $Al_{0.14}Ga_{0.86}N/GaN$ HEMT.

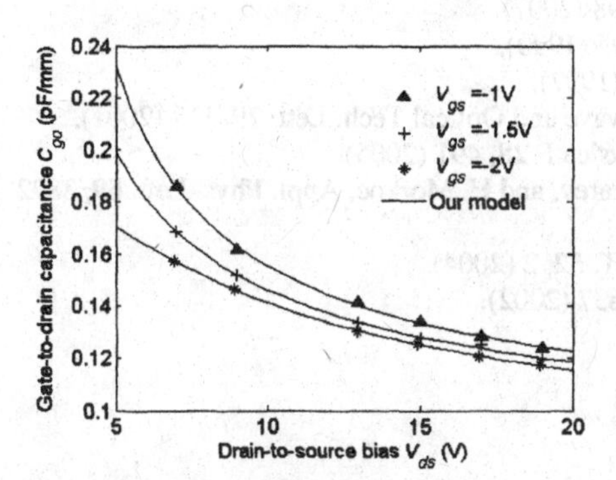

Figure 6. Simulated (solid line) and experimental (symbol) [14] gate-to-drain capacitance for an $Al_{0.14}Ga_{0.86}N/GaN$ HEMT.

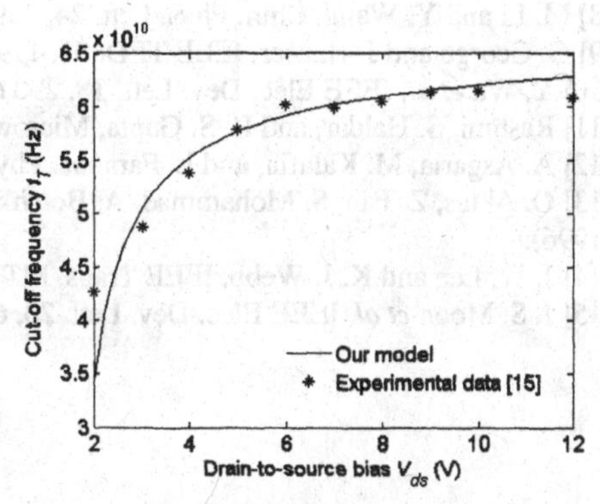

Figure 7. Simulated and measured [15] cut-off frequency for an $Al_{0.3}Ga_{0.7}N/GaN$ HEMT. The gate-to-source bias is -5.25V.

CONCLUSIONS

An analytical model is presented to describe the DC and CV characteristics of AlGaN/GaN HEMTs. We develop a compact gate charge model considering the contribution of donor neutralization and free electron in the AlGaN layer. The gate-to-source and gate-to-drain capacitances are obtained analytically and the cut-off frequency is predicted. The model has been implemented in the HSPICE and verified by experiments and numerical results.

ACKNOWLEDGMENTS

This work was subsidized by Special Funds for Major State Basic Research Projects 2002CB311907.

REFERENCES

[1] P. Ramvall, Y. Aoyagi, A. Kuramata, P. Hacke, and K. Horino, Appl. Phys. Lett. **74,** 3866 (1999).
[2] V. Kaper, R. Thompson, T. Prunty, and J. R. Shealy, IEEE Trans. MTT. **53,** 55 (2005).
[3] V. Kaper, R. Thompson, T. Prunty, and J. R. Shealy, 11th GAAS Symposium-Munich (2003).
[4] J. B. Shealy, J. Smart, and J. R. Shealy, IEEE Microwave and Wireless Components Letters **11**(6), 244 (2001).
[5] P. Gangwani *et al*, Solid-State Elec. **51**, 130 (2007).
[6] M. Chattopadhyay and S. Tokekar, Solid-State Elec. **50,** 220 (2006).
[7] M. Li and Y. Wang, IEEE TED. **55**, 261 (2008).
[8] M. Li and Y. Wang, Chin. Phys. Lett. **24**, 2998 (2007).
[9] G. George and J. Hauser, IEEE TED. **37**, 1193 (1990).
[10] Y. Wu *et al*, IEEE Elec. Dev. Lett. **18,** 290 (1997).
[11] Rashmi, S. Haldar, and R. S. Gupta, Microwave and Optical Tech. Lett. **29**, 117 (2001).
[12] A. Asgaria, M. Kalafia, and L. Faraone, Physica E **28**, 491 (2005).
[13] O. Aktas, Z. Fan, S. Mohammad, A. Botchkarev, and H. Morkoc, Appl. Phys. Lett. **69**, 3872 (1996).
[14] J. W. Lee and K. J. Webb, IEEE Trans. MTT. **52,** 2 (2004).
[15] J. S. Moon *et al*, IEEE Elec. Dev. Lett. **23**, 637 (2002).

Mater. Res. Soc. Symp. Proc. Vol. 1068 © 2008 Materials Research Society 1068-C03-02

Enhancement Mode GaN MOSFETs on Silicon Substrates with High Field-effect Mobility

Hiroshi Kambayashi, Yuki Niiyama, Shinya Ootomo, Takehiko Nomura, Masayuki Iwami, Yoshihiro Satoh, Sadahiro Kato, and Seikoh Yoshida
The Furukawa Electric Co., LTD., Yokohama, 220-0073, Japan

ABSTRACT

In this report, we have demonstrated enhancement-mode n-channel GaN MOSFETs on silicon (111) substrates. We observe a high field-effect mobility of 115 cm^2/Vs, the best report for GaN MOSFET fabricated on a silicon substrate to our knowledge. The threshold voltage was estimated to be +2.7 V, and the maximum operation current was over 3.5 A. This value is the largest which have ever been reports.

INTRODUCTION

GaN-based devices are expected candidates for high-power switching systems since GaN has excellent figures of merit. [1] Furthermore, normally-off operation is required for fail safe operation and noise margin. Recently, normally-off AlGaN/GaN HFETs have been demonstrated by using several techniques [2]-[4]. However, these threshold voltages are below 1 V. GaN MOSFET is one of the candidates to obtain further high threshold voltage. The some groups have reported on their work of GaN MOSFETs [5]-[11]. Most of these MOSFETs have been fabricated on sapphire or GaN substrates. However, these substrates are very expensive and it is difficult to fabricate these with a large diameter.

In this report, we have demonstrated enhancement-mode n-channel GaN MOSFETs on silicon (111) substrates with high field-effect mobility and a large output current operation. In order to realize a larger output current operation, we have investigated the activation annealing condition of ion implanted silicon, which was implanted into GaN to form n^+ regions.

EXPERIMENT

Figure 1 shows the schematic cross-sectional view of the GaN MOSFET we have fabricated. A heterostructure of a 2.0 μm-thick GaN-based buffer layer and a 1.5 μm-thick GaN epilayer with Mg acceptor doping at a concentration of 1×10^{17} cm^{-3} was grown on a silicon (111) substrate using metal-organic chemical vapor deposition (MOCVD). Then silicon ions were implanted into GaN layer to fabricate n^+ source and drain regions. The dose of implanted silicon was 3×10^{15} cm^{-2} with a maximum energy of 160 KeV to achieve a junction depth of 300 nm. After 500-nm-thick SiO_2 deposition as a capping layer, we performed rapid thermal annealing in an Ar ambient to activate the implanted silicon. After removing the SiO_2, 60-nm-thick SiO_2 was deposited by plasma-enhanced chemical vapor deposition (PE-CVD) as a gate oxide, and then annealed at 800°C for 30 min in a N_2 ambient. This annealing is effective to decrease the SiO_2/GaN interface-state density (Dit) near the GaN conduction band [12]. Ohmic electrodes were formed by using lift-off technique of Ti/Al, and the electrodes were annealed at 600°C for 10 min in a N_2 ambient. The gate electrode was also defined by lift-off of Ti/Au. The channel widths of the MOSFETs are 1.1 mm, 8.7 mm, and 16 mm with the cannel length of 4 μm.

RESULTS AND DISCUSSION

We have investigated the condition of activation annealing of ion implanted silicon into the GaN layer. It is important for n-channel GaN MOSFET to be fabricated with the n^+ region. However, the characteristics of activation annealing of ion implanted silicon into a GaN layer grown on a silicon substrate are unclear. Figure 2 shows the dependency of the sheet resistance at the activation annealing temperature and time. The sheet resistances were measured by the Van der Pauw method. As shown in this figure, the sheet resistance has is lower as the annealing temperature is higher and the annealing time is longer. However, the GaN layers peeled off from the silicon substrates below 100 ohm/square. Therefore, we decided to apply 1200°C for 10 sec for activation annealing of the implanted silicon. In this condition, the sheet resistance was 125 ohm/square. The contact resistance between Ti/Al ohmic electrode and n^+ region was 7.5×10^{-7} ohm-cm^2.

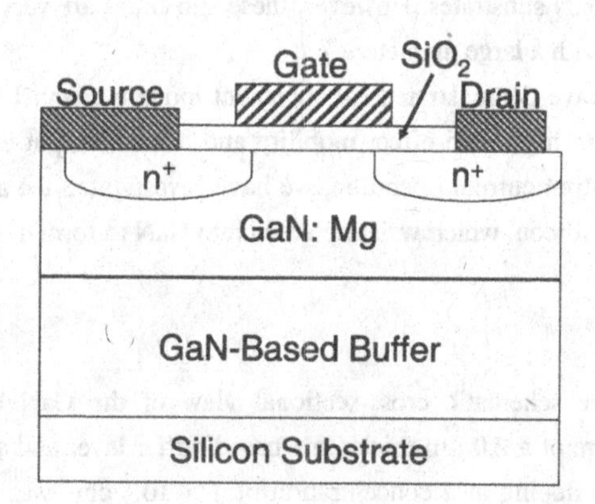

Figure 1. A schematic cross-section of an n-channel GaN MOSFET.

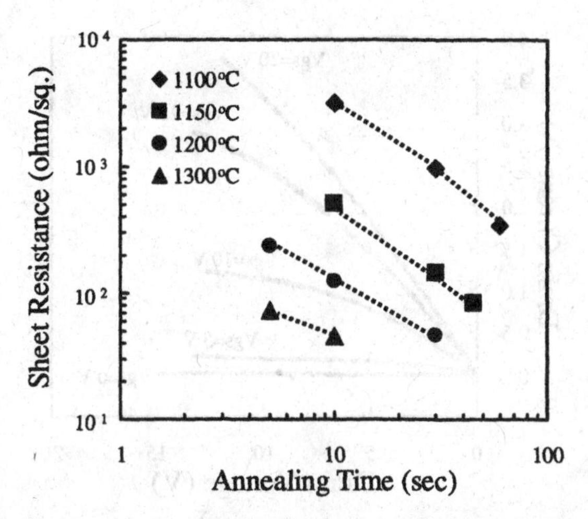

Figure 2. Sheet resistance versus activation annealing temperature and time.

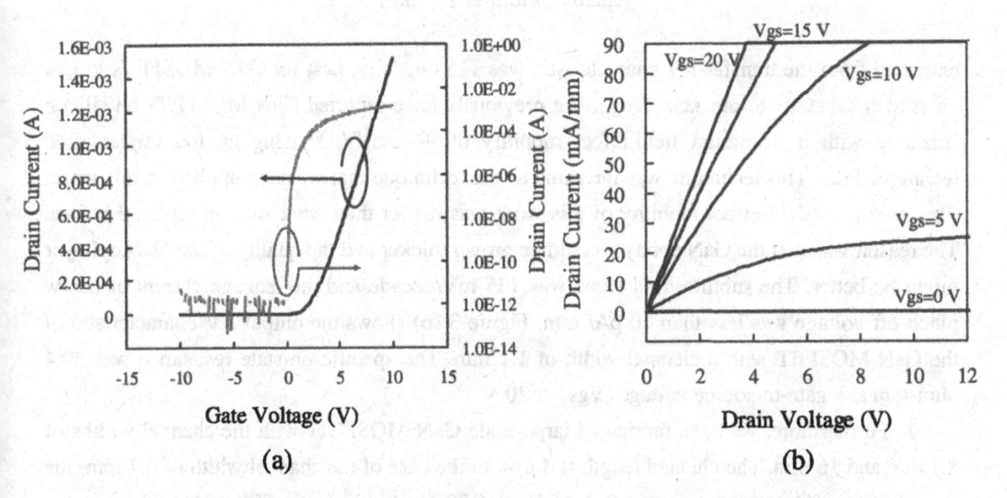

Figure 3. A typical transfer I-V characteristics at Vds of 0.1 V (a) and a typical output I-V characteristics (b) of the GaN MOSFET with the channel width of 1.1 mm.

Figure 3 (a) shows the transfer I-V characteristic of the GaN MOSFET with a channel width of 1.1 mm. The drain-to-source voltage was 0.1 V. We have realized good normally-off operation on GaN MOSFETs with threshold voltages of +2.7 V. The maximum field-effect mobility

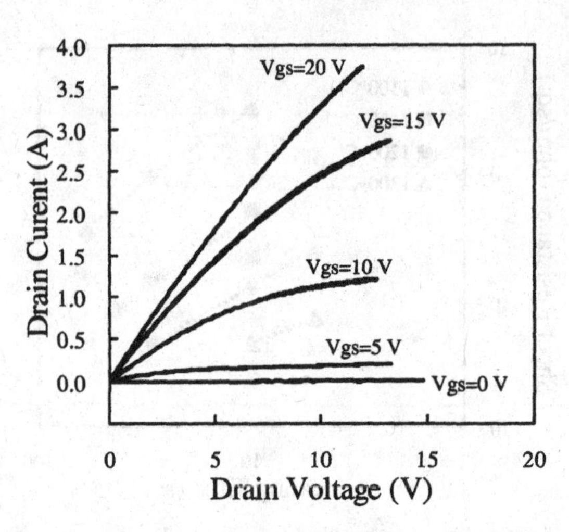

Figure 4. A typical output I-V characteristics of the GaN MOSFET.
(channel width of 16 mm)

extracted from the transfer I-V characteristic was 115 $cm^2/V \cdot s$, best for GaN MOSFETs grown on silicon substrate to our knowledge. We previously have reported GaN MOSFETs on silicon substrate with a maximum field-effect mobility of 94 $cm^2/V \cdot s$ using an ion implantation technique [12]. This technique was the same as the technique that we have applied in this work. The maximum field-effect mobility of this work was higher than what we had reported before. The reason was that the GaN epilayer could be grown thicker and the quality of the GaN epilayer might be better. The subthreshold slope was 135 mV/decade and the leakage current at below pinch-off voltage was less than 10 pA/ mm. Figure 3 (b) shows the output I-V characteristic of the GaN MOSFET with a channel width of 1.1 mm. The specific on-state resistance was 39.4 ohm-mm at a gate-to-source voltage (Vgs) of 20 V.

Furthermore, we have fabricated large-scale GaN MOSFETs with the channel widths of 8.7 mm and 16 mm. The channel length is 4 μm. In the case of the channel width of 8.7 mm, the maximum output current was over 2 A at Vgs of 20 V. As shown in Figure 4, the maximum output current with the channel width of 16 mm was over 3.5 A at Vgs of 20 V. The current is the highest in reported GaN MOSFETs. Figure 5 shows the channel width dependency of the specific on-state resistance. This figure shows that the specific on-state resistance has hardly changed even though channel length becomes longer. This is because the fabrication process we applied might be effective for GaN MOSFET on silicon substrates.

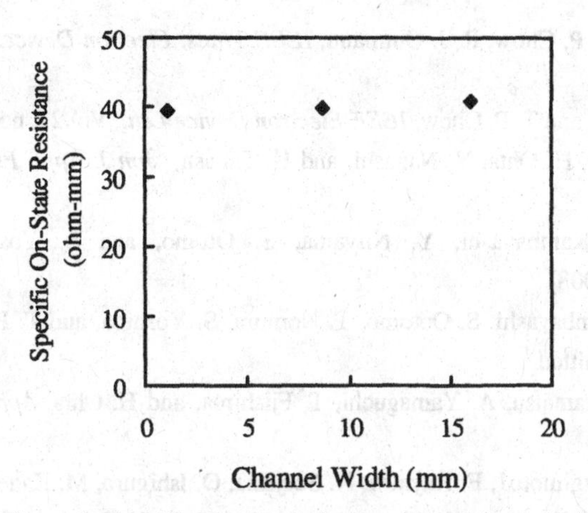

Figure 5. Specific on-state resistance versus channel width.

CONCLUSIONS

We have demonstrated enhancement-mode GaN MOSFETs on silicon (111) substrates with a high field-effect mobility of 115 cm^2/Vs. This is best for GaN MOSFETs grown on silicon substrate to our knowledge. Threshold voltage was +2.7 V. A maximum output current of over 3.5 A was obtained with a channel length of 16 mm. This is the highest current of all known reports in the literature. In order to perform a larger output current, we have investigated the activation annealing conditions of ion implanted silicon, which was implanted into GaN to form n^+ regions. As a result, the sheet resistance was 125 ohm/square. These results suggest that the fabrication process we applied is effective for GaN MOSFET on silicon substrates.

REFERENCES

1. T.P. Chow, and R. Tyagi, *IEEE Trans. Electron Devices*, Vol.41, 1481 (1994).
2. W. Saito, Y. Tanaka, M. Kuraguchi, K. Tsuda, and I. Ohmura, *IEEE Trans. Electron Devices*, Vol.53, 356 (2006).
3. S. Jia, Y. Cai, D. Wang, B. Zhang, K. M. Lau, and K. J. Chen, *IEEE Trans. Electron Devices*, Vol. 53, 1474 (2006).
4. Y. Umemoto, M. Hikita, H. Ueno, H. Matsuo, H. Ishida, M. Yanagihara, T. Ueda, T. Tanaka, and D. Ueda, *IEDM Technical Digest*, 2006.

5. K. Motocha, and T. P. Chow, R. J. Gutmann, *IEEE Trans. Electron Devices*, Vol.52, no. 1, 6 (2005).
6. W. Huang, T. Khan, and T. P. Chow, *IEEE Electron Device Lett.*, Vol.27, no. 10, 796 (2006).
7. H. Otake, S. Egami, H. Ohta, Y. Nanashi, and H. Takasu, *Jpn J. Appl. Phys. Lett.*, Vol.46, L599 (2007).
8. T. Nomura, H. Kambayashi, Y. Niiyama, S. Otomo, and S. Yoshida, Solid-State Electronics ,**52** 150 (2008).
9. Y. Niiyama, H. Kambayashi, S. Ootomo, T. Nomura, S. Yoshida, and T. Paul Chow, *Jpn J. Appl. Phys. Lett.*, submitted.
10. H. Otake, K. Chikamatsu, A. Yamaguchi, T. Fjishima, and H. Ohta, *Appl. Phys. Expr.*, **1**, 011105-1 (2008).
11. M. Kodama, M. Sugimoto1, E. Hayashi, N. Soejima, O. Ishiguro, M. Kanechika, K. Itoh, H. Ueda, T. Uesugi, and T. Kachi, *Appl. Phys. Expr.*, **1**, 021104-1, (2008).
12. H. Kambayashi, Y. Niiyama, S. Ootomo, T. Nomura, M. Iwami, Y. Satoh, S. Kato, and S. Yoshida, *IEEE Electron Device Lett.*, Vol.28, no. 12, 1077, (2007).

Mater. Res. Soc. Symp. Proc. Vol. 1068 © 2008 Materials Research Society 1068-C03-06

Integrated GaN/AlGaN/GaN HEMTs with Preciously Controlled Resistance on Silicon Substrate Fabricated by Ion Implantation

Kazuki Nomoto, Tomo Ohsawa, Masataka Satoh, and Tohru Nakamura
Department of EECE and Research Center for Micro-Nano Technology, Hosei University, Tokyo, 1840003, Japan

ABSTRACT

Multiple ion-implanted GaN/AlGaN/GaN high electron-mobility transistors (HEMTs) and preciously controlled ion-implanted resistors integrated on silicon substrate are reported. Using ion implantation into source/drain (S/D) regions, the performances were significantly improved. On-resistance reduced from 10.3 to 3.5 Ω·mm. Saturation drain current and maximum transconductance increased from 390 to 650 mA/mm and from 130 to 230 mS/mm. Measured transfer curve shows that I/O gain of 4.5 can be obtained at Vdd = 10 V.

INTRODUCTION

GaN is a wide bandgap semiconductor material, with a high breakdown electric field, high saturation drift velocity, and good thermal conductivity as compared to GaAs. In addition, one of the most important features of AlGaN/GaN system is its large piezo-electric polarization in the strained AlGaN, which generates highly dense two-dimensional electron gas (2DEG) with concentrations exceeding 10^{13} /cm^2. Due to their high power handling capability at high frequencies, AlGaN/GaN HEMTs are emerging as the promising candidates for radio frequency and microwave frequency power amplifiers used in advanced wireless communication systems. Performance of these devices has been thus far limited by self-heating. Most of the AlGaN/GaN HEMTs structures have been grown on sapphire or SiC substrates [1]-[3]. The sapphire substrates are low-cost, but feature inefficient heat dissipation due to their poor thermal conductivity. Epi-layers with excellent crystal quality and devices with the highest output power densities have been obtained on SiC substrates [4]-[7]. Recently, there have been increasing activities in AlGaN/GaN structures on silicon substrates, which offer several advantages, including large-size substrates, low cost, and good thermal conductivity.

We demonstrate novel technology for the integration of GaN/AlGaN/GaN HEMTs with extremely low source resistance fabricated by multiple ion implantation and precious controlled ion-implanted resistors on the silicon substrate. This technology will be a key process for future high-power, high-speed and high-temperature operation devices.

EXPERIMENT

The GaN (5 nm)/$Al_{0.25}Ga_{0.75}N$ (25 nm)/GaN (2 μm) HEMT structure was grown on silicon substrate by MOVPE. The sheet resistance of 2 DEG layer was 415 Ω/sq. Silicon ions were implanted into S/D regions in GaN/AlGaN/GaN HEMTs and resistor regions at the energy of 30 keV with a dose of 1.5×10^{15} /cm^2 through a 25 nm thick SiN_x layer to obtain low contact resistance for ohmic contact to the GaN layer with high surface carrier concentration. This was followed by additional ion implantation at the energy of 80 keV with ion dose of 1.0×10^{15} /cm^2 to decrease source parasitic resistance, as shown in Figure 1. After re-deposition of SiN_x, the samples were annealed at the temperature of 1200 °C for 2 min in N_2 ambient to activate implanted Si. Mesa isolation was carried out by ICP dry etching using CF_4 and Ar gas mixture. A SiN_x layer for the surface passivation was deposited by plasma enhanced CVD. S/D ohmic contacts were formed by depositing Ti/Al(30/200 nm) layers, followed by an annealing at 550 °C. Finally, gate contacts were formed by depositing Ni/Al (50/200 nm) layers. In order to clear the effect on the source resistance by ion implantation, multiple ion-implanted HEMTs and conventional HEMTs without ion implantation were fabricated on the same substrate. Devices with 1.9 μm in gate length and 12 μm in gate width were tested. The distances from the gate edge to the ion-implanted source edge and from the gate edge to the drain edge were both 1.5 μm, respectively. The distances between the gate and the source/drain contacts with Si ion implantation are both 3.0 μm, whereas without Si ion implantation was 1.5 μm. The width and length of resistors were 6.6 μm and 50 μm, respectively, as shown in Figure 2.

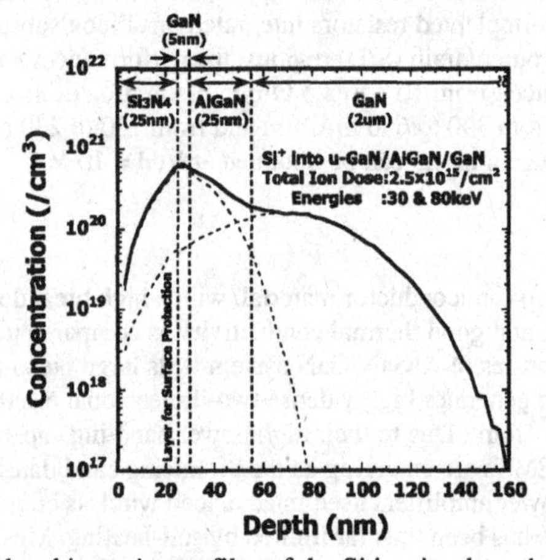

Figure 1. Simulated impurity profiles of the Si ion-implanted source/drain regions.

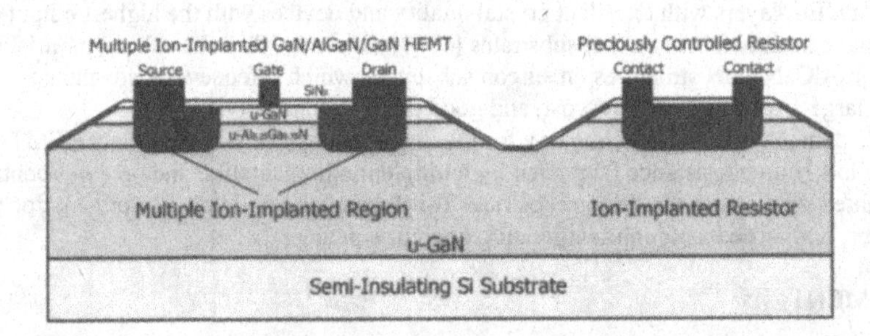

Figure 2. Device structures of multiple ion-implanted HEMTs and preciously controlled ion-implanted resistors on the silicon substrate.

RESULTS AND DISCUSSION

Sheet resistance and contact resistance were measured by means of the transmission line model (TLM) method for Si ions implanted GaN/AlGaN/GaN layers at energies of 30 and 80 keV. The length of the contact metals is 80 μm for TLM structure. The sheet resistance is reduced from 415 Ω/sq. for the as-grown sample to 160 Ω/sq. for Si ions implanted at energy of 80 keV, which is in good agreement with the results of sheet resistance evaluated by Van der Pauw measurements. The contact resistances of 0.37 Ω·mm (9.3×10^{-6} Ω·cm^2) was obtained for Si ions implanted GaN/AlGaN/GaN layers at energies of 30 and 80 keV. It is found that the contact resistance depends on the implanted Si concentration at the surface. In these material systems, the major element in source resistance is considered to be energy barriers at heterointerfaces. The peak position in the Si concentration for the ion implantation at 80 keV coincides with the heterointerface. The heavy implantation followed by high temperature annealing was speculated to have resulted in a disordered heterointerface and reduced the barrier height [8].

DC and RF characteristics of multiple ion-implanted (MI/I) HEMTs and conventional (C) HEMTs without ion implantation were measured. Id-Vds and gm-Vg characteristics of both devices are shown in Figures 3 and 4. Maximum drain current of 650 mA/mm at Vg=0 V and maximum transconductance of 230 mS/mm were obtained for MI/I HEMTs because of extremely low contact and source resistance, while those of C-HEMTs were 390 mA/mm and 130 mS/mm, respectively. From the analysis of on-resistance components, contact resistance (R_c) was reduced from 8.2 to 0.9 Ω·mm. The improvement of DC characteristics of MI/I GaN/AlGaN/GaN HEMTs was caused by a remarkable reduction of contact resistance. At an operation temperature of 200 °C, MI/I GaN/AlGaN/GaN HEMTs operated with a maximum drain current of 320 mA/mm. MI/I GaN/AlGaN/GaN HEMTs stably operated at high temperature atmosphere. Maximum cutoff frequency of 4 GHz and maximum oscillation frequency of 5.7 GHz were obtained for MI/I HEMTs with 80 μm bonding pads.

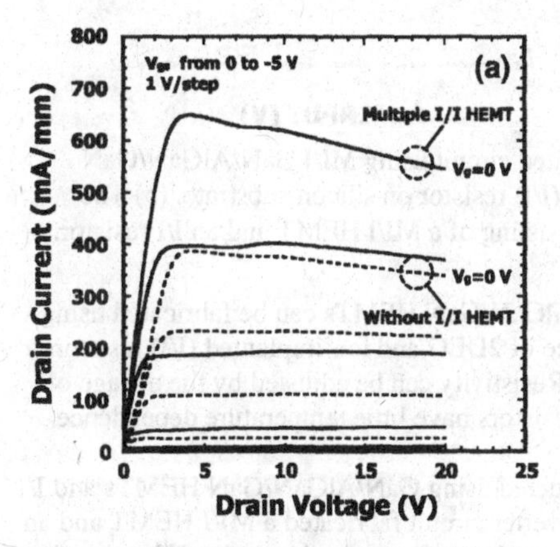

Figure 3. Id-Vds characteristics of MI/I HEMTs and C HEMTs.

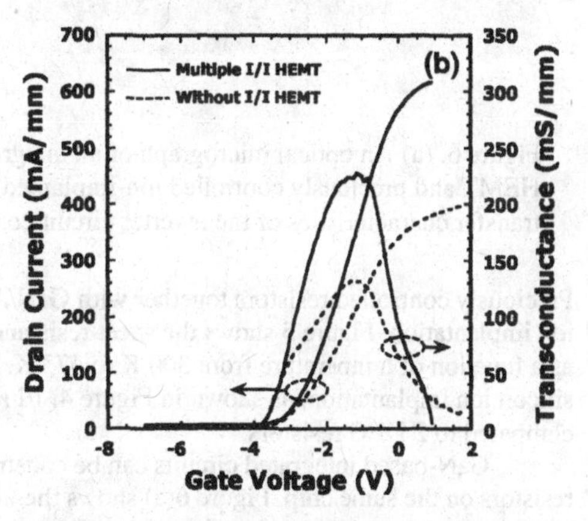

Figure 4. Drain current and transconductance as a function of gate voltage for MI/I HEMTs and C HEMTs.

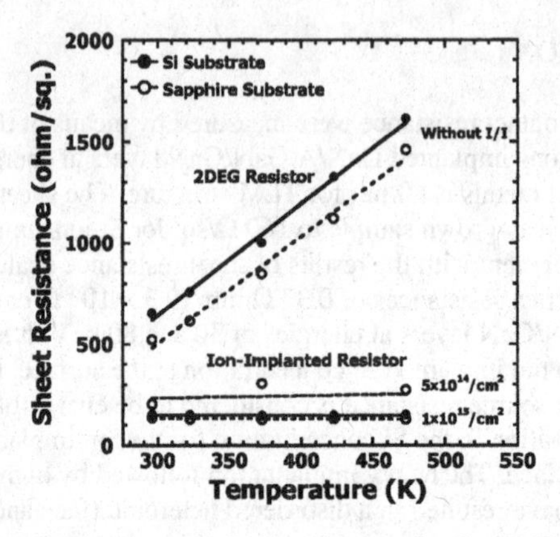

Figure 5. Temperature dependence of sheet resistance for the 2DEG and ion-implanted region.

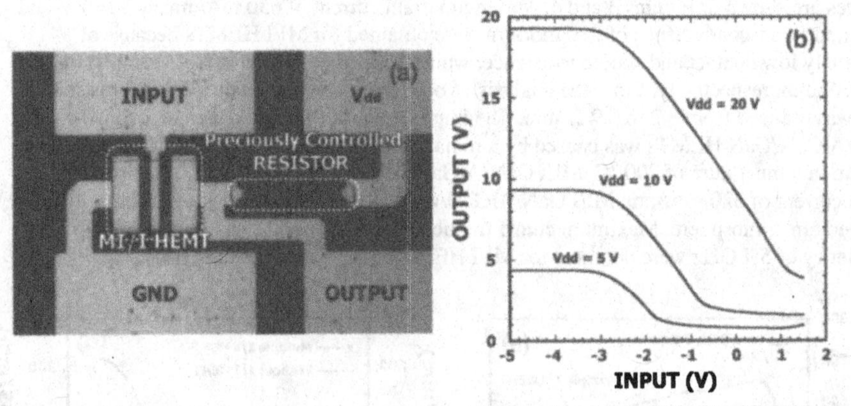

Figure 6. (a) An optical micrograph of an integrated circuit using MI/I GaN/AlGaN/GaN HEMT and preciously controlled ion-implanted (I/I) resistor on silicon substrate. (b) The transfer characteristics of the inverter circuit consisting of a MI/I HEMT and an I/I resistor.

Preciously controlled resistors together with GaN/AlGaN/GaN HEMTs can be fabricated using ion implantation. Figure 5 shows the sheet resistance of 2DEG and ion-implanted (I/I) resistors as a function of temperature from 300 K to 473 K. Resistivity can be adjusted by the dosage of silicon ion implantation, as shown in Figure 4. I/I resistors have little temperature dependence compared to 2 DEG resistors.

GaN-based integrated circuits can be constructed using GaN/AlGaN/GaN HEMTs and I/I resistors on the same chip. Figure 6(a) shows the inverter circuit fabricated a MI/I HEMT and an I/I resistor without any level shift circuits. The measured transfer curve shows that I/O gain of 4.5 can be obtained at Vdd = 10 V, as shown in Figure 6(b).

CONCLUSIONS

We fabricated multiple ion-implanted GaN/AlGaN/GaN HEMTs and resistors integrated on silicon substrate. Ion implantation is indispensable technology for fabricating high-performance integrated circuits including GaN HEMTs and preciously controlled resistors on silicon substrate.

REFERENCES

[1] Wataru Saito, Nasahiko Kuraguchi, Yoshiharu Takada, Kunio Tsuda, Ichiro Omura, and Tsuneo Ogura, "High Breakdown Voltage Undoped AlGaN-GaN Power HEMT on Sapphire Substrate and Its Demonstration for DC-DC Converter Application", IEEE Transaction on Electron Devices Vol. 51, No. 11, pp.1913-1917, 2004.
[2] Vipan Kumar, D. H. Kim, A. Basu, and I. Adesida, "0.25 μm Self-Aligned AlGaN/GaN High Electron Mobility Transistors", IEEE Electron Device Letters Vol. 29, No. 1, pp.18-20, 2008.
[3] Kazukiyo Joshin and Toshihide Kikkawa, "High-Power and High-Efficiency GaN HEMT Amplifiers", Radio and Wireless Symposium, 2008 IEEE, 22-24 Jan. 2008 Page(s):65 - 68
[4] Eduardo M. Chumbes, A. T. Schremer, Joseph A. Smart, Y. Wang, Noel C. MacDonald, D. Hogue, James J. Komiak, Stephen J. Lichwalla, Robert E. Leoni, III, and James R. Shealy, "AlGaN/GaN High Electron Mobility Transistors on Si(111) Substrates", IEEE Transaction on Electron Devices Vol. 48, No. 3, pp.420-426, 2001.
[5] Shuo Jia, Yilmaz Dikme, Deliang Wang, Kevin J. Chen, Kei May Lau, and Michael Heuken, "AlGaN-GaN HEMTs on Patterned Silicon (111) Substrate", IEEE Electron Device Letters Vol. 26, No. 3, pp.130-132, 2005.
[6] Young Chul Choi, Milan Pophristic, Ho-Young Cha, Boris Peres, Michael G. Spencer, and Lester F. Eastman, "The Effect of an Fe-doped GaN Buffer on OFF-State Breakdown Characteristics in AlGaN/GaN HEMTs on Si Substrate", IEEE Transaction on Electron Devices Vol. 53, No. 12, pp.2926-2931, 2006.
[7] Mark Yu, Robert J. Ward, and Gamal M. Hegazi, "High Power RF Switch MMICs Development in GaN-on-Si HFET Technology", Radio and Wireless Symposium, 2008 IEEE, 22-24 Jan. 2008 Page(s):855 - 858
[8] C. Ronning, E. P. Carlson and R. F. Davis, "Ion Implantation into Gallium Nitride", Physics Reports vol. 351, pp.349-385, 2001.

Mater. Res. Soc. Symp. Proc. Vol. 1068 © 2008 Materials Research Society 1068-C04-02

Power Performance of AlGaN/GaN HEMT's Grown on 6" Si Substrates

Joff Derluyn, Jo Das, Kai Cheng, Anne Lorenz, Domenica Visalli, Stefan Degroote, Marianne Germain, and Staf Borghs
RDO-PT-NEXT-NEXT35, IMEC, Kapeldreef 75, Leuven, B3001, Belgium

ABSTRACT

In this paper we report on our latest results on crack-free AlGaN/GaN HEMT devices grown on 6" Si Substrates with low sheet resistance. By using in situ SiN passivation, which solves the dispersion problem and avoids the relaxation of the AlGaN layer by inhibiting the mobility of Ga we obtain a perfectly strained AlGaN layer without cracking. The devices were grown on 6" Si <111> substrates and passivated using IMEC's in situ SiN technique. The epitaxial growth was optimized for minimum RF losses while at the same time maintaining high buffer resistivity and low trap density. To assess RF losses of the epitaxial layer structure, coplanar waveguides were defined on places of the wafer where the top in-situ SiN and AlGaN has been etched away. The attenuation of the RF signals on the coplanar waveguides remained below 0.3dB/mm for frequencies up to 6GHz. The devices' RF power performance was characterised on-wafer which is the worst case scenario due to severe thermal limitations. It is observed that the maximum gate current remains below 50μA/mm even at power density levels of 7.9W/mm, which promises very good reliability. The larger 20-finger devices of 5mm total gate periphery reach a maximum absolute output power of 20W at a bias of 40V, which represents the limit of IMEC's on-wafer measurement system. These results prove that the use of large area Si substrates is the only viable route forward for AlGaN/GaN HEMTs.

INTRODUCTION

Even though impressive results have been obtained with AlGaN/GaN HEMT's grown on SiC substrates [1], the high cost and limited availability of high quality and large area substrates prevent this technology from being used in mainstream consumer products. A few groups have reported good results on 4" Si <111> substrates and commercial products are available today [2]. In this paper we show that we can scale up such AlGaN/GaN HEMT technology to 6" Si substrates which doubles the available area, while maintaining high device performance.

EXPERIMENT AND DISCUSSION

Epitaxy

We have reported the growth of AlGaN/GaN HEMTs on highly resistive 6" Si <111> by MOCVD before [3, 4], demonstrating crack-free epitaxial layers that show at the same time very uniform and low sheet resistance values. A typical layer stack and sheet resistivity mapping is shown in figure 1.

When designing the layer stack, we have focused on two issues. First, we emphasise the importance of the in-situ SiN that is deposited as a passivation layer on top of the AlGaN cap layer in the MOCVD reactor in the same growth run as the rest of the layer stack. Given the sensitivity of AlGaN/GaN HEMTs to charge trapping in surface states, depositing very high quality SiN in-situ before exposing the sample to the atmosphere avoids the creation of new surface states (e.g. by oxide formation) while at the same time the SiN layer passivates existing ones [5]. We have also shown elsewhere that the in-situ SiN can even prevent relaxation of the AlGaN cap layer by preventing the migration of Ga that causes the formation of strain-mitigating grooves in the AlGaN [4, 5].

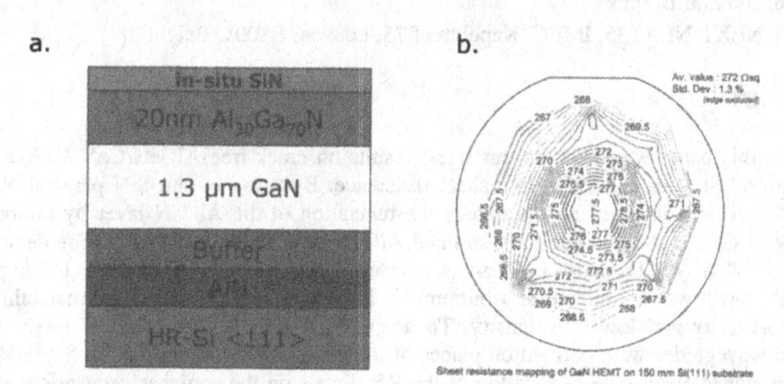

Figure 1. Epitaxial layer stack (a.) and sheet resistivity mapping (b.) of AlGaN/GaN HEMTs grown on 6" Si <111> substrate.

Secondly, to obtain good RF performance, one needs to optimize the interface between the Si substrate and the AlN layer. During the start of the MOCVD epitaxial growth, Al and N species may react with the Si substrate resulting in the formation of a doped layer. Also, an inversion layer may be created at the interface of Si and any high bandgap material. Both phenomena cause a conductive layer to be formed which causes capacitive losses for RF signals. We have characterised this losses by measuring transmission line losses in coplanar waveguides deposited on the buffer layer stack (i.e. the HEMT layer stack with the AlGaN cap layer and SiN passivation etched off). As a reference, we made waveguides on a blank HR-Si wafer.

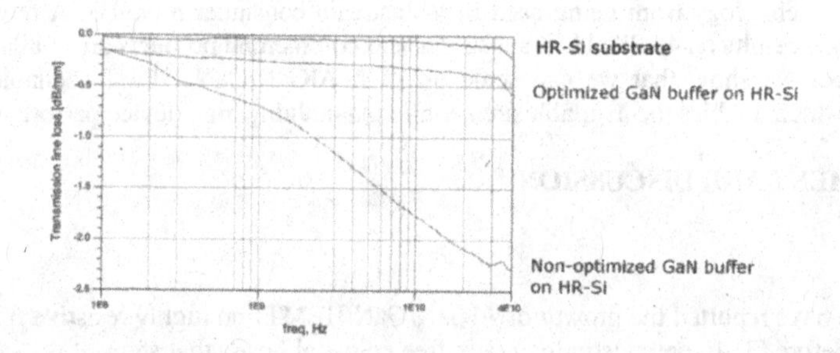

Figure 2. Transmission line losses of coplanar waveguides on different GaN buffers and on highly resistive Si substrate.

As is clearly depicted in figure 2, optimised growth conditions reduce RF losses to a level that comes very close to the best-case reference situation. The attenuation of the RF signals on the coplanar waveguides on the optimized buffer remains below 0.3dB/mm for frequencies up to 6GHz.

Processing and characterisation

Subsequently, devices were processed on these layers. A fairly standard process order was used consisting of: a mesa isolation step using an ICP plasma etch based on Cl_2 chemistry; the deposition of Ti/Al/Mo/Au ohmic contacts that were consecutively rapid thermally annealed at 850°C; the deposition of an interconnect metallisation; the definition of the gates by e-beam lithography producing 500nm long Mo/Au gates; the deposition and subsequent patterning of a second SiO_2 dielectric layer on top of which source-connected field-plates are defined and finally the fabrication of air-bridges using Au-electroplating. The basic device configuration has 6 fingers, each 250μm wide, resulting in a total gate periphery of 1.5mm. A different configuration has 20 gate fingers, totaling 5mm gate periphery.

We performed pulsed power measurements at 2GHZ and at varying bias conditions with a load that was optimized to obtain maximum output power. All measurements were performed on-wafer and without any substrate thinning. Obviously, this is the worst-case scenario for the thermal management. As a result, even while performing pulsed measurements, we had to limit the DC power dissipation by operating the devices in deep class AB bias conditions. The pulse period was 100μs and the pulse width 10μs. The drain bias was increased gradually from 30V to 60V.

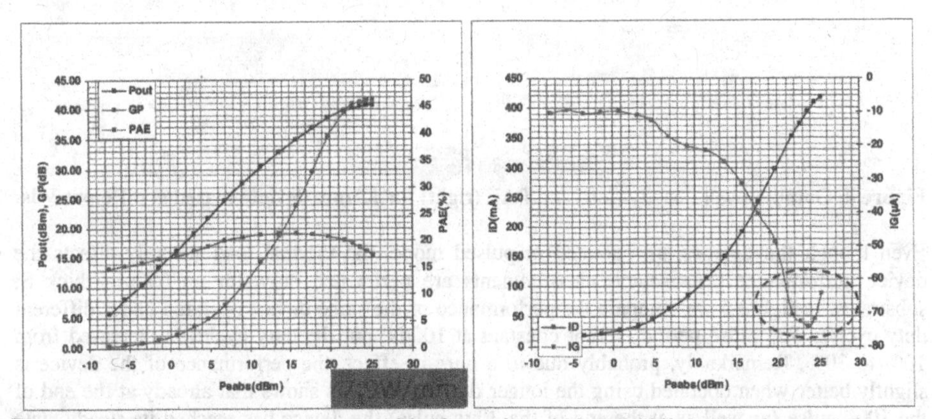

Figure 3. Pulsed power measurements with 10μs pulse width and 10% duty cycle on a 1.5mm wide device at a drain bias of 60V, operated in deep class AB with a load optimised for maximum output power. Power, gain and PAE are shown on the left, drain and gate currents on the right.

In figure 3 we show the measurement performed at a drain bias of 60V. Due to the fact that the device is biased in deep class AB, the gain at low input power is severely reduced. However, severe self-biasing causes the gain to increase steadily until compression is reached. Under these conditions, a maximum saturated power density of 7.9W/mm with an associated power added efficiency of 47% was obtained. Remarkably, the gate current under these high drain bias and input power remains below 50μA/mm. This should prove beneficial for the long-term reliability of the device.

The evolution of the power density and PAE as a function of bias voltage is summarized in figure 4. The fact that the increase of the power density with drain bias is linear and that the PAE increases slightly, is a proof of low DC-to-RF dispersion in the devices. We attribute this to the quality of the in-situ SiN surface passivation, as well as to the excellent crystal quality we obtain for these layers. This is also reflected in the FWHM values of the XRD rocking curves which give values of 650arcsec for the symmetric reflection (around 0002) and 1200arcsec for the asymmetric reflection (around 10-12). Unfortunately, 60V is the maximum drain bias we can apply to our devices as their breakdown voltage lies between 120V and 150V. Currently we are spending great effort on improving these values by adopting a double heterostructure configuration.

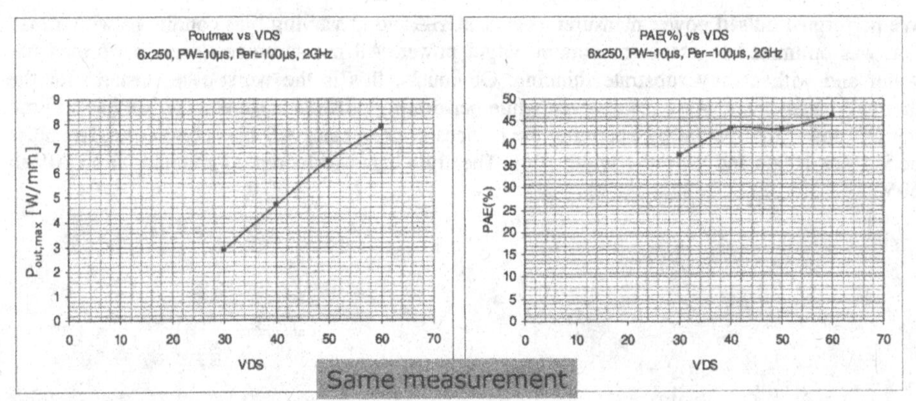

Figure 4. Output power density (left) and PAE (right) for 1.5mm devices biased at different Vds

Even though the devices are operated in pulsed mode, the thermal load severely affects the device performance because the measurements are performed on-wafer on 650μm thick Si substrates. In figure 5 we compare the performance of the same device operated under different duty cycles. The pulse period is kept constant at 100μs, but the duty cycles is increased from 10% to 30%. Remarkably, probably due to a burn-in effect, the performance of the device is slightly better when operated using the longer duty cycle. This shows that already at the end of the 10μs pulse (as well as at the end of the 30μs pulse) the device has reached its steady state temperature and thus that its performance is a good indication of what could be obtained in CW operation.

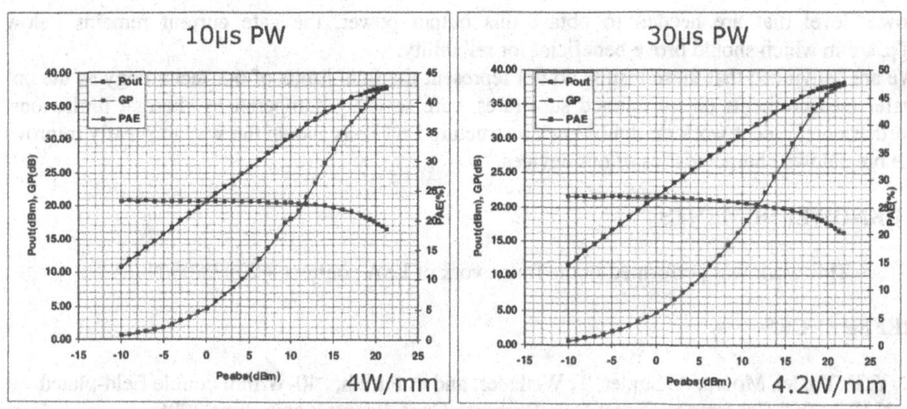

Figure 5. **Comparison of pulsed power performance of the same device with different duty cycle**

Finally in figure 6 we show the pulsed power performance of a larger device with 5mm gate periphery, biased at a drain voltage of 40V. The saturated output power reaches 20W which is the limit of our on-wafer measurement system, with a flat gain curve at 15dB and a PAE of 40%.

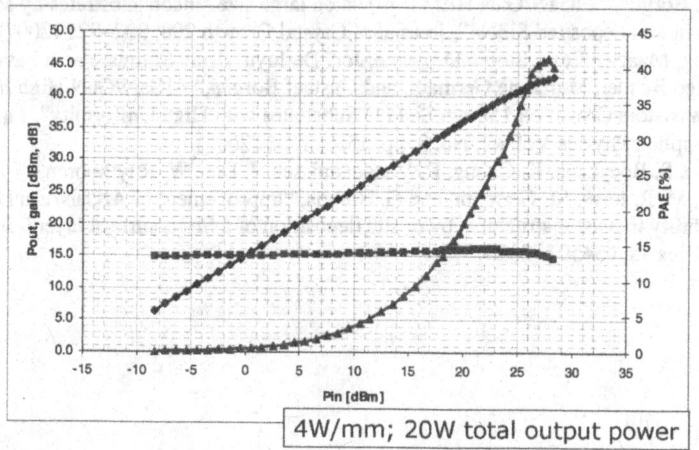

Figure 6. **Pulsed power measurement of 5mm device with 10μs pulse width and 10% duty cycle at a drain bias of 40V**

CONCLUSIONS

We have demonstrated very promising power performance of AlGaN/GaN HEMTs grown on 6" Si substrates. By minimizing the RF loss at the Si/AlN interface while at the same time maintaining low buffer trap density, it is possible to achieve a linear gain of 22dB and a power density as high as 7.9W/mm at a frequency of 2GHz. At the bias voltage (60V) and input

power level that are needed to obtain this output power, the gate current remains below 50μA/mm which should prove beneficial for reliability.
We are convinced that these results do not represent the final limits of our technology as the on-wafer measurements on un-thinned substrates were severely hampered by thermal limitations. Furthermore, recent work on double-heterostructure FETs has shown the way to greatly improve the breakdown characteristics of our devices.

ACKNOWLEDGMENTS

This work was performed in the framework of ESA contract # 20073/06/NI/IA

REFERENCES

1. Y. F. Yu, M. Moore, A. Saxler, T. Wisleder, and P. Parikh, "40-W/mm double field-plated GaN HEMTs," Conference Digest Dev. Research Conf. Pennsylvania, June 2006
2. Therrien, B., Singhal, S., Johnson, J.W., Nagy, W., Borges, R., Chaudhari, A., Hanson, A.W., Edwards, A., Marquart, J., Rajagopal, P., Park, C., Kizilyalli I.C., Linthicum, K.J. (2005). "A 36mm GaN-on-Si HFET Producing 368W at 60V with 70% Drain Efficiency", IEDM, Washington, DC (2005)
3. Kai Cheng, M. Leys, J. Derluyn, S. Degroote, D.P. Xiao, A. Lorenz, S. Boeykens, M. Germain, G. Borghs, "AlGaN/GaN HEMT grown on large size silicon substrates by MOVPE capped with in-situ deposited Si3N4", Journal of Crystal Growth 298, 822–825 (2007)
4. Kai Cheng, Maarten Leys, Stefan Degroote, Joff Derluyn, Brian Sijmus, Paola Favia, Olivier Richard, Hugo Bender, Marianne Germain, and Gustaaf Borghs, "AlGaN/GaN High Electron Mobility Transistors Grown on 150mm Si(111) Substrates with High Uniformity", Japanese Journal of Applied Physics Vol. 47, No. 3, pp. 1553–1555 (2008)
5. J. Derluyn, S. Boeykens, K. Cheng, R. Vandersmissen, J. Das, W. Ruythooren, S. Degroote, M. R. Leys, M. Germain, and G. Borghs, "Improvement of AlGaN/GaN high electron mobility transistor structures by in situ deposition of a Si_3N_4 surface layer", Journal of Applied Physics 98, 054501 (2005)

Mater. Res. Soc. Symp. Proc. Vol. 1068 © 2008 Materials Research Society 1068-C04-03

MOVPE Growth and Characterization of AlInN FET Structures on Si(111)

Christoph Hums[1], Aniko Gadanecz[1], Armin Dadgar[1], Jürgen Bläsing[1], Hartmut Witte[1], Thomas Hempel[1], Annette Dietz[1], Pierre Lorenz[2], Stefan Krischok[2], Jürgen Alois Schaefer[2], Jürgen Christen[1], and Alois Krost[1]
[1]Institute of Experimental Physics, Otto-von-Guericke-University Magdeburg, Universitätsplatz 2, Magdeburg, 39106, Germany
[2]Institute of Micro- and Nanotechnologies and Institute of Physics, Technical University Ilmenau, Weimarerstr. 32, Ilmenau, 98684, Germany

ABSTRACT

In this work metalorganic chemical vapor phase epitaxy (MOVPE) growth and characterization of AlInN in the whole compositional range and the impact on the development of field effects transistors (FET) structures will be presented. Due to the large difference in the lattice parameters of the binaries AlN and InN the growth of AlInN with high indium concentrations is ambitious, and first the growth conditions for the alloy will be discussed. An experimental phase diagram and corresponding theoretical calculations will be displayed. The critical layer thickness of AlInN on GaN has been experimentally determined. Relaxation of the AlInN-layer has a strong influence on the sample morphology. At indium concentration exceeding 30% an polarization induced hole gas is expected at the AlInN/GaN interface from theoretical calculations, but no p-channel conductivity could be confirmed. The absence of the two dimensional hole gas will be discussed.

INTRODUCTION

Because of their wide band gaps, wurtzite-structure AlN (6.2eV) [1], and GaN (3.44eV) [2] are widely used for the fabrication of optoelectronic and electronic devices. Alloying with InN (0.68eV [3]) is used to tune the band gaps and lattice constants, but is limited for InGaN and AlInN due to the immiscibility of the binaries in a wide concentration range. In contrast to InGaN which is widely explored because of its importance for light emitters, basic properties of AlInN (band gap, miscibility, optical properties) are still unknown. Only $Al_{1-x}In_xN$ nearly lattice matched (LM) to GaN (x ~ 17.5 % [4]) has been investigated in detail, since it is an excellent material for DBRs and VCSELs [5] because of its high refractive index contrast to GaN. In addition LM-AlInN has been successfully applied for strain free AlInN/GaN FET structures [6] with the advantage of higher sheet carrier concentrations compared to conventional AlGaN/GaN FETs (FIG.1b) originating in a high spontaneous polarization of the alloy. MOVPE growth of AlInN is quite challenging because of the different growth conditions of the binaries AlN and InN: Growth of AlN requires a high temperature because of the low Al-adatoms mobility and a low V-III-ratio whereas InN requires low temperature and a high V-III-ratio due high indium desorption. In this work the miscibility of the alloy is determined for the whole compositional range. Additionally the critical layer thickness $d_{critical}$ for pseudomorphic growth of AlInN on GaN is presented. Caused by the wurtzite structure, the nitrides exhibit a polarization at heterointerfaces [7]. An n-channel AlInN/GaN FETs has already been realized, but a new approach is the realization of a p-channel FET which requires indium contents higher than 30% (FIG.1b).

EXPERIMENT

AlInN/GaN heterostructures were grown in an AIX200/4 RF-S MOVPE research reactor with standar precursors ($Al(CH_3)_3$, $Ga(CH_3)_3$, $In(CH_3)_3$, and NH_3). As carrier gas for the AlInN layers only nitrogen wa used. The sample structure is displayed in FIG.1a. The GaN-template was grown on Si(111) substrate wit an AlN-seed and a strain engineering low temperature AlN-interlayer. Details of the GaN-template growt on Si(111) can be found elsewhere [8]. High resolution X-ray diffraction (XRD) measurements around th (0002) reflection were performed using a Seifert 3003 diffractometer equipped with a Cu *K*α source, Bragg mirror, a fourfold Ge(220) monochromator, and a single Ge(220) analyzer crystal. Grazing incidenc measurements (GID) around the (10-10) reflection were taken using a Seifert URD6 diffractomet equipped with Cu *K*α source and a primary and secondary Soller collimator giving a resolution of 0.06 Electron micrograph images have been recorded with a Hitachi S4800 FESEM and 15kV – 25k accelerating voltage. The heterojunction band offsets have been measured by X-ray photoelectr spectroscopy (XPS) measurements with a monochromatic Al *K*α source (1486.7eV) and an energ resolution below 0.6eV (Ag($3d_{5/2}$) at 15eV pass energy). Hall-Effect measurements have been performed i van-der-Pauw geometry with a RH2010 Hall measurement system. For this reason ohmic indium contac where applied on the surface at 350°C.

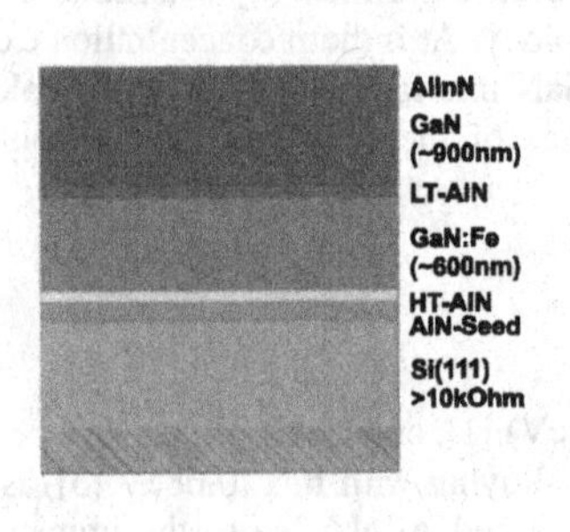

FIG.1a: Sample structure. The AlInN layer is grown on a GaN-template on Si(111). Various AlN-seed and strain engineering AlN interlayers are necessary for a high quality GaN-template.

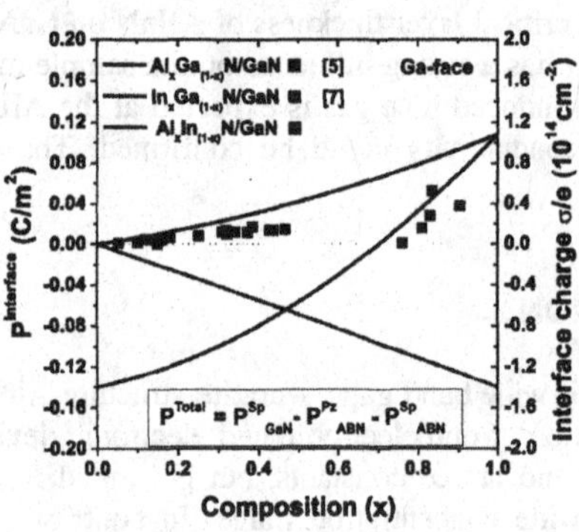

FIG.1b: Theoretical polarizations and th corresponding interface charge density at th interfaces of an ABN/GaN heterostructure [7 The rectangles represent experimental interfac charge densities measured [7,9].

DISCUSSION

Growth of AlInN is challenging: Linked to the enormous difference in the band gap of the binary crystal the bond energy of AlN (2.88eV) and InN (1.93eV) [10] show large variations. This results in contra growth conditions for the binaries AlN and InN. To guarantee a sufficient high Al-adatom mobility hig temperatures (> 1000°C) are necessary for step by step growth of AlN. The V-III-ratio has to be low prevent parasitic pre-reactions of TMAl. In contrary a low temperature and a high V-III-ratio is essential f the growth of InN, since the nitrogen partial pressure over InN increases steep above T = 630°C [11]. Th poor NH_3 pyrolysis at low temperatures [12] intensifies the need of a high NH_3 flow. For the growth AlInN the temperature is the most influential growth parameter, since the alloy composition is driven by th chemical potential. With decreasing temperature the indium content increases (FIG.2a). The indiu oncentration has been determined in all cases by XRD measurements of the c- and a- lattice parameters via e (0002) and (10-10) reflections, respectively. Both lattice parameters are necessary to calculate the orrect indium content. If the growth temperature is lower than 700°C the alloy is multi-phased (red ctangles connected with a line correspond to one sample).

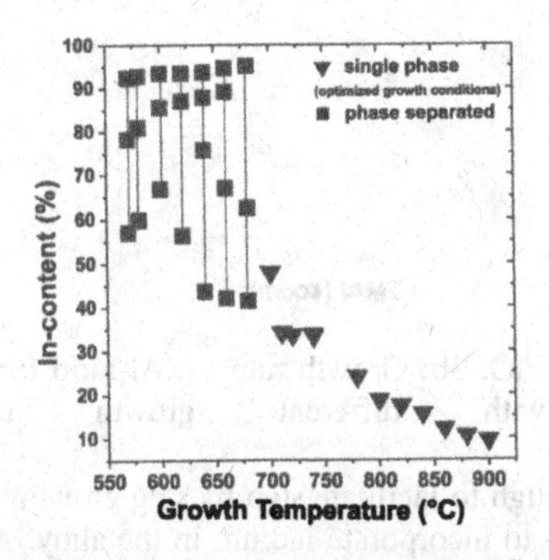

IG. 2a: In-content vs. growth temperature. With creasing temperature the In-content is ecreasing. Blue points mark single phase amples. All red points connected by a line orrespond to one sample which is phase eparated.

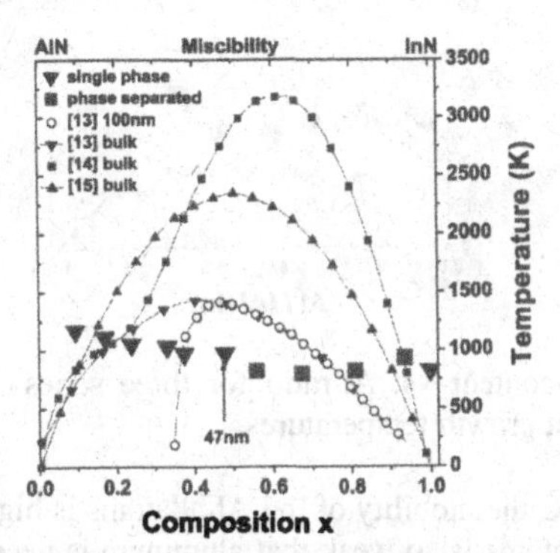

FIG. 2b: T-x phase diagram: Spinodal curves from Ref. 13,14,15 plotted as solid lines. The experimental data are plotted as blue rectangles for single phase and red rectangles for phase separated samples, respectively.

espite prognoses of theoretical calculations [13,14,15], $Al_{1-x}In_xN$ up to an indium content of 49% has been rown without phase separation (FIG.2b). The remarkable difference of experiment and theory can be xplained by the theoretical presumptions: Spinodal phase diagrams have been calculated for cubic material t equilibrium conditions (bulk & 100nm thick pseudomorphic to GaN). Figure 3a shows the indium content s. the Al/(Al+In) flow for three series grown at temperatures from 700°C to 820°C and Figure 3b the orresponding growth rates. Both series grown at 740°C and 820°C exhibit a similar growth behavior: The MAl flow does not alter the indium content but the growth rate, which increases linearly. Since the indium ontent depends only on the chemical potential the temperature determines the In-content. The growth ehavior of the 700°C series is diverse: With increasing TMAl-flow the In-content decreases linearly, hereas the growth rate is constant. Additionally field emission scanning electron microscope (FESEM) nages of the samples (not displayed here) reveal that the samples grown at 700°C have rough surfaces grains), whereas samples grown at T ≥ 740°C have smooth surfaces.

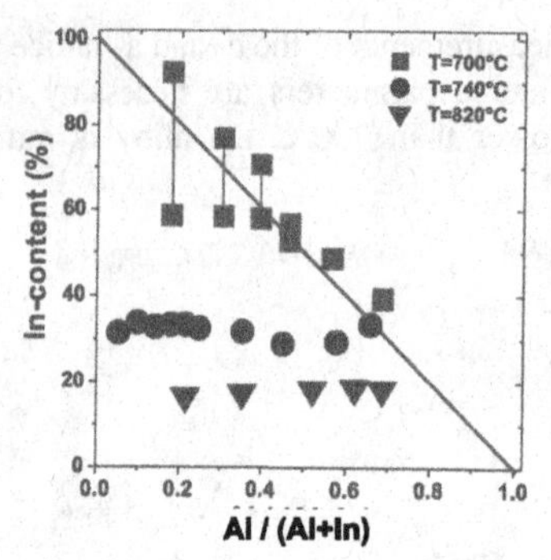

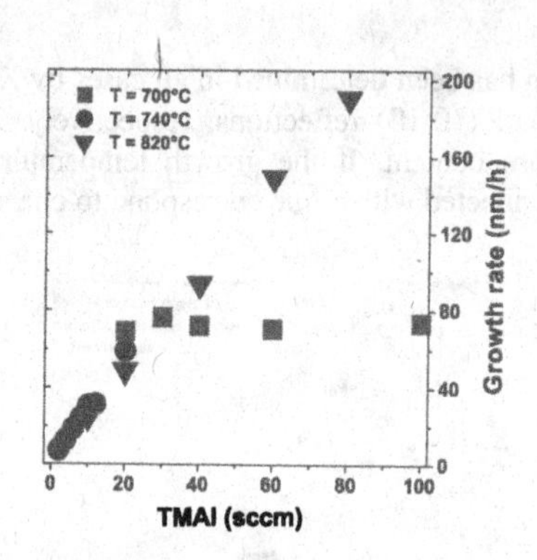

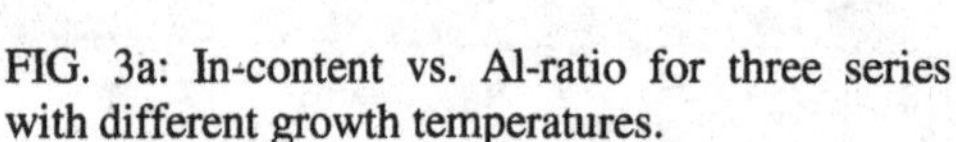

FIG. 3a: In-content vs. Al-ratio for three series with different growth temperatures.

FIG. 3b: Growth rate vs. Al-ratio for three seri with different growth temperature

At T ≥ 740°C the mobility of the Al-adatoms is high enough to facilitate step by step growth. The sticki of the In-adatoms is so weak that aluminum is necessary to incorporate indium in the alloy. At 700°C t aluminum mobility is too low and island growth is observed. Similar observations have been made for lo temperature AlN-layers on GaN [16]. The composition is only determined by proportion of the indium a aluminum flow to the reactor. Another important parameter for smooth surfaces is the critical lay thickness ($d_{critical}$) for pseudomorphic growth on GaN. To determine the strain state of the AlInN-layer G measurements have been performed (not shown here). If the $d_{critical}$ is exceeded, the surface of the samp roughens. In the electron micrographs of four samples (Fig. 4a) with different layer thickness, this behavi is visible: The fully strained sample A exhibits a very smooth surface. With increasing AlInN-lay thickness from sample B to D, relaxation starts and the surface roughens. The black spots in the FESE images are unidentified structural defects. Additionally a honeycomb like structure is visible which may caused by lateral phase separation [17]. To clarify the origin of the observed features TEM measuremen are in progress.

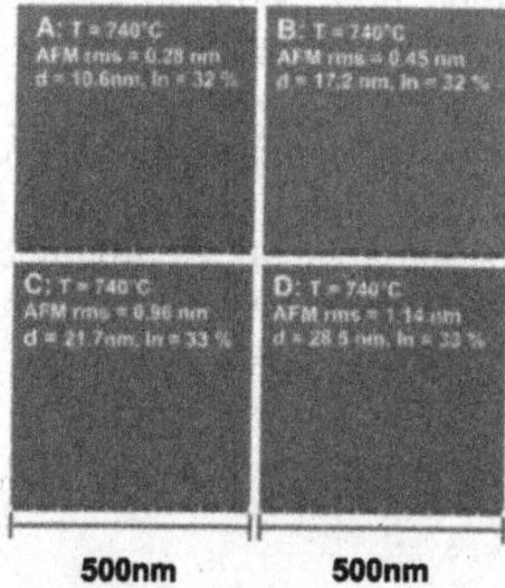

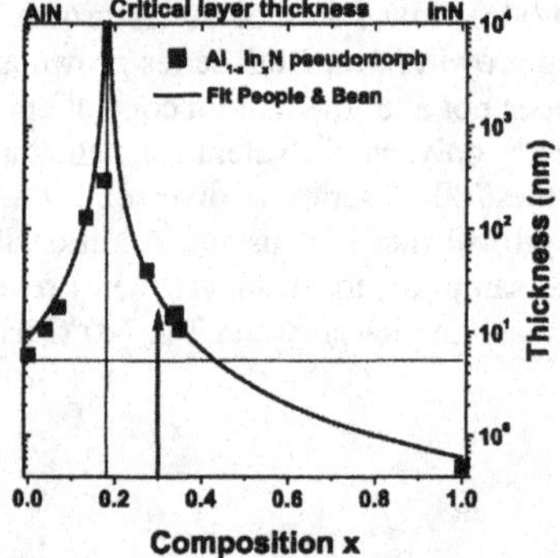

FIG. 4a: 30° to plan-view FESEM micrographs of four samples with different layer thicknesses. A roughening of the surface is visible with increasing layer thickness.

FIG. 4b: Critical layer thickness for pseudomorph growth of AlInN on GaN (blue line). Black poin correspond to experimental values and the red li is a fit [18].

he critical layer thickness has been fitted with the model of People and Bean [18], as seen in Fig. 4b. For e realization of a p-channel FET it is essential that $d_{critical}$ of $Al_{1-x}In_xN$ with x > 0.3 is thick enough since a rface depletion region has been found. This surface depletion region has been measured to be ~ 6nm for $l_{0.66}In_{0.34}N$ with sheet resistance measurements. Since $d_{critical}$ = 15nm for x = 0.34 all premises for a p-annel should be fulfilled. However, so far a 2DHG has not been found by Hall-Effect measurements at eudomorphic AlInN/GaN heterostructures with x = 0.34 and d = 14nm. The lack of the 2DHG cannot be plained by the polarization charges (Fig.1b), but by the band line up at the heterojunction. A type-I-terojunction with a negative valence band offset from AlInN to GaN is the prerequisite for the successful rmation of a p-channel at the interface (Fig. 5a). A type-II heterojunction does not work since carrier combination at the interface is likely to destroy the channel. To gain information about the band line up, e valence band offset (VBO) at the $Al_{1-x}In_xN$/GaN-interface has to be determined. For this purpose XPS easurements have been performed at a series of $Al_{1-x}In_xN$-layers with indium concentrations ranging from % to 33%. All AlInN-layers investigated where grown fully strained on the GaN. The XPS method to etermine the VBO has been described by Waldrop and Grant [19]. Details on these XPS measurements ill be presented elsewhere. Some of the results of the XPS measurements are displayed in Fig. 5b (black ctangles): A negative VBO has been found for all samples (valence band of $Al_{1-x}In_xN$ below the valence and of GaN).

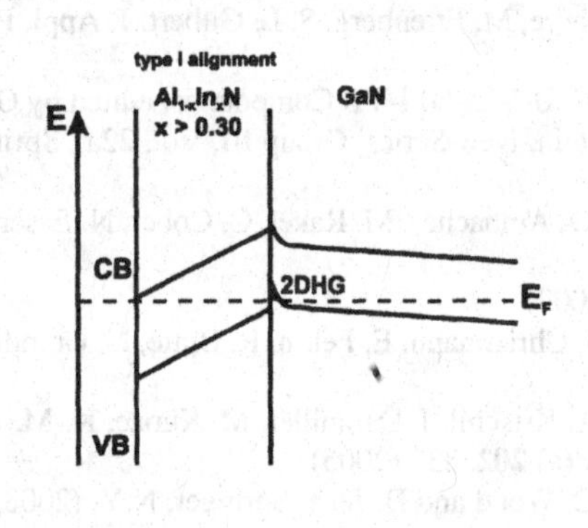

IG. 5a: Schematic band structure of AlInN/GaN eterostructure with a type-I alignment. The ormation of a 2DHG is expected at the interface.

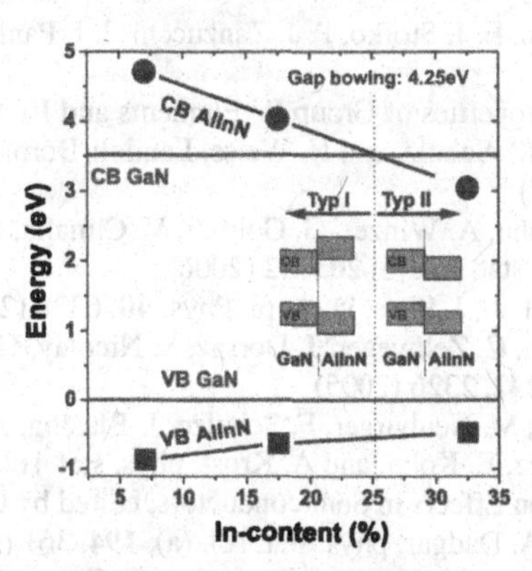

FIG. 5b: Band alignment at the AlInN/GaN interface. At ~ 25% a crossing from a type-I to a type-II heterojunction has been found

o complete the band lineup the band gap energy (E_{gap}) was added to the valence band. A bowing for E_{Gap} f 4.25eV was chosen [3]. As visible a change from a type-I to a type-II heterojunction occurs at x > 0.26. ence no n-channel exists for 0.26 < x < 0.3. This finding secondary explains the low sheet carrier density or the sample with x = 0.24 (Fig. 1b) [4]. Additionally the population of the p-channel for x > 0.3 is revented by the staggered band line up since the carriers recombine in an indirect transition at the interface. he indirect transition is under investigation but will be subject of another publication [20].

:ONCLUSIONS

he MOVPE growth conditions of AlInN in the whole compositional range have been explored. It has been emonstrated that the indium content of high quality AlInN is limited by the Al-adatom mobility and indium desorption. Two different growth modes have been identified for temperatures below and above 740°C. phase diagram was presented with single phase AlInN layers up to 49% InN content. Up to an indiu content of 34% smooth and pseudomorphic layers have been achieved. The critical layer thickness in range from 0 to 34% indium has been determined experimentally. If the critical layer thickness is exceed 3D growth is observed. At an indium concentration of 34% the critical layer thickness was identified to 14nm. No p-channel conductivity has been found at samples with this high indium content and lay thickness. The staggered band lineup at the $Al_{0.66}In_{0.34}N$/GaN has been identified as the most likely reas for the absence of a 2DHG. The p-channel at the interface is not populated since indirect recombination likely to occur cross the AlInN/GaN interface.

ACKNOWLEDGMENTS

The authors gratefully acknowledge the funding by the Deutsche Forschungsgemeinschaft under contract No. DA 466/6-2.

REFERENCES

1. W. M. Yim, E. J. Stofko, P. J. Zanzucchi, J. I. Pankove, M. Ettenberg, S. L. Gilbert, J. Appl. Phys. **44**, 292 (1973)
2. Intrinsic Properties of Group IV Elements and III-V, II-VI, and I-VII Compounds, edited by O. Madelung, M. Schulz, and H. Weiss, Landolt-Bornstein, New Series, Group III, Vol. 22a , Springer, Berlin, (1987)
3. R. Goldhahn, A. Winzer, G. Gobsch, V. Cimalla, O. Ambacher, M. Rakel, C. Cobet, N. Esser, H. Lu, W Schaff, phys. stat. sol. (a) **203**, 42 (2006)
4. R. Butté et al., J. Phys. D. Appl. Phys. **40**, 6328 (2007)
5. J.-F. Carlin, C. Zellweger, J. Dorsaz, S. Nicolay, G. Christmann, E. Feltin, R. Butté, N. Grandjean, phys. stat. sol. (b) **242**, 2326 (2005)
6. A. Dadgar, M. Neuburger, F. Schulze, J. Bläsing, A. Krtschil, I. Daumiller, M. Kunze, K.-M. Günther, H Witte, A. Diez, E. Kohn, and A. Krost, phys. stat. sol. (a) **202**, 832 (2005)
7. Polarization Effects in Semiconductors, edited by C. Wood and D. Jena, Springer, N.Y. (2008)
8. A. Krost, A. Dadgar, phys. stat. sol. (a), **194**, 361 (2002)
9. M. Neuburger, M. Kunze, I. Daumiller, T. Zimmermann, G. Koley, M. G. Spencer, A. Dadgar, A. Krtschil, A. Krost, and E. Kohn, Electronics Letters, **39**, 1614 (2003)
10. J.H. Edgar (Ed.), „Properties of group-III nitrides", INSPEC (1994)
11. O. Ambacher ,M. S. Brandt, R. Dimitrov, T. Metzger, M. Stutzmann, R. A. Fischer, A. Miehr, A. Bergmaier, G. Dollinger, J. Vac. Sci. Technol. B **14**, 3532 (1996)
12. M. Kamp, M. Mayer, A. Pelzmann, K.J. Ebeling, Mat. Res. Soc. Symp. Proc. **449**, 161 (1997)
13. V.G. Deĭbuk , A.V. Voznyĭ, Semiconductors, **39**, 623 (2005)
14. S. Y. Karpov, N. Podolskaya, I.A. Zhmakin, A.I. Zhmakin, Phys. Rev. B **70**, 235203 (2004)
15. M. Ferhat, F. Bechstedt, Phys. Rev. B **65**, 075213 (2002)
16. S. Keller, S.P. DenBaars, J. Crystal Growth **248**, 479 (2002)
17. L. Zhou, D.J. Smith, M.R. McCartney, D.S. Katzer, D.F. Storm, Appl. Phys. Lett. **90**, 081917 (2007)
18. R. People, J. Bean., Appl. Phys. Lett. **47**, 322 (1985)
19. J.R. Waldrop, R.W. Grant, Appl. Phys. Lett. **68**, 2879 (1996)
20. C. Hums, B. Bastek, F. Bertram, A. Franke, A. Dadgar, J. Bläsing, J. Christen, A. Krost (to be published)

Mater. Res. Soc. Symp. Proc. Vol. 1068 © 2008 Materials Research Society 1068-C04-04

Growth of AlGaN/GaN HEMTs on Silicon Substrates by MBE

Fabrice Semond[1], Yvon Cordier[1], Franck Natali[1], Arnaud Le Louarn[1], Stéphane Vézian[1], Sylvain Joblot[1,2], Sébastien Chenot[1], Nicolas Baron[1,3], Eric Frayssinet[1], Jean-Christophe Moreno[1,2], and Jean Massies[1]

[1]CRHEA, CNRS, rue Bernard Gregory, Valbonne, 06560, France

[2]STMicroelectronics, Crolles, France

[3]Picogiga International, Courtaboeuf, France

ABSTRACT

During the last ten years, we have developed an efficient growth process of nitrides on silicon substrates by molecular beam epitaxy. In collaboration with partners AlGaN/GaN HEMTs on Si having promising performances have been fabricated. Focusing on the growth aspect and underlying some of the key issues, we present in this paper an overview of our contribution in the field of AlGaN/GaN HEMTs on Si substrates.

INTRODUCTION

Since the first demonstration of the existence of a two-dimensional electron gas (2DEG) at an AlGaN/GaN interface in 1992 [1], tremendous progress has been realized in the field of AlGaN/GaN HEMTs. Nowadays AlGaN/GaN HEMTs are seriously considered for next generation microwave power transistors due to their high voltage and high power density capabilities [2]. Although AlGaN/GaN HEMTs are mostly grown on SiC substrates for thermal management issues, growth on lower cost Si substrates has received a lot of attention in the past few years and exciting results have been demonstrated [3-21]. Since the first AlGaN/GaN HEMT on silicon has been reported by Chumbes *et al.* in 1999 [3], it was early recognized that 2DEG mobilities as high as those reported on SiC or sapphire could be achieved on Si [4-6]. During the last years, DC and RF performances have been continuously improved and the capability of AlGaN/GaN HEMTs fabricated on silicon for microwave power applications has been demonstrated [5-19]. More recently, targeting the use of AlGaN/GaN HEMTs on Si for low-cost high efficiency switching application, very promising results have also been reported [20,21].

GROWTH

The AlN/Si(111) interface

As far as the growth of group-III nitrides in their stable wurtzite crystallographic form is concerned, the (111) surface orientation of Si with its hexagonal symmetry is used. While many different materials have been proposed as a nucleation layer, AlN layers are mostly used. In the early days, it was thought that a good AlN/Si interface could be very difficult to build up. Several potential problems were expected like the Si surface nitridation, the Al diffusion into the substrate and the Si out-diffusion from the substrate into the nitride epilayer. Surprisingly, it turned out that well defined AlN/Si(111) interfaces were reported by many groups using many different nucleation schemes. Rather sharp interfaces with no or almost no amorphous nitridation interlayer and a good epitaxial relationship are mostly reported (Fig. 1). Also, despite the fact that silicon is a non-polar substrate, only one polarity is usually obtained (the Al-polarity). Clearly this is the first reason of the success of nitride growth on Si and this is due to the strongest binding energy of Al-N (11.5 eV) compared to Si-N (3.5 eV) bonds. As a consequence, AlN nucleation takes place at the expense of the SiN interlayer. Using MBE, we followed in-situ by RHEED surface reconstruction changes during the very first atom exposures [5,22,23]. We found out that even when the silicon surface is firstly covered by SiN then a subsequent Al exposure reduces the SiN layer and a sharp AlN-Si interface is formed. Usually the SiN layer formed on Si(111) is very thin (unless the Si surface is intentionally exposed to NH_3 at high temperature and for a long time) and crystalline as checked by RHEED (Fig. 1). On the other hand, exposing the clean silicon surface to Al leads to the formation of a (Al-Si) surface alloy which appears to be very detrimental, especially in terms of strain in the subsequent grown layer [22,23]. Based on these observations, starting from a well developed Si(111)-7x7 surface reconstruction (top RHEED diagram, Fig. 1), our optimized process consists first, in a NH_3 exposure (very low amount) in order to intentionally form a very thin crystalline SiN layer (middle RHEED diagram, Fig. 1). As shown in Fig.1 (bottom RHEED diagram) about one monolayer of Al is sufficient in order to convert the crystalline SiN layer into AlN. Then growth of the AlN nucleation layer is initiated by opening simultaneously Al and NH_3 shutters increasing the temperature from 650 to 900°C. Typically, the growth rate is 0.1 µm/h and the buffer layer thickness is about 40 nm. From the RHEED pattern one can observe that both the structural quality and the surface morphology improve as a function of the AlN nucleation layer thickness. At the end of the buffer layer growth, the RHEED pattern is streaky and decreasing the temperature an intense 2x2 reconstruction (under NH_3) is seen indicating a smooth Al-polarity surface. Actually the layer consists of grains about 20-40 nm in diameter meaning that the dislocation density is huge.

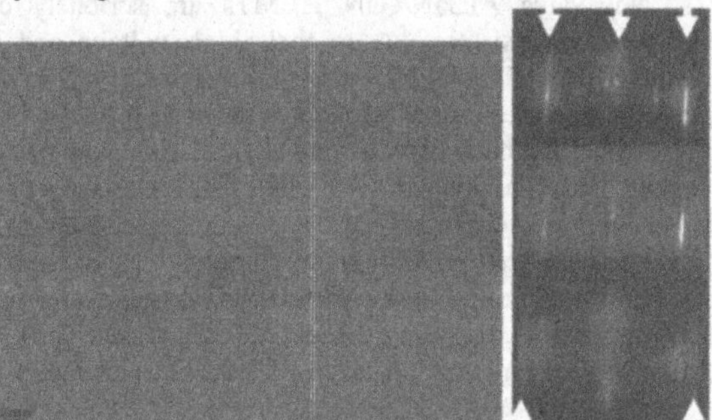

Figure 1. Left, HRTEM cross-section image of a typical AlN/Si(111) interface. Right, RHEED patterns -(Top) The clean Si(111)-7x7 surface reconstruction. -(Middle) after a shallow nitridation showing a faint 8x8-like surface reconstruction. -(Bottom) after a first Al monolayer deposition, both the pattern of the Si substrate and the AlN fractional layer are seen.

GaN growth, strain engineering and dislocation reduction

Unfortunately, growing GaN films on top of a single AlN buffer layer results in highly dislocated films because most of the dislocations inside the AlN buffer layer ($\geq 10^{11}$ cm^{-2}) are propagating vertically through the AlN/GaN interface. Another key issue dealing with the growth of group-III nitrides on silicon substrates is the cracking problem. Due to the very high thermal expansion coefficient mismatch between Si and GaN, epitaxial layers are tensely strained upon the post growth cooling and above a critical thickness the tensile stress accumulated is so high that GaN layers have a strong tendency to crack. Typically, on top of a single AlN buffer layer, one micron thick GaN layers are cracked. Therefore, new strategies have been developed in order to improve the layer quality and to prevent cracking. We choose to take advantage of the 2.5% lattice mismatch between AlN and GaN to compressively stress GaN layers during the growth. Of course relaxation occurs along the GaN growth but full relaxation is never reached. Actually, we found out that, using proper growth conditions and starting with a fully relaxed AlN thick layer, the amount of residual compressive strain in GaN layers could be high enough to noticeably mitigate the thermal mismatch strain upon cool down and then prevent cracking [4,5,24]. The amount of residual compressive strain in GaN layers depends on several parameters. Most importantly the number of pre-existing dislocations in the underlying AlN layer affects significantly the relaxation profile [25,26]. The lower the pre-existing dislocation density in the AlN layer is, the slower the GaN relaxation. This explains why we have developed the layer structure shown Fig. 2 [5,27], actually the second thick AlN layer has a better quality than the AlN nucleation layer and should be fully relaxed at the growth temperature. Also, the growth mode related to both the growth temperature and the V/III flux ratio is going to affects strongly the relaxation profile. While a 3D growth mode leads to a faster relaxation, a 2D growth mode slows down the relaxation mechanism [24]. So to prevent cracking, it is absolutely mandatory that growth parameters are adjusted in order to get a 2D growth mode all along the growth. To find out how looks like the relaxation profile of the GaN layer grown above the AlN thick layer we followed in-situ by RHEED the evolution of the surface lattice parameter as a function of the GaN thickness [28] (Fig. 3).

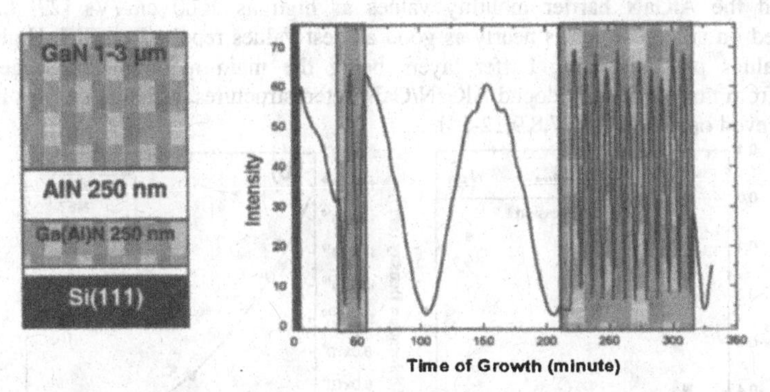

Figure 2. Left- Schematic diagram of the layer structure developed, Right- In-situ reflectometry recorded along the growth showing the different growth sequences (λ=640 nm).

At the growth temperature, it is shown on the relaxation profile that most of the strain is released in the first 500 nm and then the profile begins to saturate and a significant amount of residual compressive strain is still present for GaN up to 2-3 µm. We are still currently studying influences of various parameters on the relaxation profile and interesting results have been obtained recently which are very promising in order to further extend the cracking limit [26,29]. Now looking at the defect density using our growth scheme, we observed that the dislocation density decreases exponentially as a function of the GaN film thickness [24,28]. Actually, the dislocation density is typically around 2 to $5x10^9$ cm^{-2} for crack-free film thicknesses ranging in between 2-3 μm.

AlGaN/GaN HETEROSTRUCTURES

We are now going to see why it has been so successful to fabricate and to study AlGaN/GaN HEMTs heterostructures on Si. As soon as we started to grow unintentionally doped AlGaN/GaN heterostructures on Si [4], reproducible high mobility values ranging in between 1400 and 1700 cm^2/Vs were measured at RT [4]. Interestingly, we found out at that time that it was more difficult to achieve such high mobility values for similar quality unintentionally doped AlGaN/GaN heterostructures grown on Al_2O_3. Actually, in GaN on Al_2O_3 the residual free carrier concentration deduced from capacitance voltage measurements was 2 to 3 orders of magnitude higher than the onc in GaN on Si. Due to a parallel conduction path in the buffer, the Hall mobility values measured at RT on structures grown on sapphire were always much lower. Surprisingly, for unintentionally doped GaN on Si, the background free carrier concentration is typically in between 1 to $5x10^{14}$ cm^{-3} (Fig. 4). The reason for that is still not clear. The high dislocation density is usually believed to be responsible for this low free carrier concentration through a compensation mechanism but GaN layers grown on sapphire having dislocation densities in the same range exhibit significantly higher free carrier concentrations. Anyway, after the surprisingly good quality of the AlN/Si(111) interface, the lower residual free carrier concentration measured in GaN layers grown on Si is clearly the second reason of the success to grow AlGaN/GaN HEMTs on silicon. Meanwhile, using an AlN spacer layer in between the channel and the AlGaN barrier mobility values as high as 2000 cm^2/Vs [29] have been demonstrated on silicon which is nearly as good as best values reported on SiC. High electron mobility values plus insulating buffer layers being the main requirements to get HEMT transistors from unintentionally doped AlGaN/GaN heterostructures, efficient devices have been rapidly achieved on silicon [4,5,7,8,9,12-14].

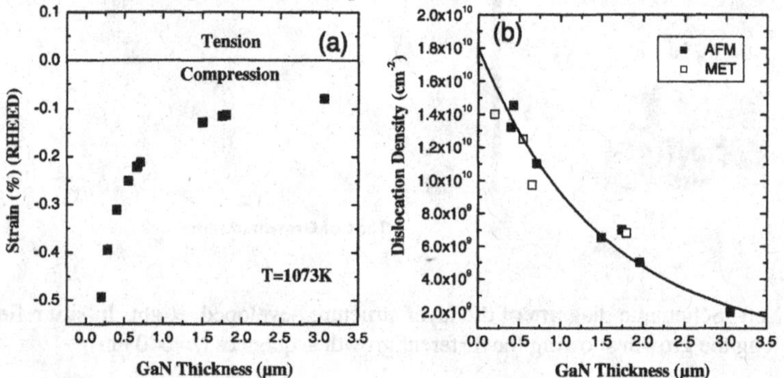

Figure 3. a) Residual strain vs GaN thickness measured at the growth temperature. b) Total threading dislocation density vs GaN thickness measured by AFM and TEM.

In collaboration with both IEMN and TRT-THALES, our first AlGaN/GaN transistors on reisitive Si were fabricated and studied in 2001 [5,7]. Already it was reported that the high sheet carrier density into the channel and the insulating buffer layer result in high static power capability (1 W/mm at 4 GHz [8]) in the devices, combined with high frequency performances. The absence of charge coupling with the substrate was also noticed and a f_{max}/f_t ratio as high as 2.35 was obtained indicating that devices were suitable for RF power applications. High linearity as well as low low-frequency noise performances were also reported [9,11,12]. Then the high potential at microwave frequencies of AlGaN/GaN HEMTs on highly resistive Si substrate for power applications has been demonstrated [12,16,13]. For instance, in collaboration with Daimler-Chrysler, an output power density of 6.6 W/mm at 2 GHz was achieved in 2003 [14]. Since then several groups working on AlGaN/GaN epilayers on Si provided by Picogiga International (Picogiga is basically using our patented growth process [27] which has been exclusively licensed to Picogiga in 2004 [30]), have reported output power density up to 10 GHz, and a value of 7 W/mm have been obtained and promising RF reliability performances were reported [15]. More recently in a joint work between IEMN-THALES (TIGER) and Picogiga, an output power density of 5.1 W/mm at 18 GHz has been reported reinforcing clearly the good potentialities of AlGaN/GaN HEMTs fabricated on Si [18].

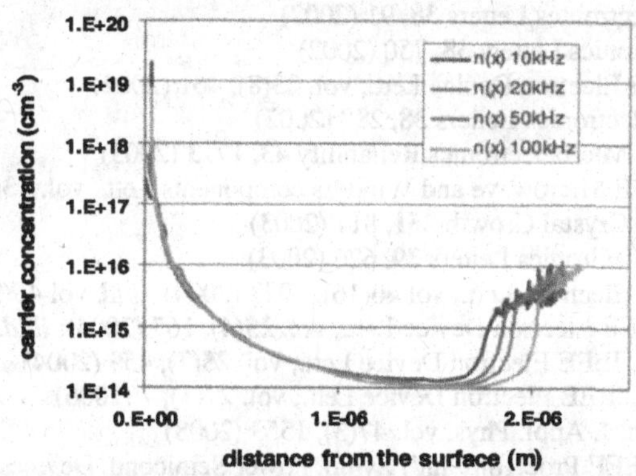

Figure 4. Typical carrier concentration profile of AlGaN/GaN heterostructures grown on Si. Profile deduced from C-V measurements with a Hg probe operating at 10, 20, 50 and 100 KHz.

CONCLUSIONS

Since now almost ten years, we have intensively investigated the growth by MBE of AlGaN/GaN HEMTs on Si(111) substrates. An original buffer layer stack has been developed in order to prevent cracking and to improve the quality of GaN layers [24]. Also, the AlN/Si(111) interface formation has been thoroughly studied and despite the fact that chemical bonds are energetically favorable to form an abrupt good quality interface, we found out that the nucleation step is going to have a significant impact on the quality and the strain of the final heterostructure [22]. Thanks to a very fruitful collaborative work with people involved in device processing, AlGaN/GaN HEMTs on Si have shown excellent DC and RF performances demonstrating the feasibility of those heterostructures for microwave power applications. However there are still some issues like the reduction of the dislocation density, the relation between threading dislocations and the electrical isolation of buffers, the electrical isolation as function of the temperature and so on. We are currently investigating these issues [31] and promising results are still to come.

More recently, addressing the integration issue we have focused our attention on the Si(100) surface orientation [33-34] and amazingly very promising DC and RF performances have also been obtained. Considering this work and other latest developments [19-21,26,29], GaN on Si will certainly plays an important role in the production of future electronic devices.

REFERENCES

1. M. Asif Khan *et al.*, Appl. Phys. Lett. **60**, 3027 (1992)
2. U. K. Mishra *et al.*, IEEE Proceedings **96**(2), 287 (2008)
3. E. M. Chumbes *et al.*, IEEE Proceedings of Int. Electron Devices Meeting, p. 397 (1999); A.T. Schremer *et al.*, Appl. Phys. Lett. **76**, 736 (2000); E.M. Chumbes *et al.*, IEEE Transactions on Electron Devices, **48**(3), 420 (2001)
4. F. Semond *et al.*, Appl. Phys. Lett. **78**, 335 (2001)
5. F. Semond *et al.*, Phys. Stat. Sol.(a) **188**, 501 (2001)
6. J. D. Brown *et al.*, Solid-State Electron. **46**, 1535 (2002)
7. Y. Cordier *et al.*, Electronics Letters **38**, 91 (2002)
8. V. Hoël *et al.*, Electronics Letters **38**, 750 (2002)
9. N. Velas *et al.*, IEEE Electron Device Lett., vol. **23**(8), 461 (2002)
10. P. Javorka *et al.*, Electronics Letters **38**, 288 (2002)
11. A. Curutchet *et al.*, Microelectronics Reliability **43**, 1713 (2003)
12. N. Velas *et al.*, IEEE Microwave and Wireless components Lett., vol. **13**(3), 99 (2003)
13. Y. Cordier *et al.*, J. Crystal Growth **251**, 811 (2003)
14. R. Behtash *et al.*, Electronics Letters **39**, 626 (2003)
15. D.C. Dumka *et al.*, Electron. Lett., vol **40**(16), 1023 (2004); *ibid.* vol **40**(24), (2004)
16. A. Minko *et al.*, IEEE Electron Device Lett., vol. **25**(4), 167 (2004); *ibid.* vol. **25**(7), 453 (2004)
17. J.W. Johnson *et al.*, IEEE Electron Device Lett., vol. **25**(7), 459 (2004)
18. D. Ducatteau *et al.*, IEEE Electron Device Lett., vol. 27(1), 7 (2006)
19. K. Cheng *et al.*, Jpn. J. Appl. Phys. vol. 47(3), 1553 (2008)
20. S. Yoshida *et al.*, IEEE Proc.18th Int'l Symp. Power Semicond. Devices and IC's, p.4 (2006)
21. S. Iwakami *et al.*, Jpn. J. Appl. Phys. vol. 46(24), L587 (2007)
22. A. Le Louarn, PhD thesis Université Nice Sophia-Antipolis, CRHEA-CNRS (2006)
23. S. Vézian *et al.*, J. Crystal Growth **303**, 419 (2007)
24. F. Natali, PhD thesis Université Nice Sophia-Antipolis, CRHEA-CNRS (2003)
25. R. Langer *et al.*, J. Crystal Growth **205**, 31 (1999)
26. E. Frayssinet *et al.*, to be published
27. Patent filed by CRHEA-CNRS, PTC/FR 01/01777, June 2001
28. S. Vézian *et al.*, Phys. Rev. B **69**, 125329 (2004)
29. N.Baron *et al.*, MRS 2008, Symposium C
30. http://www.soitec.com/picogiga/research-development/
31. N. Baron, PhD thesis Université Nice Sophia-Antipolis, CRHEA-CNRS (2009)
32. S. Joblot, PhD thesis Université Nice Sophia-Antipolis, CRHEA-CNRS (2007)
33. S. Joblot *et al.*, Electronics Letters 42, 117 (2006)
34. S. Joblot *et al.*, Superlattices and microstructures 40, 295-299 (2006)

Mater. Res. Soc. Symp. Proc. Vol. 1068 © 2008 Materials Research Society 1068-C04-05

Growth of AlGaN/GaN HEMTs on 3C-SiC/Si(111) Substrates

Yvon Cordier[1], Marc Portail[1], Sébastien Chenot[1], Olivier Tottereau[1], Marcin Zielinski[2], and Thierry Chassagne[2]

[1]CNRS-CRHEA, rue Bernard Gregory, Sophia Antipolis, Valbonne, 06560, France

[2]Novasic, Savoie Technolac, Arche Bât. 4, BP 267, Le Bourget du Lac, 73375, France

ABSTRACT

In this work, we study cubic SiC/Si (111) templates as an alternative for growing GaN on silicon. We first developed the epitaxial growth of 3C-SiC films on 50mm Si(111) substrates using chemical vapor deposition. Then, AlGaN/GaN high electron mobility transistors were grown by molecular beam epitaxy on these templates. Both the structural quality and the behavior of transistors realized on these structures show the feasibility of this approach.

INTRODUCTION

Gallium Nitride (GaN) based epitaxial films are widely used in light emission applications, but other applications like high power electronics benefit from the electron transport properties and the high critical electrical field of this wide band gap material. If high frequency electronics are mainly based on AlGaN/GaN High Electron Mobility Transistor heterostructures (HEMT) grown on highly resistive or semi-insulating substrates (silicon carbide (SiC) [1,2], silicon (Si) [3,4], Sapphire, free-standing GaN [5]), applications like power switching often rely on vertical structures grown on conductive substrates. Both of these applications necessitate to sustain high operating voltages and to dissipate the heat generated under high current density. For these reasons, hexagonal silicon carbide (4H-SiC or 6H-SiC) is often preferred due to a superior thermal conductivity as compared to sapphire or silicon substrates. Nevertheless, silicon is inexpensive and available in large size, high volume and the realization of GaN based high performance RF devices has been demonstrated on such a substrate. However, the high reactivity of silicon with the different compounds frequently used for the growth of nitrides (Ga, Al, H_2, NH_3...) makes the substrate preparation and the nucleation more delicate [6] than on a substrate like SiC, and GaN on Si suffers from a risk of crack generation due to the tensile stress induced by the large lattice mismatch (17%) and thermal expansion coefficient difference between GaN and Si.

Cubic SiC (3C-SiC) presents a closer lattice mismatch with GaN (3% for the hexagonal phase of GaN) and a thermal expansion coefficient of 4.5×10^{-6} K^{-1} intermediate between the ones of GaN 5.6×10^{-6} K^{-1} and Si (3.6×10^{-6} K^{-1}). (001) oriented self supported 3C-SiC substrates have been used for growing cubic GaN based heterostructures [7]. On the other hand, (111) oriented cubic SiC can be employed as an intermediate layer for the growth of hexagonal phase GaN on Si. The suppression of crack generation in GaN grown on 3C-SiC/Si [8,9] as well as an enhancement of GaN crystal quality as compared with similar structures grown on Si [10,11] have been reported.

Schottky diodes with a reverse breakdown voltage exceeding -250V have been realized on a GaN layer grown on 1 µm thick 3C-SiC intermediate layer on Si [8]. In the present work, our aim is to show the feasibility of field effect power devices like AlGaN/GaN HEMTs on such an alternative substrate.

EXPERIMENT

In this work, up to 2.2 µm thick 3C-SiC films were grown on 2 in. nominal Si(111) substrates in a resistively heated hot wall chemical vapor deposition (CVD) reactor [12]. A classical two stages process was used to deposit 3C-SiC films on silicon substrates [13]. The first carbonization stage was achieved with propane at 1100°C and was immediately followed by the growth step at 1350°C with silane and propane as precursors. Hydrogen was used as vector gas. Prior to the growth of GaN based structures, the 3C-SiC/Si substrates were out-gassed under high vacuum in the molecular beam epitaxy (MBE) chamber until a clear 3x1 reflection high-energy diffraction pattern was observed at 940°C. The regrown structure was exactly the same as the ones typically grown on Si(111) [14], the only difference being the nucleation conditions of the 42 nm thick AlN nucleation layer (920°C in the present case). Ammonia was used as nitrogen precursor. After growing a 0.25 µm GaN / 0.25 µm AlN stack, a 1.7 µm thick GaN buffer was grown before the realization of the HEMT structure. The HEMT active layers consisted in 1 nm AlN / 21 nm AlGaN barriers capped with 5 nm GaN layers. In particular, two HEMTs structures were studied with Al contents in the AlGaN barrier of 25% and 29% on 0.85µm and 0.6µm thick SiC/Si(111) templates respectively. The crystal quality of the SiC and III-Nitride films was studied using X-ray diffraction technique. rocking curves were recorded for (111) and (113) reflection planes of SiC, and (002) and (302) reflection planes of GaN. The residual stress was evaluated by measuring the 3C-SiC/Si(111) substrate bowing with a profilometer and by measuring the strain sensitive low-temperature photoluminescence (PL) energy of GaN. Scanning electron microscopy (SEM) and tapping-mode atomic force microscopy (AFM) were used to study surface morphology. The charge control inside the two-dimensional electron gas (2DEG) in the HEMT structures was studied using Capacitance-Voltage measurements at frequencies ranging from 1 kHz to 100 kHz, with a Mercury probe or evaporated NiAu Schottky contacts. Hall effect devices and field effect transistors were realized to assess the electrical quality of the samples using a simple 3 level photolithographic process. Device isolation was achieved using Cl_2/Ar reactive ion etching. Ohmic contacts with a specific resistance of 0.8 Ω.mm were obtained after deposition of Ti/Al/Ni/Au metals and annealing at 730°C. A Ni/Au sequence was evaporated for Schottky gates and devices were not passivated. The buffer resistivity and the current-voltage output characteristics were evaluated for transistors with 3 µm x 150 µm nominal gates in 12 µm source-drain spacings.

RESULTS AND DISCUSSION

Structural quality of 3C-SiC and GaN

X-ray diffraction ϕ scan precession around asymmetric reflection lines of Si and 3C-SiC confirms the epitaxial relationships between SiC and Si: (111) 3C-SiC // (111) Si and {110} 3C-SiC // {110} Si. Figure 1 shows the full width at half of maximum (FWHM) of X-ray diffraction rocking curves for (111) 3C-SiC and (002) GaN reflections: a trend appears with the film thickness. The quality of the SiC film is enhanced with its thickness; the widths of rocking curves range from 2600 arcsec (0.72°) for 0.5 μm SiC films down to 1150 arcsec (0.32°) for 2.2 μm thick films. Nevertheless, the lattice and thermal mismatch leads to further wafer bowing and generation of cracks in films thicker than 1 μm (but thicker crack free films can be obtained using 4° off silicon substrates). Scanning electron microscopy and atomic force microscopy reveal that the surface of the films is characterized by well coalesced three fold symmetric grains (Fig.2-left). A resulting rms roughness of about 10 nm is measured by AFM. AlGaN/GaN layers exhibit the typical morphology of thick GaN layers grown by molecular beam epitaxy using ammonia [15]. The rms roughness of 5 nm and the disappearing of the underlying SiC pattern attest the quality of GaN regrowth (Fig.2-right). A power density as high as 6.6 W/mm at 2GHz has been demonstrated with GaN on silicon having such a roughness [4]. X-ray diffraction ϕ scans around (302) asymmetric reflection line of GaN confirm the epitaxial relationships: (0001) GaN // (111) 3C-SiC and {100} GaN // {110} 3C-SiC [10]. Widths of 1040 and 987 arcsec are measured for the (002) reflection line of GaN grown on 0.6μm and 0.85μm thick 3C-SiC on Si(111) while 700 arcsec is typically obtained on bulk Si(111). Furthermore, the width of (302) GaN lines and (113) SiC lines are in the range of 2210-2508 arcsec and 1598-1864 arcsec respectively while the (302) GaN line width on bulk Si(111) is typically 1900 arcsec. The widths of GaN X-ray lines correlate with the ones of SiC lines and are superior to those measured on bulk Si(111) but the GaN threading dislocation density measured by plan-view transmission electron microscopy is below $5x10^9$ cm^{-2}, that is very similar to the one measured in GaN structures we have grown on Si(111). This indicates that the broadening is more due to a memory effect of the tilt of 3C-SiC crystal grains (X-ray line width superior to 1800 arcsec) rather than induced by additional defects in GaN. Indeed, the dislocation density is effectively reduced within the GaN buffer grown with a low relaxation rate [14], but X-ray diffraction probes the entire structure, especially this GaN layer with a significant resulting strain gradient that is responsible for the broadening of the lines.

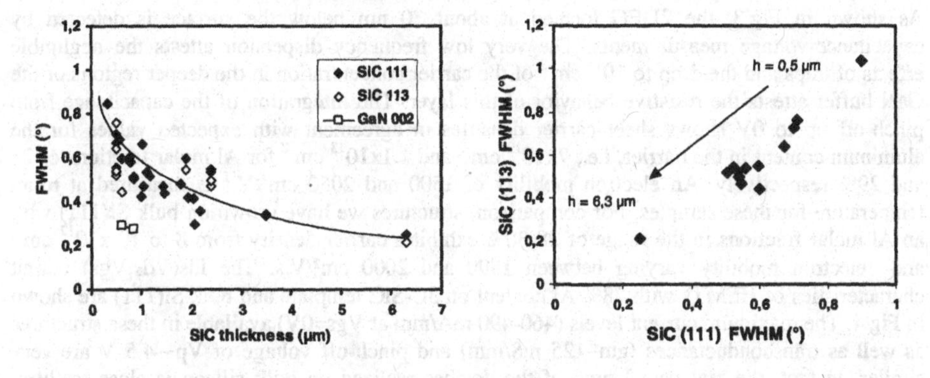

Fig. 1: Dependence of the full width at half of maximum (FWHM) of the 3C-SiC(111), (113) and GaN (002) X-ray reflection line with the thickness of the film deposited on Si(111).

Furthermore, due to the lower thermal mismatch between GaN and the 3C-SiC intermediate layer, no concave additional bowing of the 3C-SiC/Si substrate was noticed after the growth of the GaN structure. This is confirmed by the low-temperature (10K) photoluminescence energy of GaN E=3.471 eV for both structures that is comparable to the ones of our best GaN-on-silicon layers which have a bow indicating a significant residual compressive strain and a PL energy ranging from E=3.468 eV to E=3.473eV.

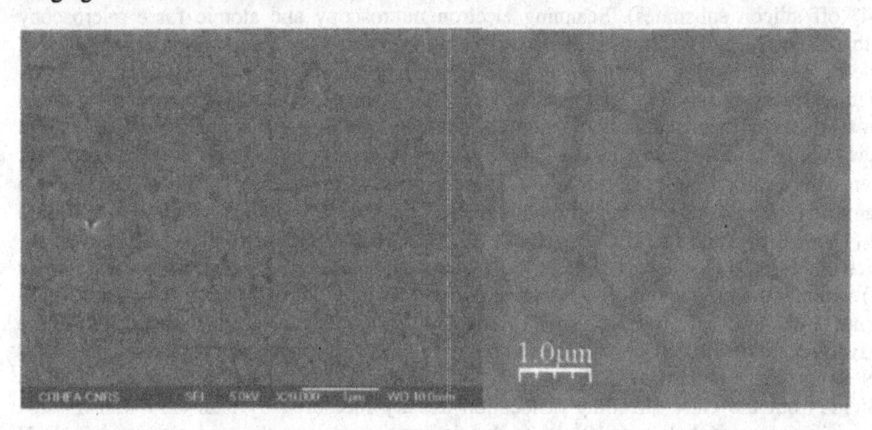

Fig. 2: Scanning electron micro-graph of the surface of a 0.5 μm thick 3C-SiC film (left) and AFM view of the surface of a GaN based HEMT grown on 3C-SiC/Si substrate (right)

Electrical characterizations of AlGaN/GaN HEMTs

The Gallium polarity orientation of the GaN films is confirmed by the presence of a two dimensional electron gas (2DEG) at the interface between the barrier and the GaN channel [16]. As shown in Fig.3, the 2DEG located at about 30 nm below the surface is detected by capacitance-voltage measurements. The very low frequency dispersion attests the negligible effects of traps and the drop to 10^{14} cm^{-3} of the carrier concentration in the deeper regions of the GaN buffer attests the resistive behavior of this layer. The integration of the capacitance from pinch-off up to 0V shows sheet carrier densities in agreement with expected values for the aluminum content in the barrier, i.e., $7x10^{12}$ cm^{-2} and $1.1x10^{13}$ cm^{-2} for Al molar fractions of 25 and 29% respectively. An electron mobility of 1500 and 2050 $cm^2/V.s$ is measured at room temperature for these samples. For comparison, structures we have grown on bulk Si(111) with an Al molar fractions in the range of 28-30% exhibit a carrier density from 8 to 10 $x10^{12}$ cm^{-2} and electron mobility varying between 1590 and 2000 cm²/V.s. The Ids(Vds,Vgs) output characteristics of HEMTs with 28% Al content on 3C-SiC template and bulk Si(111) are shown in Fig.4. The maximum current levels (460-490 mA/mm at Vgs=0V) available in these structures as well as transconductances (gm~125 mS/mm) and pinch-off voltage of Vp~-4.5 V are very similar. In fact, the real gate length of the devices realized on bulk silicon is close to 2μm, whereas it is close to 4 μm on 3C-SiC template, so that one can conclude that the electrical behavior on 3C-SiC template is slightly superior. Furthermore, the buffer leakage currents between isolated devices or measured at pinch-off of the transistors are in the same range (10-100 μA/mm at Vds=10V) for both kind of substrates which attest the quality of the devices.

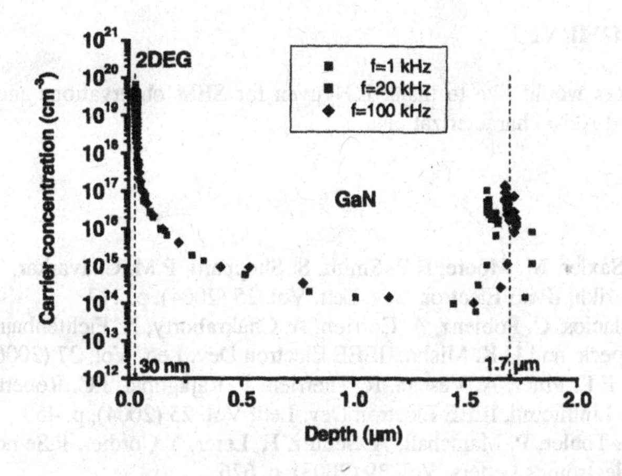

Fig.3: Carrier concentration profile extracted from capacitance-voltage measurements.

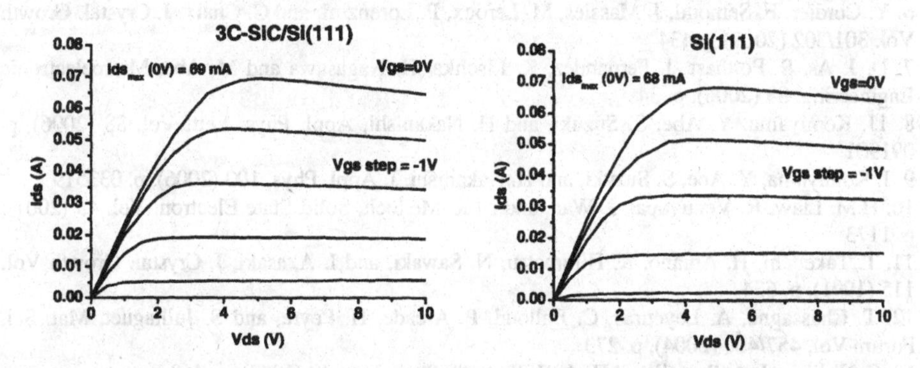

Fig. 4: I-V output characteristics of a 3 µm x 150 µm nominal gate transistor realized on the GaN based HEMT grown on 3C-SiC/Si(111) (left) and on bulk Si(111) (right).

CONCLUSIONS

AlGaN/GaN HEMT heterostructures have been grown on Si(111) with intermediate 3C-SiC layers. These 3C-SiC layers allow to simplify the growth of GaN structures on Si. The dislocation density and the characteristics of the transistors attest the quality of the heterostructures. It is a first step in the demonstration of the potentialities of the 3C-SiC/Si(111) substrate approach for the growth of high quality GaN based heterostructures.

ACKNOWLEDGMENTS

The authors would like to thank L.Nguyen for SEM observations and N.Baron for his help for X-ray and AFM characterizations.

REFERENCES

1. Y.-F. Wu, A. Saxler, M. Moore, R.P. Smith, S. Sheppard, P.M. Chavarkar, T. Wisleder, U.K. Mishra, and P. Parikh, IEEE Electron Dev. Lett. Vol. 25 (2004), p. 117
2. L. Shen, T. Palacios, C. Poblenz, A. Corrion, A. Chakraborty, N. Fichtenbaum, S. Keller, S.P. DenBaars, J.S. Speck, and U.K. Mishra, IEEE Electron Dev. Lett. Vol. 27 (2006), p. 214.
3. J.W. Johnson, E.L. Piner, A. Vescan, R. Therrien, P. Rajagopal, J.C. Roberts, J.D. Brown, S. Singhal, and K.J. Linthicum, IEEE Electron Dev. Lett. Vol. 25 (2004), p. 459
4. R. Behtash, H. Tobler, P. Marschall, A. Schurr, H. Leier, Y.Cordier, F.Semond, F.Natali, and J.Massies, IEE Electronics Letters, Vol. 39 (2003), p. 626
5. D.F. Storm, D.S. Katzer, J.A. Roussos, J.A. Mittereder, R. Bass, S.C. Binari, D. Hanser, E.A. Preble, and K.R. Evans, J. Crystal. Growth. Vol. 301/302 (2007), p. 429
6. Y. Cordier, F. Semond, J. Massies, M. Leroux, P. Lorenzini, and C. Chaix, J. Crystal. Growth Vol. 301/302 (2007), p.434
7. D. J. As, S. Potthast, J. Fernandez, K. Lischka, H. Nagasawa and M. Abe, Microelectronic Engineering. 83 (2006), p. 34
8. 1J. Komiyama, Y. Abe, S. Suzuki, and H. Nakanishi, Appl. Phys. Lett. Vol. 88 (2006), p. 091901
9. J. Komiyama, Y. Abe, S. Suzuki, and H. Nakanishi, J. Appl. Phys. 100 (2006), p. 033519
10. H.M. Liaw, R. Venugopal, J. Wan, and M.R. Melloch, Solid-State Electron. Vol. 45 (2001), p. 1173
11. T. Takeuchi, H. Amano, K. Hiramatsu, N. Sawaki, and I. Akasaki, J. Crystal. Growth Vol. 115 (1991), p. 634
12. T. Chassagne, A. Leycuras, C. Balloud, P. Arcade, H. Peyre, and S. Juillaguet, Mat. Sci. Forum Vol. 457/460 (2004), p. 273
13. S. Nishino, J. A. Powell and H. A. Will, Appl. Phys. Lett. 42 (1983), p. 460
14. F. Semond, Y. Cordier, N. Grandjean, F. Natali, B. Damilano, S. Vezian and J. Massies, Phys. Stat. Sol. (a) Vol. 188 (2001), p. 501
15. S. Vézian, F. Natali, F. Semond and J. Massies, Phys. Rev. Vol. B 69 (2004), p. 125329
16. O. Ambacher, B. Foutz, J. Smart, J.R. Shealy, N.G. Weimann, K. Chu, M. Murphy, A.J. Sierakowski, W.J. Schaff, L.F. Eastman, R. Dimitrov, A. Mitchell and M. Stutzmann, J. Appl. Phys. Vol. 87 (2000), p. 334

Mater. Res. Soc. Symp. Proc. Vol. 1068 © 2008 Materials Research Society 1068-C08-02

Epitaxial Growth of High-κ Dielectrics for GaN MOSFETs

Jesse S. Jur[1], Ginger D. Wheeler[1], Matthew T. Veety[2], Daniel J. Lichtenwalner[1], Douglas W. Barlage[2], and Mark A. L. Johnson[1]
[1]Materials Science and Engineering, North Carolina State University, 3000 Engineering Building I, 911 Partners Way, Raleigh, NC, 27695-7907
[2]Electrical and Computer Engineering, North Carolina State University, Raleigh, NC, 27695

ABSTRACT

High-dielectric constant (high-κ) oxide growth on hexagonal-GaN (on sapphire) is examined for potential use in enhancement-mode metal oxide semiconductor field effect transistor (MOSFET). Enhancement-mode MOSFET devices ($n_s > 4x10^{13}$ cm^{-2}) offer significant performance advantages, such as greater efficiency and scalability, as compared to heterojunction field effect transistor (HFET) devices for use in high power and high frequency GaN-based devices. High leakage current and current collapse at high drive conditions suggests that the use of a high-κ insulating layer is principle for enhancement-mode MOSFET development. In this work, rare earth oxides (Sc, La, etc.) are explored due to their ideal combination of permittivity and high band gap energy. However, a substantial lattice mismatch (9-21%) between the rare earth oxides and the GaN substrate results in mid-gap defect state densities and growth dislocations. The epitaxial growth of the rare earth oxides by molecular beam epitaxy (MBE) on native oxide passivated-GaN is examined in an effort to minimize these growth related defects and other growth-related limitations. Growth of the oxide on GaN is characterized analytically by RHEED, XRD, and XPS. Preliminary MOS electrical analysis of a 50 Å La_2O_3 on GaN shows superior leakage performance as compared to significantly thicker Si_3N_4 dielectric.

INTRODUCTION

GaN-based electronic devices are attractive due to their potential use in high-temperature, high-power and high-frequency operation [1]. The combination of a high bandgap energy, electron mobility and thermal conductivity as compared to traditional semiconductors provide a basis for use in such applications. Extensive research has been conducted toward developing AlGaN/GaN heterojunction field effect transistors (HFETs), normally-on (depletion-mode) devices. However, the progression toward a normally-off (enhancement-mode) GaN power device is preferred toward meeting the goal of highly linear, low leakage devices with simplified circuit design, similar to standard Si-based designs [2]. Such devices could potentially be scalable to THz frequencies. To realize an enhancement-mode device with optimal capacitance density and current, an effective gate dielectric is required. Dielectrics such as HfO_2, Al_2O_3, Si_3N_4, SiO_2 have been examined for use in GaN-based MOS devices [3]. In general, qualified dielectrics must have a high permittivity (ε), band gap energy (E_G), and conduction band offset (> 1 eV). Rare earth (RE) oxides (Sc_2O_3 – Lu_2O_3) offer the greatest potential as dielectrics due to their combination of high E_G and high ε [4], displayed by their growing interest in Si-based MOSFET gate stacks. While the permittivity of the bulk rare earth oxides is shown to decrease linearly with atomic number ($\varepsilon = 14.5 - 12.5$) [5], experimental determination of the permittivity of thin rare earth oxide films has been much higher ($\varepsilon = 27$ for La_2O_3) [6]. In addition, the band gap energy varies (6.5 – 3.5 eV) depending not only on the O-2p orbital filling, but the variable RE-4f orbital filling [7].

In an effort to obtain a low-leakage MOS device, epitaxial growth of rare earth oxides have been investigated [8-9]. Despite a large in-plane crystallographic lattice mismatch (Δ) between the rare earth oxide and GaN, epitaxial growth has been achieved for Sc_2O_3 ($\Delta = 9.2$ %) [8] and Gd_2O_3 ($\Delta = 20\%$) [9] on GaN. However, their potential use is limited by growth related defects in the oxide and formation of interface (oxide/GaN) defects [9], resulting in an decrease in the breakdown voltage and increase in the leakage current density. Electrically, these interface defects are energy states within the band gap typically arising as a physical result of dangling bonds or thermodynamic instabilities at the dielectric-semiconductor interface. The purpose of this work is to examine the early stages of epitaxial oxide growth of both La_2O_3 and Sc_2O_3 on GaN with a GaO_x surface passivation in order to gain a better understanding of mechanisms that will optimize dielectric functionality for future GaN-based MOS devices.

EXPERIMENTAL DETAILS

GaN films of 0.5 μm thickness were prepared on 50 mm diameter c-plane sapphire substrate by metal-organic chemical vapor deposition (MOCVD). Post deposition, the surface of the GaN was treated by a $K_2(SO_4)_2$ acid treatment followed by a submersion in H_2O in order to promote the formation of a consistent native oxide. Special care was made to reduce the atmospheric exposure time following the native oxide growth procedure and the subsequent loading the samples in the analytical characterization and MBE oxide growth chambers. La_2O_3 and Sc_2O_3 films of varying thickness were deposited by thermal evaporation of elemental high purity La (1700 °C) and Sc (1245 °C) in a ultra-high vacuum SVT Associates MBE system ($P_{background} = 5x10^{-10}$ Torr). Ultra-high purity O_2 was delivered to the system during growth to assist the oxide formation at a pressure of $5x10^{-7}$ Torr. The thickness of the oxide films reported was determined by *in situ* quartz crystal monitoring. The deposition rates of the oxide at evaporation conditions were 1.4 Å/min and 3.5 Å/min for La_2O_3 and Sc_2O_3, respectively. A GaN substrate temperature of 400 °C was used during oxide growth. *In situ* analysis was conducted by reflection high energy electron diffraction (RHEED) to examine the GaN surface prior to oxide growth and allow for real-time analysis during oxide growth. After growth of the oxide the substrate was cooled, and the oxygen flux was reduced when the substrate reached a temperature of 200 °C. For La_2O_3, a propensity for hydroxide formation [10] can be decreased by *in situ* capping of the La_2O_3 with a barrier metal such that device performance is unaffected [11].

X-ray photoelectron spectroscopy (XPS) characterization of the oxide films was performed using a Riber LAS3000 analytical system equipped with a two-stage cylindrical mirror-style analyzer. Spectra were acquired with a Mg Kα (1253.6 eV) non-monochromatic x-ray source, with a take off angle of 75° from surface, and a 0.1 eV step size. The C-1s binding energy (285.0 eV) was used as the offset reference. Chemical bonding states of the O-1s and Ga-2d were evaluated using a CASA XPS peak fitting program. X-ray diffraction (XRD) θ-2θ characterization was performed with a Philips X'Pert PRO MRD HR x-ray diffraction system equipped with a Cu Kα x-ray source (45 kV, 40 mA). In order to enhance the XRD characterization, *in situ* e-beam deposition of 150 Å Ta metal was performed after the La_2O_3 deposition in order to block the affinity for $La(OH)_3$ formation.

DISCUSSION

The initial stages of La_2O_3 and Sc_2O_3 film growth on hexagonal-GaN were examined by RHEED, XRD, and XPS. The $[01\bar{1}0]/[120]$ and $[11\bar{2}0]/[110]$ lattice planes of h-GaN prior to growth of the oxides are provided in figures 1 (a) and (b), respectively. Corresponding in-plane lattice parameter of 3.19 Å (bulk GaN = 3.189 Å) is determined by the spacing of the $[01\bar{1}0]/[120]$ lattice plane, calibrated against a Si (001) standard. The elongated streaks in the RHEED pattern is representative of a thin GaO_x passivation on the GaN surface, induced by the $K_2(SO_2)_4$ pretreatment. La_2O_3 and Sc_2O_3 grown 50 Å thick on the GaN substrates are characterized by RHEED in figures 1 (c)-(d) and (e)-(f), respectively. These patterns show epitaxial La_2O_3 and Sc_2O_3 films. As has been shown previously in literature, the Sc_2O_3 preferential grows in an fcc bixbyite structure on the (0001) surface of GaN [12]. As observed in the RHEED patterns, the symmetry of the (111) fcc array of Sc_2O_3 is identical to that of the (0001) hexagonal array. An in-plane lattice parameter of 3.42 Å is obtained for the 50 Å Sc_2O_3 film, yielding $\Delta = 9$ %, with the GaN substrate. The additional diffraction streaks in the RHEED pattern a surface reconstruction in the Sc_2O_3 film indicates a 2D superstructure which is consistent with an atomically smooth film morphology.

The epitaxial La_2O_3 structure (figures 1 (c)-(d)) appears to be identical to that of the Sc_2O_3. It is noted that the preferred low temperature structure (< 700 °C) for La_2O_3 is hexagonal (P3m), but an equilibrium transformation to cubic (Ia3) structure does occur at high temperature anneal/growth temperatures [13]. The observed cubic nature of the La_2O_3 film suggests a metastable alignment between the La_2O_3 and the GaN, most likely induced by the high in-plane lattice mismatch at their interface. From the RHEED pattern spacing, an in-plane lattice parameter of 3.90 Å is obtained for La_2O_3 ($a_{bulk} = 3.826$ Å) and the spot-like nature of the RHEED pattern is indicative of a 3D-textured film morphology. This rough surface would also be expected due to the high lattice mismatch between the cubic-La_2O_3 and the GaN (Δ =21%).

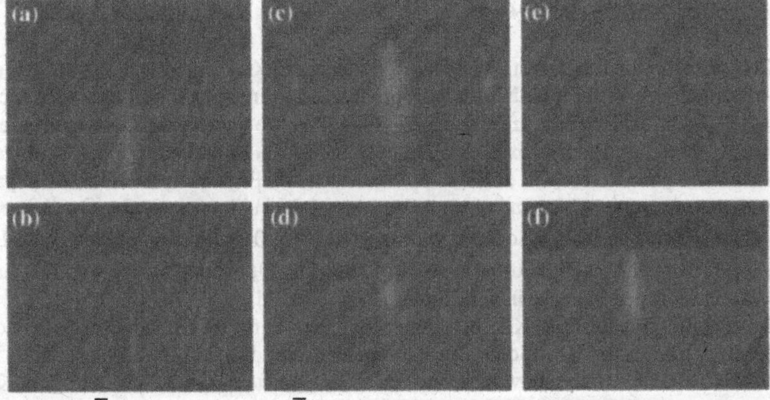

Figure 1. (a) $[01\bar{1}0]/[120]$ and (b) $[11\bar{2}0]/[110]$ lattice planes of the GaN surface along with corresponding lattice planes of epitaxial fcc bixyite La_2O_3 (c) and (d) and Sc_2O_3 (e) and (f).

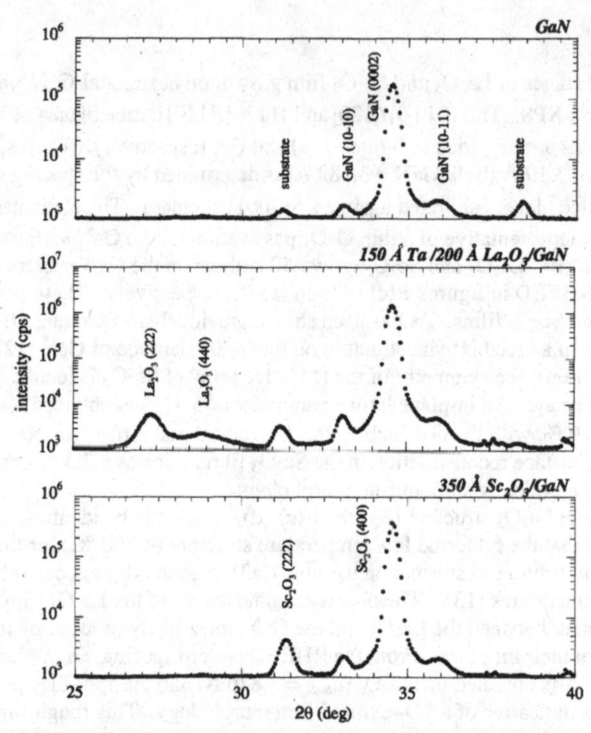

Figure 2. θ-2θ XRD patterns of the bare GaN film grown by MOCVD on a sapphire substrate, and after growth of La_2O_3 and Sc_2O_3.

XRD of the oxide films with an increased thickness (200 Å La_2O_3 and 350 Å Sc_2O_3) are shown in comparison with the GaN film in figure 2 and confirms the fcc structure for each rare earth oxide. Additionally peaks are associated with either the underlying Al_2O_3 substrate and buffer layer. To reduce the hydroxide formation of the La_2O_3 film, a Ta capping layer was deposited *in*-situ prior to removal of the sample from UHV. Peak analysis of the XRD patterns shows the presence of the (222) and (440) reflections of the cubic majority La_2O_3. A broad peak located at $2\theta = 29°$ is possibly associated with the $(10\bar{1}1)/(101)$ reflection of hexagonal-La_2O_3. For the Sc_2O_3 film, (222) and (400) reflections of cubic-Sc_2O_3 overlap the GaN (0002) substrate peaks and an associated increase in peak intensity is observed.

Figure 3 presents the O-1s XPS binding energy spectra for the MOCVD GaN film on sapphire, 50 Å La_2O_3, and 50 Å Sc_2O_3 to examine the extent of oxide surface coverage. The O-1s spectra of the GaN surface shows evidence of a GaO_x surface layer as a result of the $K_2(SO_4)_2$ acid pretreatment (followed by a H_2O immersion) conducted immediately preceding loading the samples in the XPS analysis and MBE oxide growth chambers. The thickness of the GaO_x passivation layer is believed to be only a few monolayers due to the observation of the GaN substrate peaks in the XPS analysis and ability to the observe the GaN RHEED pattern. Following the 50 Å growth of the La_2O_3 and Sc_2O_3, the binding energy peak associated with the

Ga-O-Ga is still observed. Evidence for the hydroxide formation is observed in the O-1s XPS spectra of La_2O_3 by the high intensity of the La-O-H peak at a binding energy at ~ 531 eV. This hydroxide formation could result in some surface agglomerates; however, the substantial peak for Ga-O-Ga is indicative of island growth, which would expose the GaN surface to the XPS analysis. This island-like growth mode is also suggested by the 3D character of the film growth and the high Δ observed for La_2O_3 growth on GaN [14]. In comparison, the near complete coverage of the Sc_2O_3 reduces the Ga-O-Ga peak. This corresponds to a layer-by-layer growth mode, which requires a quasi-zero lattice matching like that observed by the initial stages of hexagonal-Sc_2O_3 growth on GaN.

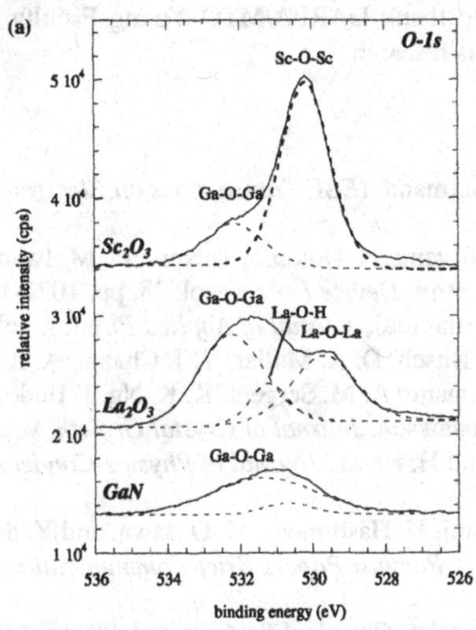

Figure 3. XPS of the O-1s binging energy spectra for GaN surface, 50 Å La_2O_3, and 50 Å Sc_2O_3.

Despite the high lattice mismatch, early results of a 50 Å La_2O_3 films (with an Ta overlayer) on a mesa-isolated GaN-based MOS capacitor (with K_2SO_4 passivation procedure) exhibits a leakage current density of $3x10^{-2}$ A/cm^2 @ Vg = 2V. Compared to a MOS capacitor with a 600 Å Si_3N_4, a 10x improvement in the leakage current density is observed. The low leakage observed is likely due to the high epitaxial quality of the film in combination with the ability of the GaO_x passivation layer to sufficiently limit the leakage pathways that would result from the high lattice mismatch. Further electrical analysis, including a determination of the density of interface traps for the surface-passivated MOS gate stack is underway. Still, it is evident that the GaO_x passivation plays an important role in achieving high device quality with epitaxial oxide growth on GaN.

CONCLUSIONS

The early-stage (~50 Å) La_2O_3 and Sc_2O_3 epitaxial oxide growth on GaN is examined by surface and structure analysis techniques. It was observed that both rare earth oxides were initially grown cubic in structure with lattice mismatch values of 21 % (La_2O_3) and 9 % (Sc_2O_3). As a result of the discrepancy in the lattice mismatch for each oxide, analysis shows island-like growth for La_2O_3 and layer-by-layer growth for Sc_2O_3 on GaN. The high lattice mismatch between the rare earth oxides and GaN might suggest an incompatible use of these oxides in GaN-based MOS device applications. However, the use of a GaO_x passivation by wet chemical treatment of the GaN, allows for a suitable growth of the epitaxial oxide on GaN and the ability to achieve low leakage current densities.

ACKNOWLEDGMENTS

The authors would to thank DARPA/MTO Young Faculty Award and NSF CAREER Award for the funding for this research

REFERENCES

[1] K. Matocha and R. Gutmann, *IEEE Transactions on Electron Devices*, vol. 52, pp. 6-10, 2005.

[2] H. Kambayashi, Y. Niiyama, S. Ootomo, T. Nomura, M. Iwami, Y. Satoh, S. Kato, and S. Yoshida, *IEEE Electron Device Letters*, vol. 28, pp. 1077-1079, 2007.

[3] J. Robertson and B. Falabretti, *Journal of Applied Physics*, vol. 100, pp. -, 2006.

[4] J. Kwo, M. Hong, B. Busch, D. A. Muller, Y. J. Chabal, A. R. Kortan, J. P. Mannaerts, B. Yang, P. Ye, H. Gossmann, A. M. Sergent, K. K. Ng, J. Bude, W. H. Schulte, E. Garfunkel, and T. Gustafsson, *Journal of Crystal Growth*, vol. 251, pp. 645-650, 2003.

[5] D. Xue, K. Betzler, and H. Hesse, *Journal of Physics-Condensed Matter*, vol. 12, pp. 3113-3118, 2000.

[6] T. Yamamoto, Y. Izumi, H. Hashimoto, M. Oosawa, and Y. Sugita, *Japanese Journal Of Applied Physics Part 1-Regular Papers Brief Communications & Review Papers*, vol. 45, pp. 6196-6202, 2006.

[7] G. Adachi and N. Imanaka, *Chemical Reviews*, vol. 98, pp. 1479-1514, 1998.

[8] A. M. Herrero, B. P. Gila, C. R. Abernathy, S. J. Pearton, V. Craciun, K. Siebein, and F. Ren, *Applied Physics Letters*, vol. 89, 2006.

[9] J. W. Johnson, B. Luo, F. Ren, B. P. Gila, W. Krishnamoorthy, C. R. Abernathy, S. J. Pearton, J. I. Chyi, T. E. Nee, C. M. Lee, and C. C. Chuo, *Applied Physics Letters*, vol. 77, pp. 3230-3232, 2000.

[10] M. Nieminen, M. Putkonen, and L. Niinisto, *Applied Surface Science*, vol. 174, pp. 155-165, 2001.

[11] D. J. Lichtenwalner, J. S. Jur, A. I. Kingon, M. P. Agustin, Y. Yang, S. Stemmer, L. V. Goncharova, T. Gustafsson, and E. Garfunkel, *Journal of Applied Physics*, vol. 98, 2005.

[12] J. Chen, B. Gila, M. Hlad, A. Gerger, F. Ren, C. Abernathy, and S. Pearton, *Applied Physics Letters*, vol. 88, 2006.

[13] N. Imanaka, T. Masui, and Y. Kato, *Journal of Solid State Chemistry*, vol. 178, pp. 395-398, 2005.

[14] H. J. Scheel and T. Fukuda, *Crystal Growth Technology*, 1 ed. Chichester, West Sussex, England: John Wiley and Sons, 2003.

GaN Based Optical Devices on Silicon

Mater. Res. Soc. Symp. Proc. Vol. 1068 © 2008 Materials Research Society 1068-C05-05

Integrated Optics Utilizing GaN-Based Layers on Silicon Substrates

Armand Rosenberg, Michael A. Mastro, Joshua D. Caldwell, Ronald T. Holm, Richard L. Henry, Charles R. Eddy, Konrad Bussmann, and Mijin Kim
Naval Research Laboratory, Washington, DC, 20375

ABSTRACT

It is now apparent that future generations of fast electronics and compact sensors may need to rely increasingly on integrated optical components. But integration of electronics and photonics in today's IC's is challenging. Silicon, the ubiquitous electronic material, is neither ideally suited for most photonic functions nor readily integrated with most of the common photonic materials, such as GaAs. The approach we describe here relies on GaN-based films, which can be grown directly on silicon substrates and hence can be potentially integrated with state-of-the-art Si-based electronics. We have demonstrated the fabrication of GaN structures on silicon wafers ranging in overall size from sub-micron to several millimeters, all containing highly accurate individual features on the nm scale. As proof of concept, we have fabricated GaN optical waveguides and photonic crystals containing optical cavities by patterning GaN membranes grown directly on Si wafers. Our optical cavities were designed to have resonant modes within the spectral region of the broad defect-induced luminescence of GaN. We have measured sharp resonant features associated with these cavities by optically pumping above the GaN band edge, and have compared the data to numerical simulations of the spectra. Our results to date demonstrate the feasibility of fabricating high-quality GaN photonic structures directly on Si wafers, thereby providing a possible path to achieving true integration of electronics and photonics in future generations of IC's.

INTRODUCTION

Recent work has demonstrated the feasibility of fabricating photonic crystal structures in GaN-based films [1-3]. These optical structures contain typical feature sizes significantly smaller than the wavelength of light, and thus qualify as true "nanophotonic" devices when the sub-micron emission wavelength typical of GaN-based materials is taken into account. Potentially, based on such photonic crystal nanocavities, LED's and lasers could be designed with enhanced light extraction efficiency and reduced lasing threshold, thus enabling optoelectronic integration on an unprecedented scale. However, although high-quality GaN layers can be grown directly on Si substrates [4], a significant roadblock to optoelectronic integration is the fact that fabrication of GaN photonic crystal membranes has only been demonstrated in GaN-based films grown on sapphire substrates [2], and hence the resulting GaN optical structures cannot be easily incorporated into Si-based devices. We have addressed this challenge, and our most recent results demonstrate that GaN-based suspended waveuides and photonic crystal membranes containing nanocavities can be successfully fabricated directly on Si surfaces [5], thus partially eliminating this remaining roadblock to widespread optoelectronic integration on the sub-micron length scale.

As is well known, the fundamental problem of light emission in Si is its indirect electronic band gap. Although numerous potential solutions are currently under investigation, a straightforward approach (which we have also been pursuing) appears to be the growth of a III-V based light emitting layer on Si [6]. However, the large epitaxial mismatch between Si and III-V semiconductors usually leads to poor emission efficiency and short lifetimes in such heteroepitaxial film structures. An exception is the case of GaN-based layers grown on Si, which have been shown to permit fabrication of efficient LED's and lasers [4]. Because the real refractive index of Si is considerably higher than the indices of GaN-based semiconductors, the optical resonator design for in-plane optical structures must include a means of confining the optical field inside the GaN-based layer (and out of the underlying Si substrate). The best way to achieve this optical confinement is to surround the GaN-based layer with a low-index region, which in turn is best accomplished by fabricating the optical devices in a GaN-based membrane suspended above Si.

FABRICATION RESULTS

The GaN films used in this study were grown on Si(111) by a metalorganic chemical vapor deposition (MOCVD) process. Growth was carried out in a modified vertical impinging flow CVD reactor. Two-inch Si wafers were cleaned via a modified Radio Corporation of America process followed by an *in situ* H_2 bake. An Al seed layer was deposited prior to the onset of NH_3 flow to protect the Si surface from nitridation. The 33-nm AlN buffer layer was deposited at 1050°C and 50 Torr. The 350-nm GaN cap layer was deposited at 1020°C and 250 Torr directly on the buffer layer. The higher deposition pressure resulted in GaN layers with larger grains and lower defect densities. *In-situ* spectroscopic interferometry monitored the layer thicknesses and reflectance in real time during deposition, to ensure that the desired thicknesses were grown. Details of this procedure and apparatus have been reported elsewhere [7,8]. The film growth conditions have not been fully optimized, and we expect that modifying the buffer layer deposition temperature and thickness can lead to improved film quality.

Following deposition, structural characterization was performed with a Panalytical X'pert x-ray diffraction (XRD) system. *Ex-situ* reflectance was measured at normal incidence using a halogen lamp as a source, and the reflected beam was dispersed through an Ocean Optics S2000 spectrometer with a 50 μm slit for the purpose of characterizing layer thicknesses. Surface characterization of the GaN films was performed with a Carl Zeiss SMT Supra 55 scanning electron microscope (SEM) and a Veeco Digital Instruments Dimension 3100 atomic force microscope (AFM) in tapping mode. Focused ion beam (FIB) milling was used to generate cross-sectional secondary electron images using a FEI Nova 600 dual-beam system.

An AFM image of a 5x5 μm area of a nominally 350 nm thick GaN film deposited on Si(111) is shown in Fig. 1. The AFM data indicate that an RMS roughness of 3.5 nm is associated with this film. This value is less than a factor of 2 greater than the value typically obtained for our highly optimized GaN films deposited on *a*-plane sapphire substrates [8]. Since, as seen in the AFM image, the defects attributed to threading screw dislocations are well localized and the local roughness of these films is not significant on the scale of this image, the morphology of these films appears suitable for fabrication of planar optical devices at visible and IR wavelengths. However, we note that GaN films deposited on sapphire show far fewer localized defects [8].

Fig. 1. AFM image of a 5x5 μm area of a 350 nm thick GaN film grown on Si(111). The vertical scale is 100 nm per division.

To further characterize the crystal quality of our GaN films grown on Si(111), we have performed XRD analysis of these films, as shown in Fig. 2(a), and compared the results to the more thoroughly studied GaN films deposited on *a*-plane sapphire substrates (not shown). The omega rocking curve of the (0002) GaN reflection displays a FWHM of 0.325° for the GaN film grown on Si(111), compared to a 0.162° FWHM for the corresponding peak of our best GaN film on sapphire (at similar film thickness). The GaN films on Si(111), comparatively, have smaller grains and a higher density of threading dislocations, such as threading screw dislocations which tilt the GaN grains and thus broaden symmetric reflections such as the (0002) peak. The photoluminescence (PL) spectrum of a 170 nm thick GaN film on Si(111), excited at 360 nm, is shown in Fig. 2(b). This spectrum is dominated by a strong nitrogen-vacancy induced emission peak near 560 nm.

We fabricated photonic crystal structures in the GaN films deposited on Si(111) as described previously for GaN films on sapphire [9]. Following the metalorganic chemical vapor deposition (MOCVD) of GaN on a Si(111) wafer, a 30 nm thick Ni film was sputter deposited to act as a Cl_2 plasma etch mask for GaN. A 100 nm thick layer of positive e-beam resist (diluted ZEP520A) was then spin coated on the wafer. The photonic crystals, consisting of high-density triangular arrays of holes with diameters in the range of 90-330 nm and lattice spacings in the range of 180-460 nm (not in all combinations), were then written using a Raith 150 e-beam lithography system at 25 kV beam voltage. After developing the exposed resist, the pattern was transferred to the Ni film using Ar^+-ion beam milling at 500 V, with continuous rotation about the sample normal which was oriented at an angle of 25° off the ion beam axis. As before, the resist and Ni layers were found to sputter at approximately the same rate. The residual resist was then removed from the Ni mask by ashing in an oxygen plasma. The GaN film was etched at a rate of approximately 250 nm/min in a Trion inductively-coupled plasma etcher using a Cl_2/Ar atmosphere (with a 2:1 ratio of etch to buffer gas). Finally, the Ni mask was removed by a brief wet etch in an acid mixture consisting of (5:1) HCl:HNO_3. The resulting photonic crystal patterns faithfully reproduced the intended geometry, which was designed to produce first-order photonic band gaps and cavity resonant modes in the UV-visible spectral region.

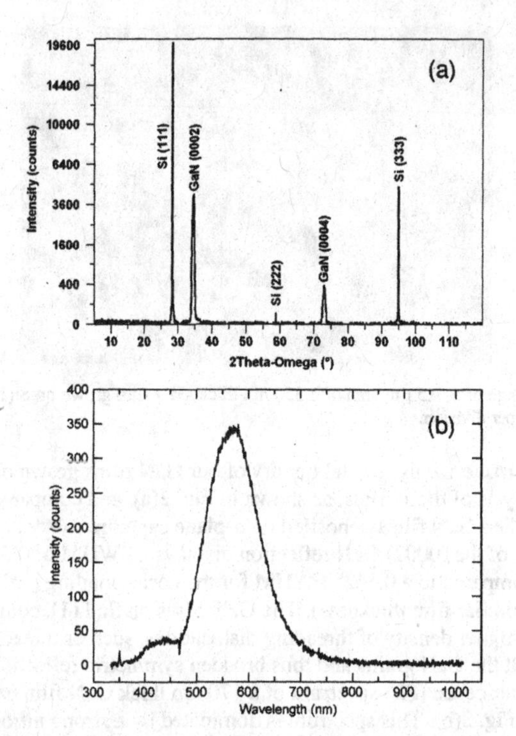

Fig. 2. (a) X-ray diffraction of a single-crystal 350 nm GaN / 25 nm AlN film on Si(111). Strong wurtzite GaN (0002) and (0004) peaks are observable at approximately 34.4° and 73.2° in the ω-2θ scan. (b) Photoluminescence spectrum of a 170 nm GaN / 25 nm AlN film on Si(111) excited at 360 nm.

To undercut the GaN film and create "bridge" (suspended) structures, an isotropic wet etch of Si was used. This etch consisted of a brief (approximately 60 s) immersion of the wafer in an acid mixture consisting of (1:99) HF:HNO_3 [10]. Due to the isotropic nature of this etch, the lateral extent of the undercutting is approximately the same as the depth. The wafer was then rinsed in DI water and isopropanol and air dried. High aspect ratio bridge structures were easily created above the Si surface (anchored to the Si surface at both ends), as shown in Fig. 3(a). These bridges could serve as conventional "ridge" optical waveguides with high index contrast (approximately 2.4:1), which would allow sharp bends with low radiation loss. In particular, large aspect ratio suspended structures with very few anchor points were demonstrated by this technique, as shown in Fig. 3(b). This "tethered" configuration has been studied in other material systems recently [11], where it was found to be a practical method of fabricating low-loss optical waveguides (in spite of the additional loss introduced by each "tether" point).

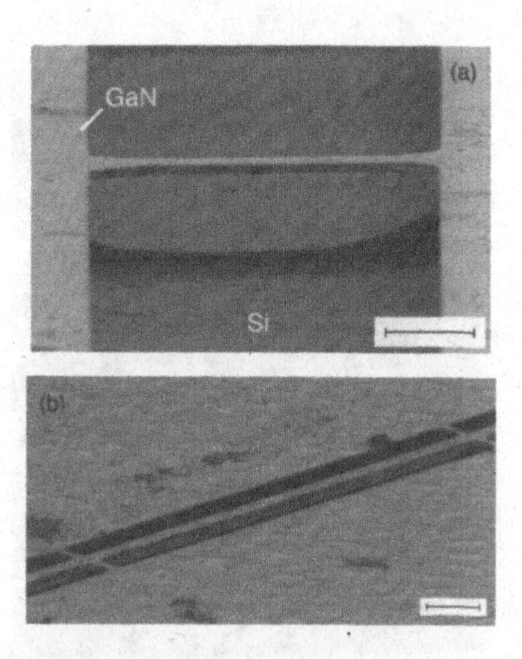

Fig. 3. SEM image of 350 nm thick GaN "bridges" suspended above Si(111). The scale bar represents 3 μm in (a) and 10 μm in (b).

For the photonic crystal patterns in the GaN films on Si(111), the wet acid etch produced photonic crystal membranes that were locally suspended several microns above the etched Si regions but were still anchored to the Si surface, as shown in Fig. 4. In this case, the acid mixture readily attacked the Si through the openings of the photonic crystal pattern itself, leading to undercutting only in the region of the photonic crystal (although the etch could continue into adjacent regions, if allowed to do so). The suspended nature of the photonic crystal regions was confirmed by using an FEI focused-ion-beam (FIB) instrument to reveal cross-sections of the photonic crystal membranes after the fabrication process was completed, as shown in Figs. 4(b) and (c). The fabricated photonic crystal regions also incorporated "defects" which act as optical nanocavities, such as the 7 missing holes in the center of the pattern shown in Fig. 4 (known as an "L7" cavity). The hole diameter and spacing of our fabricated triangular-symmetry photonic crystal patterns (70-175 nm and 200-500 nm, respectively), combined with the refractive index of the GaN membrane (approximately 2.37), ensure that the corresponding lowest-order resonant modes of these cavities will occur in the 500-1300 nm range, according to a detailed analysis of these photonic crystal nanocavities [12].

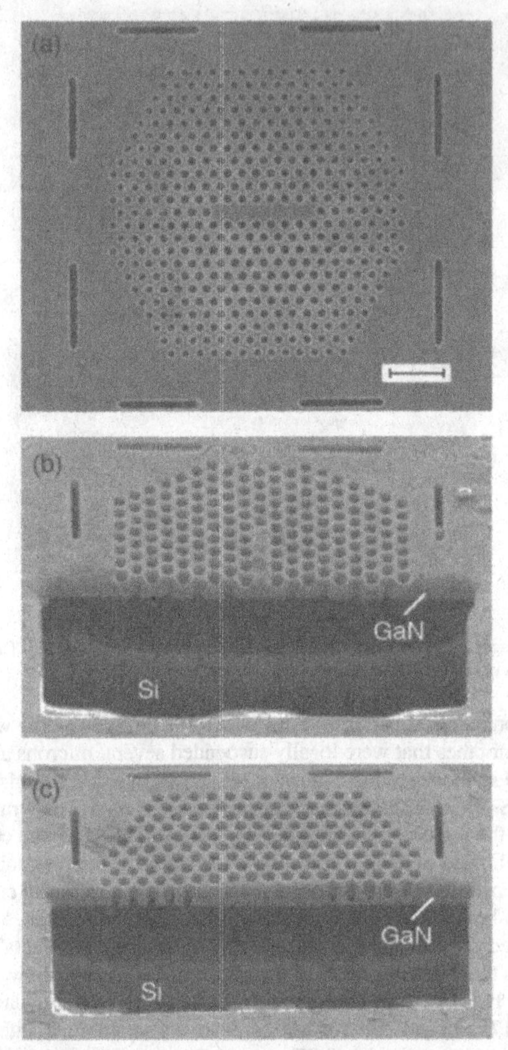

Fig. 4. SEM images of a 350 nm GaN photonic crystal membrane containing an L7 cavity suspended above Si(111). (a) Top view. (b) and (c) Views at 52° tilt after revealing the cross-section using a FIB instrument. The scale bar represents 1 μm.

In spite of the high internal tensile stress state known to exist in GaN films on Si [6,7,12] (which can be grown crack-free up to approximately 1 μm in thickness), no cracking of the GaN layer on Si(111) was observed for the undercutting conditions described above. In fact, we have found that the photonic crystal pattern (with features on the 100-nm-scale and below) acts to relax partially the internal stresses built into the as-deposited GaN film on Si, making for a structurally more stable suspended structure than patterned structures consisting only of μm-scale features, such as the bridge shown in Fig. 3. This is attributed to the ability of nm-scale GaN features to accommodate the stress that is released upon etching [14]. By contrast, we have found that KOH-based etching of Si, which is highly anisotropic, causes severe cracking of similarly patterned GaN membranes on Si, and is therefore unsuitable for this purpose.

OPTICAL MEASUREMENTS

The UV-excited broadband PL spectrum from the as-grown GaN films on Si(111) is shown in Fig. 2(b). Following fabrication of the photonic crystal membranes described above, we repeated the PL measurements in spatial regions near GaN photonic crystal L7 cavities, similar to those in Fig. 4, using a custom UV microscope set-up. This allowed us to spectrally and spatially analyze the above-gap excited PL with sub-micron spatial resolution, which in turn allowed us to isolate and analyze the PL emitted by the photonic crystal cavities. With the sample on a computer-controlled x-y translation stage, the PL was excited by the 351.0 nm laser line of an Ar^+ laser through a 100X Mitutoyo microscope objective, and the PL was collected through the same objective and analyzed by an Ocean Optics QE65000 spectrometer. Using this set-up, we obtained spatial maps of the PL intensity for the narrow features observed in the PL spectra, with a spectral resolution of 5 nm and a spatial resolution of 0.5 μm.

Qualitatively, the effects of the photonic crystal cavities are quite dramatic, as seen in Figs. 5(a) and (b). The optical image of the photonic crystal membrane shown in Fig. 5(a), obtained under broad-area white light illumination, is essentially unremarkable, revealing only a slight color change in the area of the suspended and patterned GaN membrane. This color change results from the semi-transparent quality of the suspended GaN membrane and the associated interference effects. (Outside of the suspended area, where the GaN film is directly contacting the Si surface, the light transmitted by the GaN film is absorbed by the Si surface, with the GaN film effectively acting as a quasi anti-reflection layer, thereby resulting in the observed color difference.) By contrast, under broad-area above-gap UV laser (351 nm) illumination, which is absorbed by the GaN film, a striking enhancement of the PL in the region around the photonic crystal L7 cavity is observed in Fig. 5(b). This is clearly seen as a short bright horizontal segment in the center of Fig. 5(b), which highlights the position of the L7 photonic crystal cavity corresponding to the SEM images of Fig. 4. Thus, the observed PL enhancement in Fig. 5(b) is a direct result of the resonant modes associated with the L7 photonic crystal cavity fabricated in the suspended GaN membrane.

Two PL spectra collected from a fabricated photonic crystal membrane with L7 cavity are presented in Fig. 5(c). The black curve is the spectrum collected near the position of the L7 cavity, while the red (featureless) curve is the spectrum collected from a position off the photonic crystal region. Although the shape of the broadband background spectrum (red curve) is somewhat changed from that measured on the as-grown GaN film in Fig. 2(b), the background PL intensity remains peaked around 550 nm. In particular, the fabrication process appears to induce additional broadband PL at shorter wavelengths, whereas the PL in the as-grown films decreases sharply below 550 nm. However, the narrow localized PL peaks are only observed in the spectrum collected near the center of the L7 cavity (black curve), implying that these peaks are the result of cavity resonant modes. The cavity PL spectrum (black curve) also shows an overall increase in the featureless background, which could be associated either with increased defect density resulting from the fabrication process or with enhanced extraction efficiency of the PL in the region where the GaN film is now a suspended membrane above the Si surface instead of being in direct contact with the Si surface, as is the case for the red curve spectrum in Fig. 5(c) and also for the spectrum in Fig. 2(b). Using the spatial scanning capability of our UV microscope, we obtained spatial maps of the PL intensity at various wavelengths. An example of such data is shown in Fig. 5(d), which displays a map of the peak PL intensity at 559 nm obtained with 0.5 μm spatial resolution. This spectral mapping of the region near the photonic crystal cavity clearly shows that the PL peak is strongest in the region of the L7 cavity, with the PL intensity map mirroring the elongated shape of the L7 cavity itself. Similar PL intensity maps were obtained for other narrow PL peaks visible in Fig. 5(c), confirming their origin as resonant modes of the photonic crystal cavity.

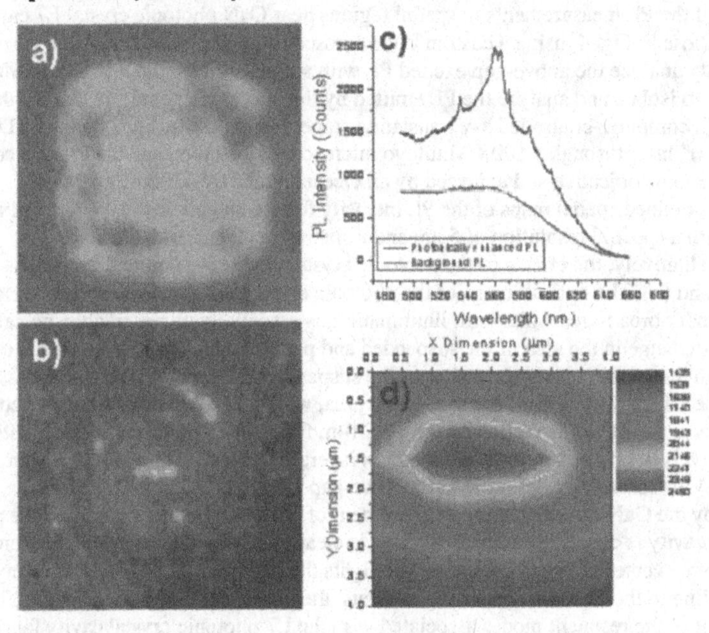

Fig. 5. (a) Optical image of the GaN photonic crystal cavity in Fig. 4(a), obtained with reflected white light. (b) Corresponding real-color PL image of the same area when excited by 351 nm laser (with cavity visible at center). (c) PL spectra collected when the laser is focused at the center of the cavity (black, top line) and away from the cavity (red, bottom line). (d) Spatial map of the intensity of the PL peak at 559 nm in the vicinity of the cavity position.

Numerical simulations using FDTD techniques were used to determine the field distributions associated with the resonant modes of these L7 photonic crystal cavities, as shown in Fig. 6. These simulations were based on the exact parameters of our fabricated structures, determined from SEM images similar to those in Fig. 4. In agreement with the measured PL intensity maps, the calculated mode patterns are highly confined to the location of the L7 cavity. We note that the numerical simulation does not appear to reproduce several of the weaker narrow peaks visible in Fig. 5(c). In fact, the measured PL intensity maps associated with these peaks (not shown here) also do not correspond to the shape of the L7 cavity, suggesting that they may be caused by some type of fabrication-induced defect. Identifying these additional defects will require additional work.

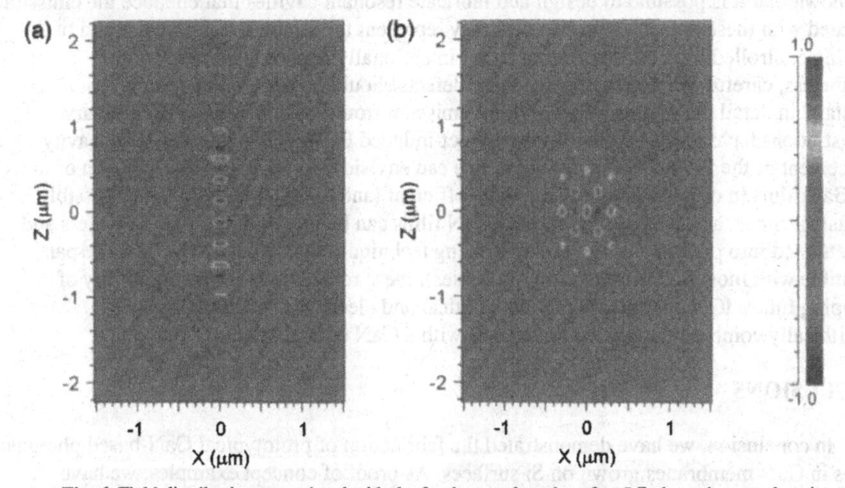

Fig. 6. Field distributions associated with the fundamental modes of an L7 photonic crystal cavity in a suspended GaN membrane. The physical parameters of the simulated cavity are the same as the fabricated structure shown in Fig. 4. (a) TM mode at 544 nm, (b) TE mode at 552 nm.

DISCUSSION

The procedure we have presented here for fabricating GaN tethered optical waveguides and photonic crystal structures in thin membranes suspended above Si differs from results in previous work, which relied on specially prepared substrate surfaces to create the suspended membranes. For example, GaN membranes with relatively large (μm-scale) features have been demonstrated by growing GaN on a previously patterned Si surface [10], in a process similar to a "lift-off" technique. In another previous demonstration, suspended photonic crystal membranes (similar to those demonstrated here) were fabricated by growing the GaN layer on top of a carefully constructed multiple-layer GaN-based film structure on a sapphire substrate [2]. In the latter case, the film stack between the GaN layer and the sapphire substrate included a sacrificial InGaN layer whose In content allowed for a selective photo-enhanced wet etching step in order to produce the locally suspended GaN membrane. Of these approaches, ours is arguably the simplest and most generally applicable, since it relies solely on standard dry and wet etching steps. Specifically, in our case, the initial patterning of the GaN film into a photonic crystal structure can proceed by any of a number of techniques demonstrated previously (we used an ion-milled Ni mask [1,9], but plasma-etched SiO_2 masks have also been demonstrated for this purpose [2], as well as more complex mask structures [3]). Unlike the other cases mentioned, no special *a priori* sample preparation is required in our fabrication procedure to enable the fabrication of suspended GaN photonic crystal membranes, consisting of high-density arrays of sub-micron scale features and anchored to the Si surface.

Our optical measurements confirm the high quality of the GaN structures fabricated on Si by demonstrating the spatial confinement of the enhanced PL associated with the fabricated cavities. While the fabrication process appears to introduce additional structural defects into the GaN films, these defects do not appear to quench the optical emission in these films. In fact, we have shown that it is possible to design and fabricate resonant cavities that enhance the emission associated with these defects, which in turn may represent the simplest possible method of achieving controlled light emission from a (not intentionally doped) GaN-based film. Nevertheless, careful characterization of these defects should be undertaken in order to understand in detail their contribution to light emission from GaN. While our preliminary demonstrations have relied on the intrinsic defect-induced PL of GaN to demonstrate cavity enhancement of the PL via resonant modes, one can envision, for example, the addition of In to these GaN films in order to obtain much more efficient (and tunable) emission, and possibly even lasing. Since, as we have shown, such GaN films can be grown directly on Si wafers and then patterned into photonic devices *in-situ* (using techniques that appear to be, in large part, compatible with most IC manufacturing processes), these results suggest the possibility of developing future IC's that incorporate both optical and electronic functionality by monolithically combining a Si electronic layer with a GaN optical layer.

CONCLUSIONS

In conclusion, we have demonstrated the fabrication of prototypical GaN-based photonic devices in GaN membranes grown on Si surfaces. As proof of concept examples, we have demonstrated optical waveguides and photonic crystal structures, including optical nanocavities designed for resonances in the visible to near-IR spectral region. These structures represent a significant step forward along the path towards monolithic integration of GaN-based photonics, including light emitting devices, with Si-based electronics. Although all the work described above was done using Si(111) wafers, recent advances [15] in the growth of GaN films on Si(100), which is the orientation most commonly used in IC manufacturing, give us reason to be optimistic about the broader applicability of the results presented above. Other than issues introduced by potential differences in GaN film quality, we expect that the results presented here could also be achieved in GaN films on Si(100) with no significant complications.

Extrapolating from these preliminary results, we envision that more sophisticated GaN-based photonic structures and devices could eventually be incorporated into future silicon-based IC's, thus adding optical functionality to these chips. Among the benefits of integrating optical devices directly into future IC's are the speed increase expected when more of the chip's functions are performed in the optical regime (rather than after conversion to an electronic signal), as well as the possibility of increased functionality and versatility within the IC itself. The latter can be achieved, for example, by directly incorporating into the IC some of the optical functions that currently require discrete optical components, such as modulators, sensors, etc.

ACKNOWLEDGMENTS

We acknowledge useful discussions with Prof. Dennis Prather (Univ. of DE), and technical assistance from Chul-Soo Kim (NRL) and J.A. Murakowski (Univ. of DE). We also acknowledge contributions by Vasgen A. Shamamian and James A. Casey (both currently at Dow-Corning Inc.) and Michael Carter (currently at the USPTO) during earlier stages of this work.

REFERENCES

1. A. Rosenberg, Michael W. Carter, J.A. Casey, Mijin Kim, Ronald T. Holm, Richard L. Henry, Charles R. Eddy, V.A. Shamamian, and K. Bussmann, Optics Express **13**, 6564 (2005).
2. Cedrik Meier, Kevin Hennessy, Elaine D. Haberer, Rajat Sharma, Yong-Seck Choi, Kelly McGroddy, Stacia Keller, Steven P. DenBaars, Shuji Nakamura, and Evelyn L. Hu, Appl. Phys. Lett. **88**, 31111 (2006).
3. D. Coquillat, J. Torres, D. Peyrade, R. Legros, J.P.Lascaray, M. Le Vassor d'Yerville, E. Centeno, D. Cassagne, J.P. Albert, Y. Chen, and R.M. De La Rue, Optics Express **12**, 1097 (2004).
4. A. Reiher, J. Bläsing, A. Dadgar, A. Diez, and A. Krost, J. Cryst. Growth **248**, 563 (2003).
5. A. Rosenberg, K. Bussmann, Mijin Kim, Michael W. Carter, M. A. Mastro, Ronald T. Holm, Richard L. Henry, Joshua D. Caldwell, and Charles R. Eddy, Jr., J. Vac. Sci. Technol. B **25**, 721 (2007).
6. S. Joblet, F. Semond, F. Natali, P. Vennegues, M. Laugt, Y. Cordier, and J. Massies, Phys. Stat. Sol. (c), **2-7**, 2187 (2005).
7. M.A. Mastro, C.R. Eddy Jr., D.K. Gaskill, N.D. Bassim, J. Casey, A. Rosenberg, R.T. Holm, R.L. Henry, and M.E. Twigg, J. Crystal Growth, **287**, 610 (2006); M.A. Mastro, R.T. Holm, N.D. Bassim, C.R. Eddy Jr., D.K. Gaskill, R.L. Henry, and M.E. Twigg, Appl. Phys. Lett., **87**, 241103 (2005).
8. C. R. Eddy, Jr., R. T. Holm, R. L. Henry, J. C. Culbertson, and M. E. Twigg, J. Electron. Mater. **34**, 1187 (2005).
9. David S.Y. Hsu, Chul Soo Kim, Charles R. Eddy, Jr., Ronald T. Holm, Richard L. Henry, J.A. Casey, V.A. Shamamian, and A. Rosenberg, J. Vac. Sci. Technol. B **23**, 1611 (2005).
10. Z. Yang, R.N. Wang, S. Jia, D. Wang, B.S. Zhang, K.M. Lau, and K.J. Chen, Appl. Phys. Lett. **88**, 41913 (2006).
11. T.H. Stievater, W.S. Rabinovich, D. Park, J.B. Khurgin, S. Kanakaraju, and C.J.K. Richardson, Optics Express **16**, 2621 (2008).
12. Se-Heon Kim, Guk-Hyun Kim, Sun-Kyung Kim, Hong-Gyu Park, and Yong-Hee Lee, and Sung-Bock Kim, J. Appl. Phys. **95**, 411 (2004).
13. A. Krost and A. Dadgar, Phys. Stat. Sol. (a), **194**, 361 (2002).
14. M.A. Mastro, C. R. Eddy Jr., N. D. Bassim, M. E. Twigg, R. L. Henry, and R. T. Holm, A. Edwards, Sol. Stat. Elec., **49-2**, 251 (2005).
15. Michael A. Mastro, Ron T. Holm, Nabil D. Bassim, Charles R. Eddy, Jr., Rich L. Henry, Mark E. Twigg, and Armand Rosenberg, Jpn. J. Appl. Phys. **31**, L814 (2006).

Mater. Res. Soc. Symp. Proc. Vol. 1068 © 2008 Materials Research Society 1068-C05-03

Effect of Si and Er Co-doping on Green Electroluminescence from GaN:Er ELDs

Rui Wang, and Andrew J Steckl
Department of Electrical & Computer Engineering, University of Cincinnati, Cincinnati, OH, 45221-0030

ABSTRACT

(Er, Si) co-doped GaN thin films were grown on Si substrates by molecular beam epitaxy (MBE) technique. Electroluminescent devices (ELDs) were fabricated and the effect of Si co-doping on the performance of GaN devices was studied. Previous results with GaN:Er ELDs reported that electroluminescence (EL) was much stronger in reverse bias than in forward bias condition, indicating that the dominant factor in EL intensity was the electric field. The results reported here show the first time GaN:Er ELDs where forward bias EL is very much larger, indicating that the dominant factor is forward bias current. The electrical properties of (Si, Er) co-doped GaN thin films are believed to be responsible for the current control mechanism.

INTRODUCTION

Rare earths are of interest in optoelectronics and photonics because of their sharp emission lines. For example, Er has emission lines at both visible (538nm and 559nm) and infrared (1.54um) wavelengths, which are very useful for display devices and optical communications. Although the emission spectrum is independent of host materials, the intensity and efficiency strongly depend on the host. It was found by Favennec[1] that thermal quenching of the RE emission is inversely dependent on the energy bandgap of host materials. Therefore, III-nitride semiconductors are excellent host materials because of their direct wide bandgap. Moreover, robust thermal and mechanical properties of nitride semiconductors enable device applications under high temperature environment.

Sapphire and SiC are typically used as substrates for GaN thin film growth because their lattice constants are close to III-nitrides. However, GaN growth on Si substrates is very attractive for integration with Si microelectronics.[2] The first report[3] of electroluminescence (EL) from GaN:Er was for diodes actually grown on Si substrates and utilized a Schottky barrier structure. Subsequently, visible and IR EL emission from GaN:Er Schottky diodes was reported[4] on Si and sapphire substrates. Zavada et al[5] reported EL emission from GaN:Er P-I-N LED on sapphire. In general, in all of these devices the EL emission was much stronger in reverse bias than in forward bias.

Here we report on Er-doped GaN ELD on Si wafer with Si co-doping. The effect of Si and Er co-doping on electrical properties and Er green electroluminescence in the GaN layer was studied.

EXPERIMENT

GaN thin films were grown with Er and Si *in situ* doping on p-type Si(111) substrates in a solid source Riber 32 MBE system. A 20nm low temperature AlN buffer layer was first grown at 500° C, followed by 1hr growth of (Si, Er) co-doped GaN layer. A very thin GaN capping layer was grown last. The substrate temperature was kept at 600° C. The Ga effusion cell temperature was kept at 875° C, while the Si and Er cell temperatures were kept at 1050° C and 860° C. The resulting thickness of GaN thin films was approximately 500nm.

Room temperature photoluminescence (PL) measurements were performed with a Melles Griot 325nm He-Cd laser. A Princeton Instruments/Acton Research Corporation SP2550i spectrometer was used to acquire the emission spectra.

Ring diode structures were patterned by photolithography with a Karl Suss MJB3 mask aligner. ITO thin films were deposited with a Denton Vacuum sputtering system. After lift-off lithography, ITO electrodes were rapid thermal annealed. GaN ELDs doped with Er only were also fabricated with the same conditions for comparison.

DISCUSSION

PL spectra from GaN:(Er+Si) and GaN:Er samples are plotted in Figure 1. Both samples show green peaks of 538nm and 559nm from Er ions, corresponding to radiative transitions (Figure 1) from Er^{3+} ions energy levels $^2H_{11}$-$^4I_{15/2}$ and $^4S_{3/2}$-$^4I_{15/2}$. The peak intensities are similar if we subtract yellow band background from the spectra. There is also a small peak around 369nm corresponding to free exciton emission from GaN bandgap for both samples. No Si-related peak was observed in the PL spectra of (Si, Er) co-doped GaN thin films.

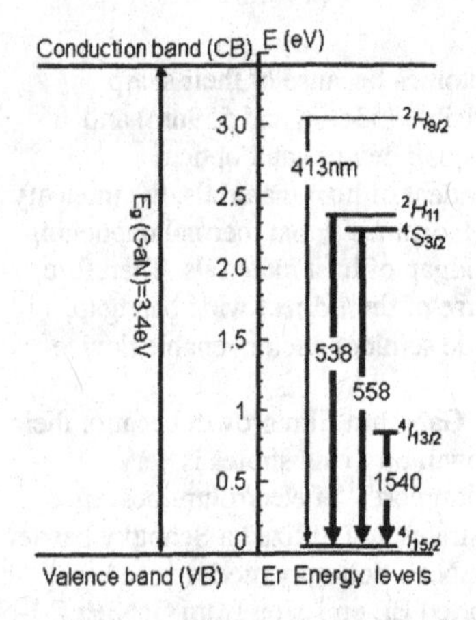

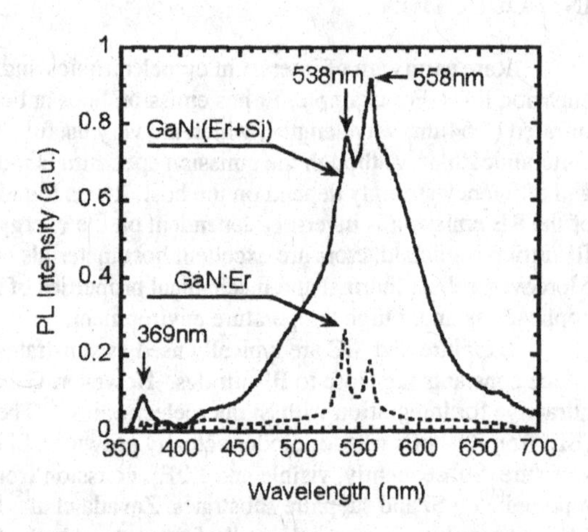

Figure 1. Energy levels of Er^{3+} ions and GaN valence and conduction bands (left); photoluminescence spectra from GaN doped with Er only, and co-doped with Er and Si (right).

Ring diode structures were fabricated with both (Si, Er) co-doped GaN samples and Er only GaN samples. The device structure is shown in Figure 2. ITO electrodes were patterned as inner and outer rings. Ag paste was used to connect the outer ring to the Si substrate. Current-voltage measurements were performed for all GaN ELDs with Labview controlled HP 6634B DC power supply. Typical results are plotted in Figure 3. Er doped GaN ELDs have similar I-V under forward and reverse bias, while (Si, Er) co-doped GaN ELDs exhibit strong I-V rectification characteristics.

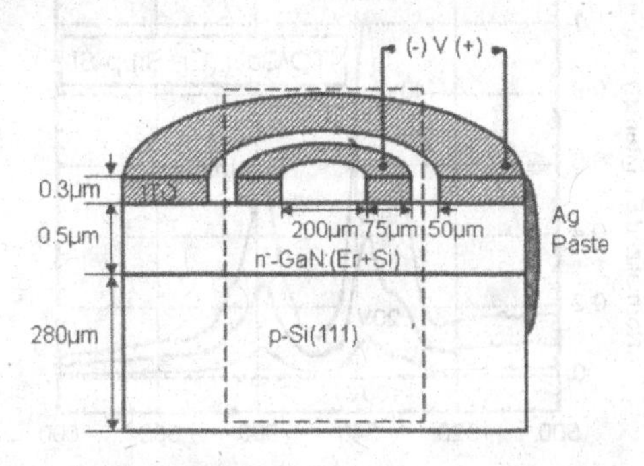

Figure 2. Diagram of (Er, Si) co-doped GaN ELDs.

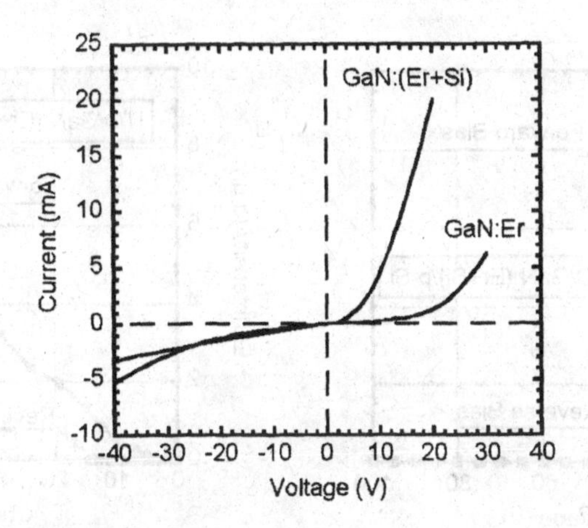

Figure 3. I-V characteristics of (Er, Si) co-doped GaN and Er doped GaN ELDs.

Electroluminescence measurements were performed on (Si, Er) co-doped GaN ELDs. Visible (green) EL spectra are shown in Figure 4 at several bias levels. The EL emission is clearly very strong under forward bias, with the intensity proportional to the current. Under reverse bias as high as 100V, little or no emission was observed. Compared to previous results on RE doped GaN ELDs, this is first time that GaN:RE ELDs were observed to have strong emission under forward bias with no emission under reverse bias. The luminescence dependence on voltage and current was also obtained and the results are shown in Figure 5.

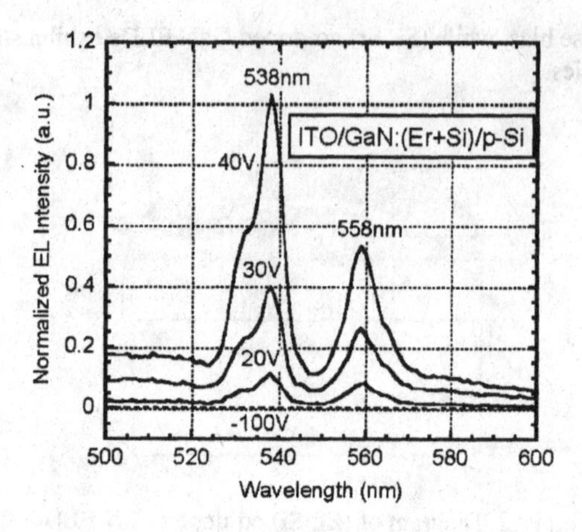

Figure 4. Visible EL spectra from ITO/GaN:(Si+Er)/p-Si(111) devices at several bias levels.

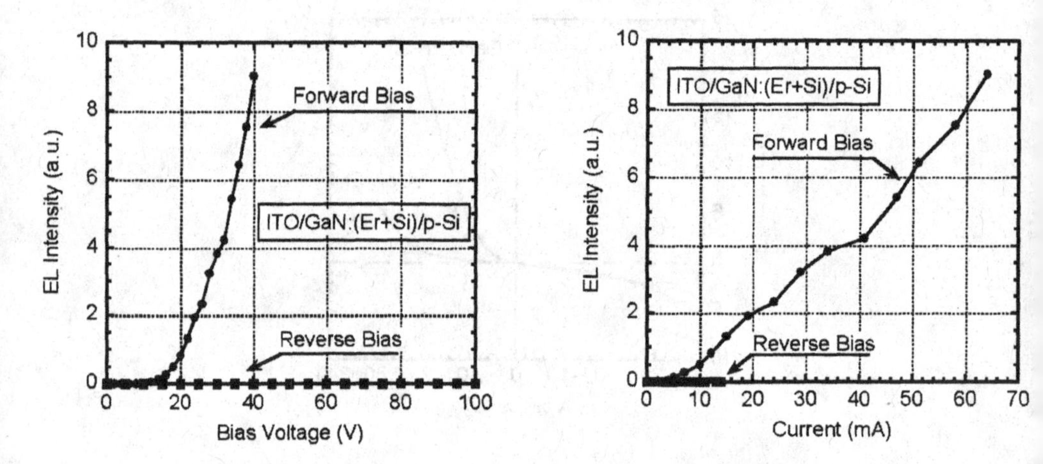

Figure 5. EL intensity dependence on voltage and current conditions from ITO/GaN:(Si+Er)/p-Si(111) device under forward and reverse bias.

Undoped GaN thin films grown in our lab typically show weak n-type conductivity with electron concentration around $10^{17}cm^{-3}$. It is believed[6] that the nitrogen vacancy (N_V) acts as the donor and accounts for such n-type conductivity. However, RE-doped GaN is generally highly insulating. The change to an insulating GaN:RE layer is not fully understood, but it is generally thought that RE ions act as electron trap centers in the GaN bandgap.

With Si co-doping in GaN:RE thin films, Si atoms act as donors and provide a significant source of electrons. The electrical properties of GaN:Er thin films have been modified from highly resistive to weakly n-type. The sheet resistance of (Si, Er) co-doped GaN thin films was measured as 1.15×10^5 Ω/□ by a six point probe meter, whereas the Er-doped-only GaN has a resistance too large to be measured (>10^{10} Ω/□). Combined with the p-type Si(111) substrate, a pn^- heterojunction is formed at the interface between the GaN layer and the Si layer, shown as the dashed rectangle area in Figure 2. The electrical properties of the pn heterostructure strongly affect the I-V characteristics of GaN:Er ELDs under different DC bias.

Generally, there are two main mechanisms for RE luminescence: electron-hole recombination and hot carriers impact excitation.[7] For EL emission, as electrons and holes are injected into the active GaN layer under forward bias, they recombine and transfer their energy to neighboring RE ions. Some of the excited RE ions relax radiatively, resulting in EL emission. Under reverse bias, hot electrons injected into the GaN layers collide with RE ions and transfer their energy to RE ions directly. For RE-doped-only GaN ELDs, the insulating GaN:RE layer acts a high barrier layer and limits the current which can go through the layer. Since the Si co-doped GaN:RE layer is much more conducting, the pn^- heterojunction experiences enhanced current flow under forward bias, but acts as a barrier for electrons under reverse bias. Although a few hot electrons can tunnel through the GaN:(Si+Er) layer, their energy probably is not sufficient for impact excitation mechanism, resulting in little or no emission.

CONCLUSIONS

In conclusion, we have reported GaN thin films co-doped with Er and Si grown by MBE. And we have fabricated GaN ELDs on Si substrates. The effect of Si and Er co-doping on the electrical and luminescent properties of (Si, Er) co-doped GaN ELDs was studied. The co-doped (Er+Si) GaN ELDs have achieved for the first time current controlled luminescence.

ACKNOWLEDGMENTS

This work is supported by an ARO grant. The authors would like to acknowledge Dr. John Zavada for his kind support and encouragement.

REFERENCES

[1] P. N. Favennec, H. L'Haridon, M. Salvi, D. Moutonnet, and Y. Le Guillou, "Luminescence of erbium implanted in various semiconductors: IV, III-V and II-VI materials," *Elec. Lett.*, vol. 25, p. 718, 1989.

[2] A. J. Steckl, J. H. Park, and J. M. Zavada, "Prospects for rare earth doped GaN lasers on Si," *Materials Today,* vol. 10, pp. 20-27, 2007.

[3] A. J. Steckl, M. Garter, R. Birkhahn, and J. Scofield, "Green electroluminescence from Er-doped GaN Schottky barrier diodes," *Appl. Phys. Lett.,* vol. 73, pp. 2450-2452, 1998.

[4] M. J. Garter and A. J. Steckl, "Temperature behavior of visible and infrared electroluminescent devices fabricated on erbium-doped GaN," *IEEE Trans. on electron devices,* vol. 49, pp. 48-54, 2002.

[5] J. M. Zavada, S. X. Jin, N. Nepal, J. Y. Lin, H. X. Jiang, P. Chow, and B. Hertog, "Electroluminescent properties of erbium-doped III-V light-emitting diodes," *Appl. Phys. Lett.*, vol. 84, pp. 1061-1063, 2004.

[6] D. W. Jenkins and J. D. Dow, "Electronic structures and doping of InN, $In_xGa_{1-x}N$, and $In_xAl_{1-x}N$," *Phys. Rev. B*, vol. 39, pp. 3317-3329, 1989.

[7] A. J. Steckl, J. Heikenfeld, M. Gartner, R. Birkhahn, and D. S. Lee, "Rare Earth Doped Gallium Nitride — Light Emission from Ultraviolet to Infrared," *Compound Semiconductor*, vol. 6, pp. 48-52, 2000.

Mater. Res. Soc. Symp. Proc. Vol. 1068 © 2008 Materials Research Society 1068-C05-04

MOVPE of m-plane InGaN/GaN Buffer and LED Structures on γ-$LiAlO_2$

H. Behmenburg[1], C. Mauder[1], L. Rahimzadeh Khoshroo[1], T.C. Wen[1], Y. Dikme[2], M.V. Rzheutskii[3], E.V. Lutsenko[3], G.P. Yablonskii[3], M.M.C. Chou[4], J. Woitok[5], M. Heuken[1,2], H Kalisch[1], and R.H. Jansen[1]
[1]Chair of Electromagnetic Theory, RWTH Aachen, Aachen, 52074, Germany
[2]AIXTRON AG, Aachen, Germany
[3]Stepanov Institute of Physics, National Academy of Sciences of Belarus, Minsk, Belarus
[4]National Sun Yat-Sen University, Kaohsiung, Taiwan
[5]PANalytical B.V., Almelo, Netherlands

ABSTRACT

We report on deposition and properties of m-plane GaN/InGaN/AlInN structures on $LiAlO_2$ substrates grown by metal organic vapor phase epitaxy (MOVPE). At first, two different buffer structures, one of them including an m-plane AlInN interlayer, were investigated concerning their suitability for the subsequent coalesced single-phase m-plane GaN growth. A series of quantum well structures with different well thickness based on one of these buffers showed absence of polarization-induced electric fields verified by room temperature photoluminescence (RT PL) measurements at different excitation intensities. Furthermore, polarization-resolved PL measurements revealed a high degree of polarization (DoP) of the emitted light with an intensity ratio of 8:1 between light polarized perpendicular and parallel to the c-axis.

INTRODUCTION

Epitaxial growth for optoelectronic devices is mostly performed in c-direction on sapphire or silicon carbide substrates. In this case, piezoelectric and spontaneous polarization occur perpendicular to the surface. The resulting electrostatic fields across the quantum well structure lead to band bending and spatial separation of electron and hole wavefunctions in c-plane light emitting diodes (LEDs) [1,2]. These effects cause an instable emission wavelength of the device with different drive currents and a reduced recombination efficiency [2,3,4].

Growth on the alternative substrate (100) γ-$LiAlO_2$ allows a deposition of group III-nitrides with the non-polar m-plane (1-100). This is possible due to a small lattice mismatch of 1.7 % for (11-20) GaN planes parallel to (001) $LiAlO_2$ and 0.3 % for the (0001) GaN planes parallel to (010) $LiAlO_2$ [5]. By growth in the m-plane mode, polarization-induced electric fields across the active region of the LED can be avoided and these non-polar devices are expected to result in stable emission wavelength with drive current and in high recombination efficiency [6,7]. γ-$LiAlO_2$ substrates can be produced inexpensively by Czochralski pulling but they have a lower thermal stability compared to sapphire and are more sensitive to hydrogen.

EXPERIMENT

All growth experiments have been performed in AIXTRON MOVPE reactors with standard precursors and process gases. Epi-ready (100)-oriented γ-$LiAlO_2$ substrates with an average surface roughness of 0.4-0.5 nm root-mean-square (RMS) were used for the experiments presented here. Details of ingot growth and polishing procedure of the substrates can be found elsewhere [8].

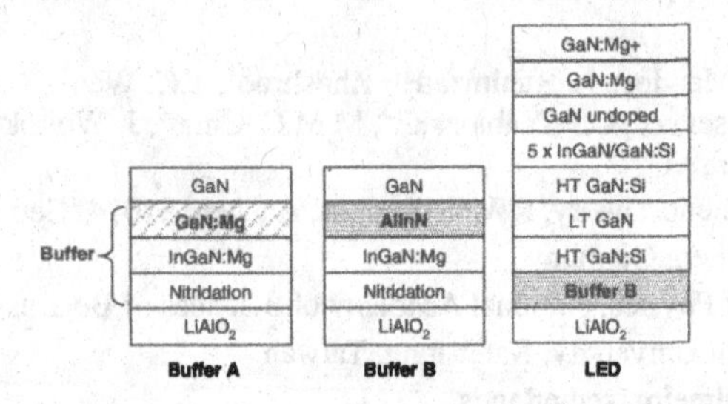

Figure 1. Schematic cross section of the investigated buffer structures A (left) and B (middle) and the LED structure (right).

A schematic cross section of the investigated buffer structures A and B is shown in Fig.1. Both structures start with an Mg-doped InGaN layer. For buffer A, the Mg doping was continued in the following 200 nm GaN layer, in buffer B, the GaN:Mg interlayer was replaced by a 150 nm thick AlInN layer. Both structures are completed with a 500 nm GaN layer. Prior to the InGaN:Mg starting layer, all samples were in-situ nitridated at temperatures between 900°C and 1050°C. While nitridation of the substrate surface is reported to result in a formation of AlN [9], growth of InGaN:Mg was found helpful to reduce the diffusion of oxygen out of the substrate into the active layer and to enable subsequent GaN growth in the m-plane mode [10]. Based on buffer A, a series of four samples with InGaN:Mg thicknesses of 25, 50, 100 and 150 nm was produced, in buffer structure B, the InGaN:Mg layer was kept at 25 nm. All buffer structures were investigated by high resolution X-ray diffraction (HRXRD) and atomic force microscopy (AFM).

Based on buffer B, a series of four LED structures with an n-type GaN:Si layer, a 5-fold multiple quantum well (MQW) and p-type GaN:Mg cap layers were grown. The high-temperature (HT) GaN:Si layer grown at approximately 1040°C surface temperature was interrupted by a thin low-temperature (LT) GaN nucleation layer grown at 560°C. A schematic cross section of this structure is also shown in Fig.1. The samples of this series have a constant GaN:Si MQW barrier thickness of 15 nm and varying InGaN well thicknesses between 1 nm and 4 nm. The samples were investigated by room temperature photoluminescence spectroscopy (RT PL) at different excitation intensities of a nitrogen laser emitting at 337 nm and by polarization-resolved PL.

DISCUSSION

In Fig. 2, HRXRD 2Theta/Omega scans (slid width 1 mm) of the samples with InGaN buffer thicknesses of 25, 50, 100 and 150 nm are displayed, respectively. Beside the (200) substrate peak at 34.69°, we only observe peaks of m-plane oriented GaN and m-plane InGaN. No c-plane material is detected. With rising InGaN thickness, we see an increasing intensity of this peak. The peaks for GaN are located around 32.27°, which indicates in-plane compressive strain and is similar to what other groups report [5,11].

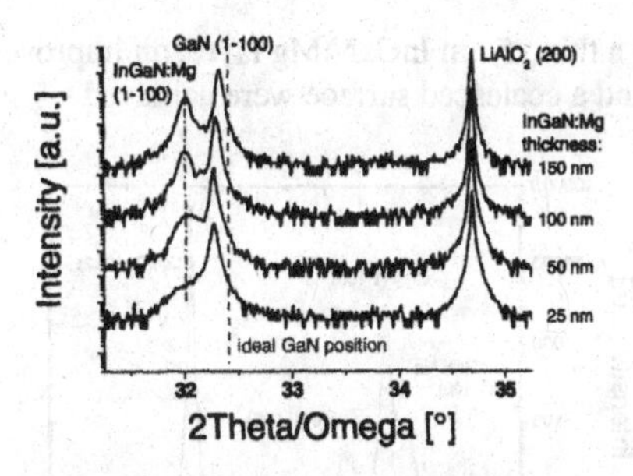

Figure 2. 2Theta/Omega scans (slid width 1 mm) of InGaN/GaN buffer A on $LiAlO_2$ with increasing InGaN:Mg thickness. The dotted line on the left side is a guide to the eye. The dashed line shows the position of unstrained m-plane GaN.

But for thicker InGaN buffers, this peak shifts to higher angles towards its nominal position at 32.39°. We suggest a relaxation of the GaN layer induced by a thicker underlying InGaN layer as possible explanation. An analysis of the relaxation status is ongoing and will be reported later. Further characterization was performed by comparing X-ray rocking curve (XRC) full width at half maximum (FWHM) values as well as the average roughnesses recorded by atomic force microscopy (AFM). These data are shown in Fig. 3. The (1-100) XRC scans with the incident beam along [0001] direction yielded an increase of the FWHM values for a thickness above 50 nm. Since basal plane stacking faults (BSF) disrupt the crystal structure parallel to the (In)GaN c-axis and thus will mainly affect the XRC FWHM along [0001], we think there might be an increase of such defects for thicker InGaN:Mg layers. In contrast, the AFM roughness decreased for higher InGaN:Mg thicknesses. This is mainly due to a better coalescence of the epilayer in the first growth stage preventing an etching effect when switching to hydrogen ambient.

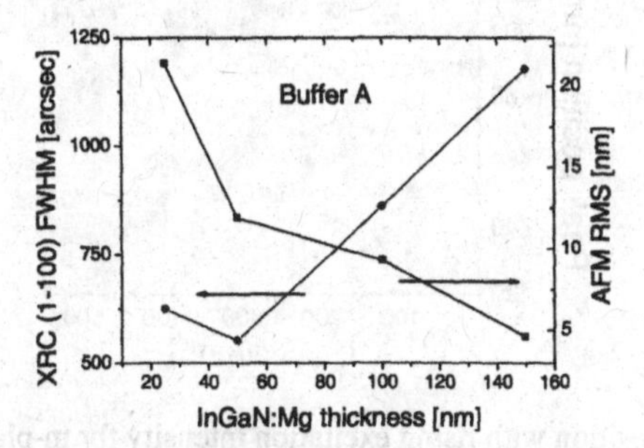

Figure 3. Development of the XRC FWHM (incident beam parallel to [0001]) and AFM RMS roughness for various InGaN:Mg buffer layer thicknesses.

In buffer B, the GaN:Mg layer was replaced by an AlInN interlayer to improve the surface coalescence for the terminating GaN layer. Reasons for introducing this interlayer are the preferable AlInN growth conditions [12] with a lower growth temperature of 860°C and maintained nitrogen ambient. These conditions prevent the thermally and H_2 sensitive substrate from damage. Despite a deterioration of the XRC FWHM of the uppermost GaN layer of this buffer with 1942 arcsec and a thin 25 nm InGaN:Mg layer, an improvement of the surface roughness to 3.5 nm RMS and a coalesced surface were achieved.

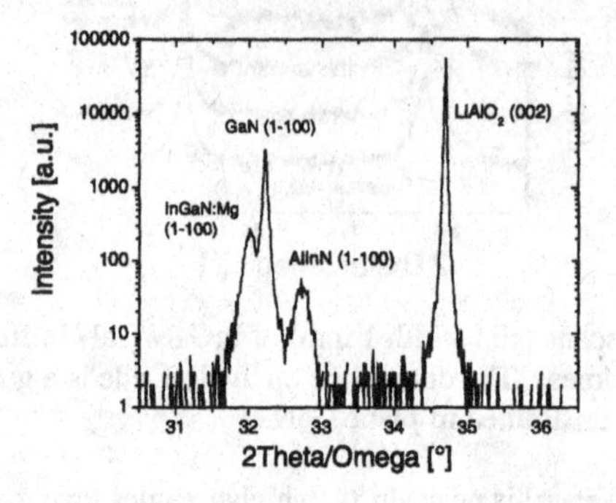

Figure 4. Triple-axis HRXRD scan of buffer structure B. Peaks of m-plane GaN, AlInN, InGaN were detected. No indication of GaN (0002), which should appear as a shoulder left to the $LiAlO_2$ peak, was found.

The 2Theta/Omega scan in Fig. 4 shows the achievement of single phase m-plane GaN growth. No reflection of c-plane GaN (0002), which should appear as a shoulder left to the $LiAlO_2$ peak, was detected. Furthermore, a new peak was found in this scan on the right side of the m-plane GaN. Correlation of appearance and angular position of the peak allows attributing this reflex to m-plane AlInN.

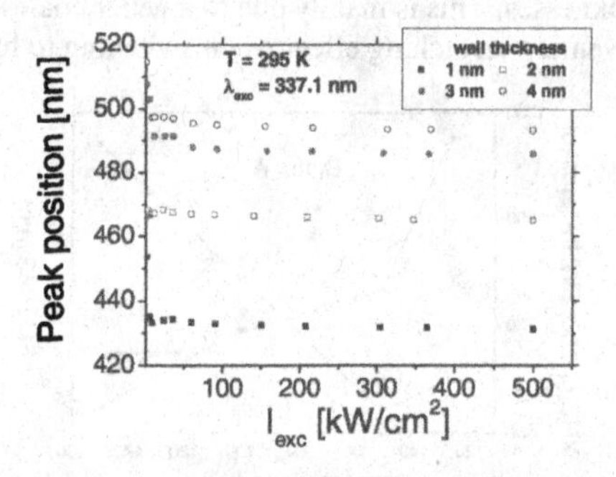

Figure 5. Stable peak position with rising excitation intensity for m-plane InGaN/GaN MQW structures with well width from 1 to 4 nm.

For common LED structures grown in the c-plane mode, a blue-shift of the emission wavelength is observed in EL or PL measurements when drive current or excitation intensity increases. This effect is usually explained by screening of the electrostatic fields inside the QW and saturation of band tails in indium clusters with rising carrier injection [3,4].
The expected absence of an electrostatic field across the heterostructures in m-plane structures was tested by RT PL with different excitation intensities on four LEDs with different well width. In Fig. 5, the peak wavelength position of all samples as a function of excitation intensity is

displayed. With rising well width, the peak position shifts to longer wavelengths. The increase of indium concentration and a decrease of the discrete energy levels inside the well with rising well width explain this red-shift. The peak position of each sample exhibited no significant shift at excitation intensities up to 500 kW/cm². The lack of polarization fields and a low account of indium clusters are a possible explanation for the spectral stability.

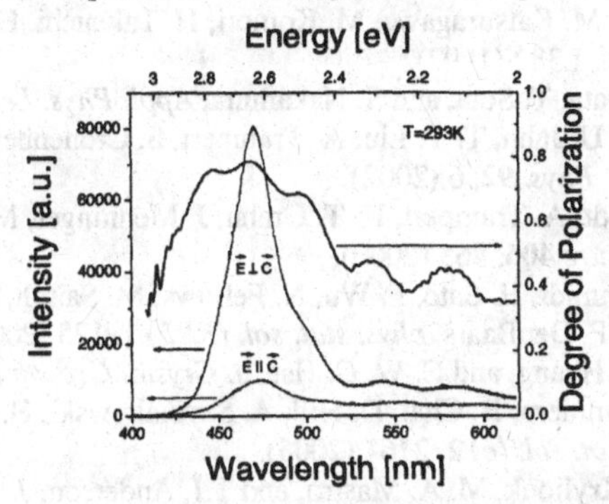

Figure 6. Intensity of spontaneous emission polarized parallel and perpendicular to the nitride c-axis. A maximum DoP of 0.78 is achieved, corresponding to an intensity ratio of 8:1.

The polarization of the RT PL emission of the LED structures was investigated by placing a polarization filter between sample and detector. By rotation of the polarization filter, alterations of the PL spectral intensity were observed. On all samples, two positions of the polarization filter, 90 degrees apart, were found, corresponding to the maximal and minimal intensity of the PL emission. This is in accordance to previous observations by Sun [13] and Domen [14], who attribute the maximal PL intensity to light polarization perpendicular to the c-axis, the minimal one to polarization parallel to the c-axis. As it can be seen in Fig. 6, a maximum DoP of 0.78 was achieved corresponding to an intensity ratio of 8:1.

CONCLUSIONS

Single-phase m-plane GaN growth was achieved on different buffer structures on γ-$LiAlO_2$ substrates. The thickness of the InGaN:Mg starting layer showed a controversial impact on surface roughness and crystal quality of the following GaN layer. The introduction of an m-plane AlInN interlayer was found helpful to improve the surface roughness and coalescence of the subsequent GaN layer. RT PL measurements at various excitation intensities on four LED structures with different well width revealed a remarkably stable emission wavelength with a shift of only a few nm for excitation intensities up to 500 kW/cm². Furthermore, polarization-resolved PL measurements showed a high degree of polarization of the emitted light with a maximum intensity ratio of 8:1. These promising results indicate the possible use of γ-$LiAlO_2$ substrates for GaN-based optoelectronic devices. Electrical characterization and processing of the LED structures are in process.

REFERENCES

1. F. Bernardini, V. Fiorentini, and D. Vanderbilt, *phys. rev. B* **56**, R10024 (1997).
2. V. Fiorentini, F. Bernardini, F. Della Salla, A. Di Carlo, and P. Lugli, *phys. rev. B* **60**, 8849 (1999).
3. T. Takeuchi, S. Sota, M. Katsuragawa, M. Komori, H. Takeuchi, H. Amano, and I. Akasaki, *Jpn. J. Appl. Phys.* **36**, L382 (1997).
4. S. Chichibu, T. Azuhata, T. Sota, and S. Nakamura, *Appl. Phys. Lett.* **69**, 4188 (1996).
5. Y. J. Sun, O. Brandt, U. Jahn, T. Y. Liu, A. Trampert, S. Cronenberg, S. Dhar, and K. H. Ploog, *J. Appl. Phys.* **92**, 6 (2002).
6. P. Waltereit, O. Brandt, A. Trampert, H. T. Grahn, J. Menninger, M. Ramsteiner, M. Reiche, and K. H. Ploog, *nature* **406**, 865 (2000).
7. K. C. Kim, M. C. Schmidt, H. Sato, F. Wu, N. Fellows, M. Saitoh, K. Fujito, J. S. Speck, S. Nakamura, and S. P. DenBaars, *phys. stat. sol. (RRL)* **1**, 125 (2007).
8. M. M. C. Chou, S. J. Huang, and C. W. C. Hsu, *J. Crystal Growth* **303**, 585 (2007).
9. Y. Dikme, P. van Gemmern, B. Chai, D. Hill, A. Szymakowski, H. Kalisch, M. Heuken, and R. H. Jansen, *phys. stat. sol. (c)* **2**, 2161 (2005).
10. M. D. Reed, O. M. Kryliouk, M. A. Mastro, and T.J. Anderson, *J. Crystal Growth* **274**, 14 (2005).
11. P. Waltereit, O. Brandt, M. Ramsteiner, A. Trampert, H. T. Grahn, J. Menninger, M. Reiche, R. Uecker, P. Reiche and K. H. Ploog, *phys. stat. sol.* (a) **180**, 133 (2000).
12. L. Rahimzadeh Khoshroo, C. Mauder, W. Zhang, M. Fieger, M. Eickelkamp, Y. Dikme, J. Woitok, P. Niyamakom, A. Vescan, H. Kalisch, M. Heuken, and R. H. Jansen, *to be published in phys. stat. sol.* (c), 2008.
13. Y. J. Sun, *PhD thesis*, HU Berlin (2004).
14. K. Domen, K. Horino, A. Kuramata, and T. Tanahashi, *Appl. Phys. Lett.* **71**, 1996 (1997).

Mater. Res. Soc. Symp. Proc. Vol. 1068 © 2008 Materials Research Society 1068-C05-06

Strong Light-Matter Coupling in GaN-Based Microcavities Grown on Silicon Substrates

Fabrice Semond[1], Ian Roberts Sellers[1], Nadège Ollier[2], Franck Natali[1], Declan Byrne[1], François Réveret[2], Flavian Stokker-Cheregi[3], Katarzyna Bejtka[1,4], Maximo Gurioli[3], Anna Vinattieri[3], Aimé Vasson[2], Pierre Disseix[2], Joël Leymarie[2], Mathieu Leroux[1], and Jean Massies[1]
[1]CRHEA, CNRS, rue Bernard Gregory, Valbonne, 06560, France
[2]UMR 6602 UBP/CNRS, LASMEA, 24 Avenue des Landais, Aubière, 63177, France
[3]Dipartimento di Fisica and LENS, University of Florence, Sesto Fiorentino, Italy
[4]Department of Physics, SUPA, University of Strathclyde, Glasgow, G4 0NG, United Kingdom

ABSTRACT

We present an overview of our work concerning the fabrication of GaN-based microcavities grown on silicon substrates dedicated to the observation of the strong light-matter coupling regime. In the view of recent promising results in the field, prospects regarding the improvement of heterostructures in order to observe room temperature polariton lasing from a GaN-based microcavity grown on a silicon substrate will be discussed.

INTRODUCTION

Although GaN-based optoelectronic devices are mostly grown on Al_2O_3 or SiC, a lot of efforts are made to develop GaN-based devices on silicon substrates. The material quality of GaN layers grown on silicon is nearly as good as the one used to fabricate commercially available GaN-based LEDs, but the light output power of LEDs grown on Si is still too low to compete with devices grown on Al_2O_3 or SiC. Knowing that, is it reasonable to think about low-threshold lasing from GaN-based heterostructures grown on Si? In this paper we are going to discuss how cavity-polaritons (CPs) could help to fabricate such a low-threshold laser on silicon.

The physics of semiconductor microcavities (MCs) operating in the strong light-matter coupling regime (SCR) has progressed rapidly in the last decade [1,2]. This regime produces new mixed quasiparticles, which are half-light and half-matter, called cavity-polaritons and having unusual and very interesting properties [1,2]. In particular, besides fundamental studies, semiconductor MCs operating in the SCR are promising candidates to develop low threshold light emitters [1,2]. GaN is a good candidate, since the large exciton binding energy and optical oscillator strengths of this materials offers the possibility to observe SCR at room temperature (RT) [3]. Since the first observation of the SCR in a GaN-based MC [4], improvements have been made and polariton emission at RT has been confirmed [5-10] and very recently, RT polariton lasing with an amazingly low-threshold has been reported [11].

FIRST OBSERVATION OF THE STRONG COUPLING IN A GaN MICROCAVITY

The first objective to achieve was to obtain the SCR in a GaN-based MC at low temperature (LT) and then at RT. Usually high finesse MCs are needed to demonstrate the SCR. For GaN-based structures it was not, and it is still not, a trivial issue to grow crack-free, highly-reflective Distributed Bragg Reflectors (DBRs). In this respect it was quickly recognized that a silicon substrate was a good candidate for the fabrication of GaN-based MCs, since the substrate can be easily removed and then high quality factor (Q) MCs using high reflectivity dielectric mirrors on both sides [12] could be fabricated. Based on simulations showing that it would be possible to observe SCR in a GaN bulk cavity as soon as the inhomogeneous broadening of the exciton is reduced below 30 meV even if the amplitude of the optical cavity field is not very high (low Q factor) [13], we decided in a very first approach to fabricate a simple low finesse GaN MC. Using the know-how we had at that time on the growth by MBE of crack-free good-quality GaN layers on Si [14,15], a λ/2 GaN bulk cavity was grown on Si. A simple three layer Bragg stack consisting of a 2λ-λ/4 layer of AlN, a 2λ-λ/4 layer of $Al_{0.20}Ga_{0.80}N$ and a λ/4 layer of AlN was grown in between the Si substrate and cavity [4]. These three layers were the buffer layers necessary to overcome the difficulties encountered during the growth of crack-free nitrides on Si [14,15] and their thicknesses were specifically chosen so as to satisfy the Bragg condition. Combined with these 3 buffer layers, the substrate forms the bottom mirror and produces an estimated reflectivity of about 30% in the near-UV range [4].

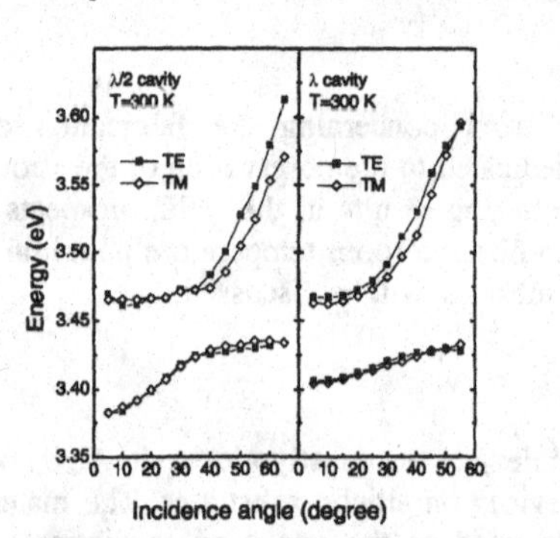

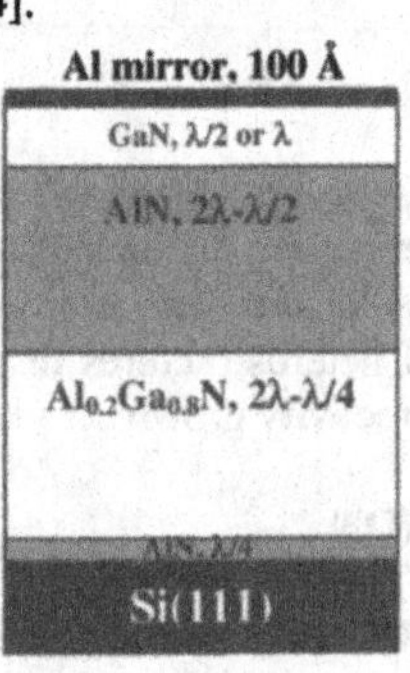

Figure 1. Right, layer structure studied. Left, cavity-polariton dispersion curves showing the anticrossing. Data deduced from angle resolved reflectivity measurements at RT. Rabi splitting values of 47 and 60 meV are measured for a λ/2 and a λ GaN bulk cavity respectively [5].

The upper mirror is a 4 period SiO_2/Si_3N_4 mirror producing a reflectivity of 83% giving a Q of about 75. This microcavity structure was studied in reflection under various angles of incidence at LT and a clear anticrossing between excitonic and photonic modes was observed indicating that the SCR was achieved [4]. A Rabi energy splitting of 31±1 meV was obtained and an overall good agreement was found between calculations and experimental results. The exciton-photon coupling has also been observed at 77K, but at RT, due to the additional thermal broadening of excitons, the polariton splitting was not resolved. Later on, using a similar MC structure (see Fig.1, right) and a thin aluminium semi-transparent top mirror instead of a dielectric top mirror, we were able to follow the SCR up to RT [5]. The CP dispersion curves, deduced from angle resolved reflectivity measurements are displayed Fig. 1. Despite the fact of a lower Q factor (~ 40), due to the top metallic mirror instead of a top dielectric mirror, a higher Rabi splitting of 47 meV was measured for a λ/2 GaN bulk cavity and it explains why the SCR was maintained up to RT. A Rabi splitting energy up to 60 meV was measured for a GaN λ cavity at RT [5].

EVIDENCE OF POLARITON EMISSION

In earlier samples, cavity and exciton modes were quite broad and the SCR was only observed by reflectivity but we could not observe the CPs dispersion curve from PL measurements. In order to improve the Q factor of our GaN-bulk MCs grown on Si we then used a bottom AlN/AlGaN epitaxial DBR [6]. DBRs having a very low surface roughness, typically 0.5 nm over a 3x3 μm^2 surface area, were grown on Si (Fig. 2). Reflectivities up to 92% were obtained using only a 12 pair DBR stack (Fig. 2). Actually, it is found out that the Si substrate helps to increase the reflectivity seen by the GaN cavity [16]. First, a bulk λ/2-GaN MC having a 7 period bottom epitaxial $AlN/Al_{0.2}Ga_{0.8}N$ DBR grown directly on Si(111) has been fabricated [6]. The microcavity was completed with the deposition of a 10 nm transparent aluminium mirror. By using a bottom epitaxial DBR instead of the 3 layer stack described above, we noticed that the PL linewidth was slightly improved (20 meV at 10K) but the Q was still low (~ 60) due to the use of a top Al mirror. Nevertheless, using angle resolved reflectivity measurements the SCR was observed on this structure up to RT with a Rabi splitting of 50 meV [6]. Interestingly, at low temperature the interaction between both the free A and B-exciton transitions with the photonic mode was clearly resolved for the first time. However there was still no evidence of the SCR by PL measurements and then no proof of the polaritonic nature of the emission.

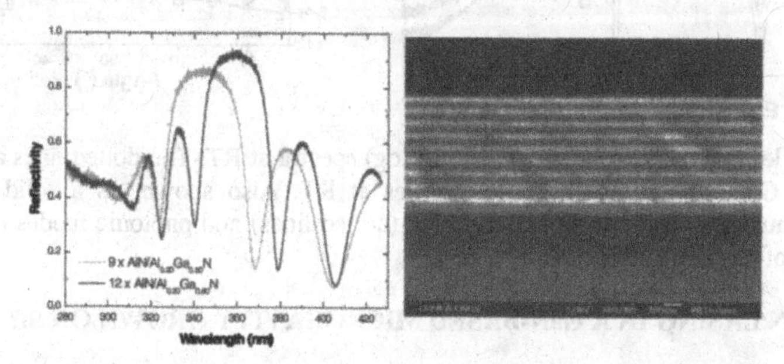

Figure 2. a) SEM image of an epitaxial AlN/AlGaN DBR grown on Si, b) RT reflectivities of AlN/AlGaN DBRs grown on silicon having 9 and 12 pairs.

Evidence of polariton emission in the nitride system was first reported by Tarawa *et al.* in 2004 [17]. However, these results remain controversial due to the very large emission linewidth of InGaN QWs (200 meV) with respect to the Rabi splitting measured (6 meV). Actually, polariton emission in a GaN MC was first reported by Butté *et al.* using a bottom lattice-matched high-reflectivity AlInN/AlGaN DBR grown on GaN/sapphire [7]. The structure was then completed by the deposition of a 10 pair SiO_2/Si_3N_4 DBR and a Q factor in excess of ~ 2000 was measured [7]. The low Q factor of microcavities grown on Si was likely the reason why we did not see so far polariton emission [18]. This motivated the growth of a λ/2-GaN MC having a 10 period bottom $AlN/Al_{0.2}Ga_{0.8}N$ DBR grown on Si closed by a 8 period SiN/SiO_2 DBR [9]. A Q of ~ 160 is measured. For the first time the SCR was observed using simultaneously reflectivity and PL measurements [9]. Due to the higher Q of this structure, the two polariton branches are clearly resolved at LT, and a Rabi splitting of ~ 30 meV is measured. Despite the thermal broadening of transitions the SCR is still observed at RT. Figure 3(a) and 3(b) show respectively angle-resolved reflectivity and PL spectra at 300K. On the 5° reflectivity spectra, the photonic mode at 3.444 eV is negatively detuned with respect to the excitonic mode at 3.499 eV. As the angle is increased an anti-crossing is observed with a resonance at about 35°. Figure 3(c) shows the dispersion of the polariton modes deduced from both reflectivity and PL measurements. Also shown in bold are the results of a (3x3) matrix calculation considering the system as the interaction between A and B excitons of constant energy and an angle dependent photonic mode. Relevant confirmations of the mixing between excitons and photons are also found in the analysis of the spectral linewidth and the time resolved kinetics [19].

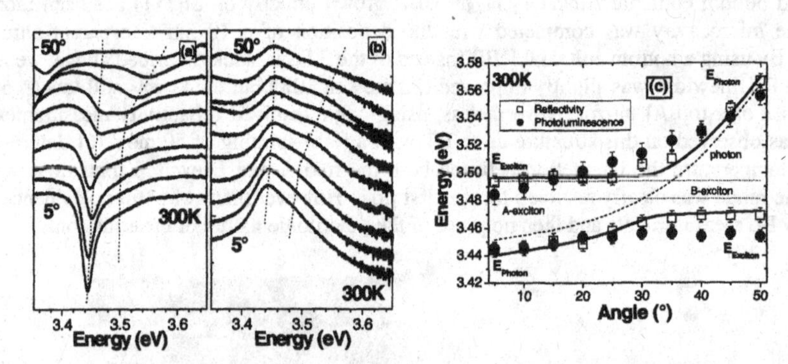

Figure 3. angle resolved a) reflectivity, b) PL (log) spectra at RT. The dotted lines are guide for eye only. c) Cavity-polariton dispersion curves at RT. Also shown, as a solid line, is the numerical simulation. The uncoupled excitonic (dotted lines) and photonic modes (dashed line) as a function of angle are also included.

POLARITON LASING IN A GaN-BASED MICROCAVITY GROWN ON Si?

Once the strong coupling regime is achieved, cavity-polaritons are formed. Having a bosonic character, these new quasiparticles may accumulate in a single quantum state which gives rise to the polariton laser effect [11,20,21]. In GaAs-based MCs a bottleneck was found to inhibit the relaxation of CPs toward the lowest state [22,23] and polariton lasing was observed only in laterally confined structures [24]. Although the polariton bottleneck problem was overcome in II-VI MCs [25], polariton lasing could not be achieved at RT in this material system [26]. In contrast, room temperature polariton lasing with a low threshold in GaN based MCs has been predicted [21] and recently experimentally observed [11]. Actually, it is reported that the threshold is one order of magnitude smaller than the best optically pumped InGaN QWs surface-emitting laser. Interestingly this much lower threshold has been observed in a microcavity containing only a simple bulk GaN cavity with no QW in the active region, supporting that polariton lasers behave very differently than conventional lasers.

The reason why we did not observe so far polariton lasing in our GaN-based microcavities grown on Si could be the lower Q and or the shorter exciton lifetime. On Si the Q factor is typically a factor 10 lower than the Q factor measured in ref.[11] and the exciton lifetime is a factor 10 shorter, likely due to the higher nonradiative defect density in MCs grown on Si. Recent simulations suggest that a Q of at least 500-1000 is needed to obtain polariton lasing [27]. Up to now, Q factor of MCs grown on Si was quite far from this value. Actually, the main drawback dealing with the growth on Si is the huge thermal expansion coefficient mismatch between nitrides and Si. This is responsible for tensile strain and cracking. Already with 10 pairs, a crack density of 100/mm is observed by optical microscopy in our samples; however in this range there is no evidence to suggest that cracks influence the optical performance of the structure but the increase of the number of epitaxial pairs will result in massively cracked structures. To enhance significantly the Q and to get rid of the cracks, we propose to fabricate fully hybrid MCs, having high-reflectivity dielectric DBRs on both sides. Before to go further in the fabrication process, let's discuss the excitonic part. The MC showing polariton lasing [11], is grown on a standard 3 μm thick GaN buffer layer on Al_2O_3 and the dislocation density is typically in the low 10^9 cm^{-2} [7]. This is typically the quality reached for thick GaN layers (~ 2-3 μm) grown on Si [14]. Unfortunately, we did not succeed so far to grow crack-free MC structures on top of such a thick GaN buffer layer grown on Si. So far the bottom epitaxial DBR was directly grown on Si and obviously this results in a higher dislocation density ($\geq 10^{10}$ cm^{-2}). Despite these differences, the suppression of the polariton bottleneck at RT has also been reported in MCs grown on Si [28] indicating that the effective relaxation, which is a key issue in order to achieve polariton lasing, is not affected by the lower quality on MCs grown on Si. Nevertheless, the poorer quality is likely the reason why polariton lasing has not been observed so far on Si.

In this last section we propose a structure in order to improve significantly both the optical mode (Q factor) and the excitonic oscillator of GaN-based MCs grown on Si. First a thick good-quality GaN layer is grown on a Si substrate. By doing so we know that in the last 100-200 nm (typically the thickness of the future cavity), the dislocation density is going to be in the low 10^9 cm^{-2}. After the deposition of a high reflectivity top dielectric mirror and bonding to a carrier substrate, the Si substrate is removed. Then, starting from the backside, the defective nitride is etched away up to an etch stop layer which would have been intentionally intercalated during the epitaxy. The use of an etch stop layer is mandatory in order to control perfectly the cavity thickness of the final MC. An AlInN layer lattice matched to GaN would be perfect [29]. Finally the microcavity will be closed by the deposition of a second dielectric mirror on the backside giving high Q cavities. Although there are some technological optimizations behind this approach, there is no major obstacle to achieve such a microcavity structure. As a preliminary result, the strong light-matter coupling regime in a double dielectric mirror GaN microcavity (only by removing the silicon substrate) has been recently observed [30].

CONCLUSIONS

During the last years, thanks to a very fruitful collaborative work, we have intensively investigated GaN-based microcavities fabricated on silicon substrates and dedicated to the observation of the strong light-matter coupling. In the early days very simple microcavity structures allowed to observe for the first time the strong coupling regime in nitrides at low and then at room temperature with Rabi splitting values up to 60 meV. Optimizing the material quality as well as the microcavity design, polariton emission at room temperature has been demonstrated and the polariton relaxation studied. Considering the good quality of GaN layers grown on silicon, the potentialities offered by the silicon substrates and the very good understanding of the behavior of our microcavities, we do think that it would be possible in a near future to achieve low threshold polariton lasing in a GaN-based microcavity grown on a silicon substrate.

ACKNOWLEDGMENTS

This work was funded by the European Community through CLERMONT (MRTN-CT-2003-503677) and STIMSCAT (FP6-517769). The work at LENS was supported by EC contract RII3-CT-2003-506350. We acknowledge discussions with Guillaume Malpuech.

REFERENCES

1. Semiconductor Microcavities, Guest Editor B. Deveaud, Phys. Stat. Sol. (b) **242**(11) (2005).
2. Microcavities, A.V. Kavokin, J.J. Baumberg, G. Malpuech, F.P. Laussy, Oxford Press, (2007).
3. A. Kavokin *et al.*, Appl. Phys. Lett. **72**, 2880 (1998).
4. N. Antoine-Vincent *et al.*, Phys. Rev. B **68**, 153313 (2003).
5. F. Semond *et al.*, Appl. Phys. Lett. **87**, 21102 (2005).
6. I.R. Sellers *et al.*, Phys. Rev. B **73**, 33304 (2006).
7. R. Butté *et al.*, Phys. Rev. B **73**, 33315 (2006).
8. E. Feltin *et al.*, Appl. Phys. Lett. **89**, 71107 (2006).
9. I.R. Sellers *et al.*, Phys. Rev. B **74**, 193308 (2006).
10. A. Alyamani *et al.*, J. Appl. Phys. **101**, 93110 (2007).
11. S. Christopoulos *et al.*, Phys. Rev. Lett. **98**, 126405 (2007).
12. J.Y. Duboz *et al.*, Phys. Stat. Sol. (a) **183**, 35 (2001).
13. N. Antoine-Vincent *et al.*, Phys. Stat. Sol. (a) **190**, 187 (2002).
14. F. Semond *et al.*, Appl. Phys. Lett. **78**, 335 (2001).
15. F. Semond *et al.*, Phys. Stat. Sol. (a) **188**, 501 (2001).
16. F. Réveret *et al.*, Phys. Rev. B accepted.
17. T. Tawara *et al.*, Phys. Rev. Lett. **92**, 256402 (2004).
18. I.R. Sellers *et al.*, Mater. Res. Soc. Symp. Proc. **892**, 0892-FF20-04.1 (2006).
19. M. Gurioli *et al.*, Superlattices and Microstructures **41**, 284 (2007).
20. A. Imamoglu *et al.*, Phys. Lett. A, **214**, 193, (1996).
21. G. Malpuech *et al.*, Appl. Phys. Lett. **81**, 412 (2002).
22. F. Tassone *et al.*, Phys. Rev. B **56**, 7554, (1997).
23. R. Butté *et al.*, Phys. Rev. B **65**, 205310, (2002).
24. D. Bajoni *et al.*, Phys. Rev. Lett. **100**, 047401, (2008).
25. Le Si Dang *et al.*, Phys. Rev. Lett. **81**, 3920, (1998).
26. J. Kasprzak *et al.*, Nature **443**, 409 (2006).
27. D. Solnyshkov *et al.*, Superlattices and Microstructures **41**, 279 (2007).
28. F. Stokker-Cheregi *et al.*, Appl. Phys. Lett. **92**, 042119 (2008).
29. F. Rizzi *et al.*, Appl. Phys. Lett. **90**, 111112 (2007).
30. K. Bejtka *et al.*, submitted.

Mater. Res. Soc. Symp. Proc. Vol. 1068 © 2008 Materials Research Society 1068-C05-07

AlGaN/GaN multiple quantum wells grown by using atomic layer deposition technique

Ming-Hua Lo, Zhen-Yu Li, Shih-Wei Chen, Jhih-Cang Hong, Ting-Chang Lu, Hao-Chung Kuo, and Shing-Chung Wang

Photonics & Institute of Electro-Optical Engineering, National Chiao Tung University, No.1001, Ta Hsueh Rd., Hsinchu City 300, Taiwan, R.O.C., Hsinchu, 300, Taiwan

ABSTRACT

We report the successful growth of high quality ultraviolet (UV) AlGaN/GaN multiple quantum wells (MQWs) structure using atomic layer deposition (ALD) technique. The AlGaN/GaN MQW sample grown on the sapphire substrate consisted of three GaN QWs and four AlGaN barriers comprised AlN/GaN superlattices (SLs). From atomic force microscope measurement, the root-mean-square value of the surface morphology was only 0.35 nm, and no crack was found on the surface. The dislocation density was estimated to be as low as 2×10^{8} cm^{-2}. X-ray and transmission electron microscope data show the grown MQW has shape interface with good periodicity. The sample has a deep UV photoluminescence emission at 334 nm (3.71 eV) with a very narrow linewidth of 45 meV at 13K. The cathodoluminescence image show fairly uniform luminescence pattern at room temperature. In conclusion, the AlGaN/GaN MQW grown by ALD technique should be useful for providing high crystalline quality AlGaN/GaN-based MQW for fabrication of AlGaN/GaN-based UV light emitting devices such as light emitting diodes and lasers.

INTRODUCTION

The AlGaN/GaN multiple quantum wells (MQWs) have attracted much attention because of their many excellent properties, such as high conduction band offset, large LO phonon energy, ultra-fast carrier and/or intersubband relaxation [1, 2] and so on. Therefore, the AlGaN/GaN MQWs are promising candidates for realizing ultraviolet (UV) light emitting diodes (LEDs), edge emitting laser diodes (LDs) [3, 4] and vertical cavity surface emitting lasers (VCSELs) [5]. Besides, the large LO phonon energy inherited in AlGaN/GaN systems could allow room-temperature operation of quantum cascade laser (QCL) possible [6]. Recent reports indicated that the optical and electrical properties of AlGaN/GaN MQWs were very sensitive to the threading dislocation density (TDD) in AlGaN/GaN epilayer suggesting high crystalline quality AlGaN/GaN MQWs could dramatically improve the device performance [7, 8]. So far most MQWs structures were grown on lattice-mismatched foreign substrates such as sapphire. As a result it is relatively difficult to grow device-quality AlGaN/GaN MQWs on sapphire substrate due to the lattice mismatch and the misfit in the thermal expansion coefficients between these two material systems. In addition, the growth rate of the epilayer using conventional metal-organic chemical vapor deposition (MOCVD) is relatively fast making the MQW thickness and quality control difficult. Recently the high quality AlGaN/GaN heterostructures using quasi AlGaN as barrier layers formed by AlN/GaN super-lattice was reported [9, 10], and their results mainly focused on the electrical properties of AlGaN/GaN HEMT. Hence, in our experiment, we used this structure to grow AlGaN/GaN MQWs, and investigated the crystalline quality and optical properties of AlGaN/GaN MQWs.

In this paper, we report the growth of high quality AlGaN/GaN MQWs by using the atomic layer deposition (ALD) grown AlN/GaN superlattices (SLs) as the AlGaN barrier. The as-grown AlGaN/GaN MQWs sample had sharp interfaces between SLs layers and QWs with good periodicity, smooth surface morphology with low root-mean-square (RMS) value, and low defect density. In addition, the optical properties of AlGaN/GaN MQWs sample were also characterized by cathodoluminescence and photoluminescence (PL).

EXPERIMENT

The AlGaN/GaN MQW structures were grown by the low-pressure metalorganic chemical vapor deposition (VEECO D75 system). The TMGa, TMAl and gaseous NH_3 were employed as the reactant source materials for Ga, Al and N, respectively and H_2 and N_2 were used as the carrier gas. The substrates used in this experiment were (0001)-oriented sapphire offset 0.2° off toward $<01\bar{1}1>$ direction. The sapphire substrate was first heated to 1000°C under an H_2 ambient for 5 min to form a clean surface. Then, about 2-μm-thick GaN epilayer was grown after the deposition of a low-temperature nucleation layer. Finally, the AlGaN/GaN MQWs structure comprising three GaN wells and four AlGaN barriers was grown at 850°C in H_2+N_2 atmosphere. Particularly, the AlGaN barriers were grown using ALD technique. The ALD process involves alternate control of mass flow of TMAl and TMGa gas during the growth of AlGaN barrier to form six pairs of AlN/GaN SLs. The TMAl and TMGa flow time of AlN and GaN layer were 6.8 and 19.8 sec, respectively under a continuous flow of the NH_3 gas at 850°C. The interruption time was 10 sec between AlN and GaN alternating layers. The growth rate of ALD AlGaN barrier was measured by an *in-situ* Filmetrics optical monitoring system. The measured growth rate was about 0.14 μm/hr which was lower than the conventional growth rate of about 0.6 ~ 1 μm/hr [11]. After AlGaN barrier was grown, only TMGa was introduced into the reactor for 34.8 second to grow the GaN well. Fig. 1 shows the growth procedure of AlGaN barrier and GaN well layers.

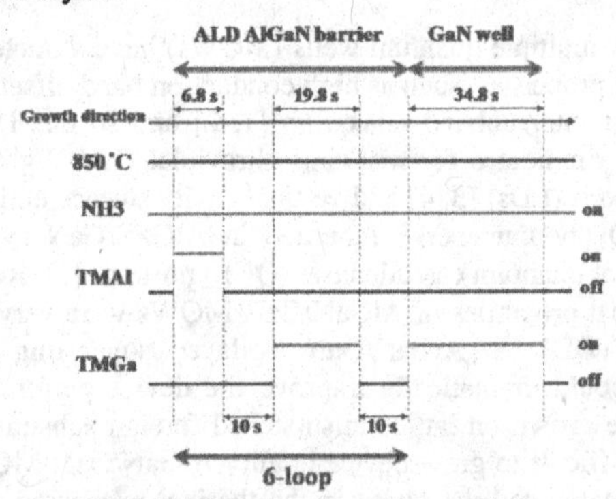

Fig. 1 The growth procedure of AlGaN barrier and GaN well layers.

The surface morphology of the top layer was observed by atomic force microscope (AFM) with a scanning area of 5 μm × 5 μm. Crystalline quality was evaluated by high resolution X-ray diffraction (HRXRD) and reciprocal space mapping (RSM), and Cu K_α radiation was used as the X-ray source. Additionally, the average thicknesses of the AlGaN barriers and the GaN wells were determined from the angular distance for (0002) plane using satellite peaks of ω/2θ-scan diffraction patterns. The optical properties were investigated by PL measurements performed at 13K. PL spectra were excited with a frequency tripled Ti: sapphire laser at wavelength of 266 nm and laser output power of 20 mW. The laser pulse width was 200 fs and the repetition rate was 76 MHz. The luminescence spectrum was measured by a 0.5 m monochromator and detected by a photomultiplier tube. The cathodoluminescence (CL) measurements were carried out at 300 K by using a MonoCL system installed on a field emission scanning electron microscope (SEM) with beam energies of 5–20 keV. The threading dislocations and the sharpness of the AlGaN/GaN structure interfaces were studied by transmission electron microscope (TEM).

DISCUSSION

The figure 2 shows that the surface morphology of the top layer was observed by atomic force microscope (AFM) with a scanning area of 5 μm × 5 μm. No cracks were found and a very small RMS value of the surface roughness of 0.35 nm was achieved. The dislocation density was estimated to be as low as 2×10^8 cm^{-2}

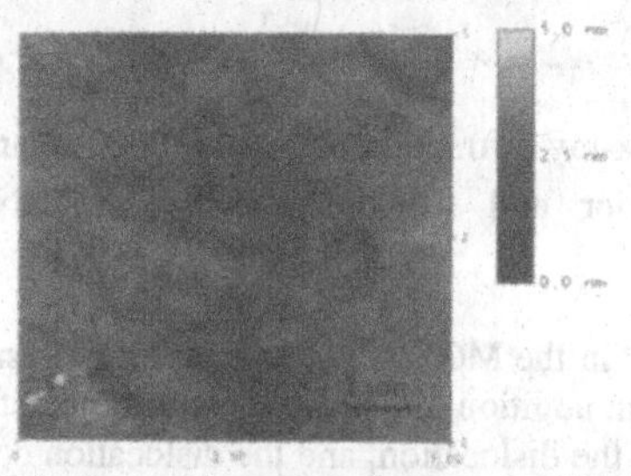

Fig. 2 The Atomic force microscope image of AlGaN/GaN MQWs

Fig. 3 (a) and (b) show the HRXRD ω/2θ diffraction pattern and the RSM of the grown AlGaN/GaN MQWs sample. As can be seen in Fig. 3 (a), the HRXRD diffraction pattern shows two periodical structures: one can be attributed to the AlGaN/GaN MQWs; another can be attributed to the AlN/GaN SLs in the AlGaN barriers. The third order satellite peak of the diffraction pattern for AlGaN/GaN MQWs can be clearly observed, suggesting that the high crystalline quality of the grown AlGaN/GaN MQWs and AlN/GaN SLs sample. Besides, from the fitting of HRXRD data indicated by the red lines in the Fig. 3 (a), we can found that the HRXRD can be fitted well by the condition of $(AlN/GaN)_6/GaN$ MQWs, suggesting that the MQWs structure consist of three GaN wells and four AlGaN barriers which was composed of six AlN/GaN SL, and furthermore we obtained the thickness of AlN and GaN in the SLs are about 0.42 nm-thick and 0.77 nm-thick, respectively, to form one AlGaN barrier layer with an overall thickness of 7.14 nm, and the GaN well has a thickness of 2.9 nm. From the RSM data shown in Fig. 3 (b), we can clearly found that the spread of RSM intensity for the AlGaN/GaN MQWs was relatively narrow indicating AlGaN epilayer with relatively small distribution of crystal orientation [12]. In addition, the reciprocal lattice points of AlGaN and GaN were lined up at the same Q_x position, i.e. the AlGaN and GaN had same lattice constant. According the earlier report [13], the degree of lattice relaxations can be estimated from the equation of $\varepsilon_{xx}=q_x^{GaN}/q_x^{MQWs}-1$, where the q_x^{GaN} and q_x^{MQWs} are the x position of GaN layer and AlGaN layer, respectively. We obtained the degree of lattice relaxation to be about 3.9×10^{-6} which is 4-5 order lower than the previous report [12], indicating the AlGaN epilayer is fully strained and pseudomorphic to the underlying GaN layer [14]. These data show that the AlGaN/GaN MQWs grown by ALD technique is viable for producing high crystalline quality with low dislocation density and crack-free surface morphology.

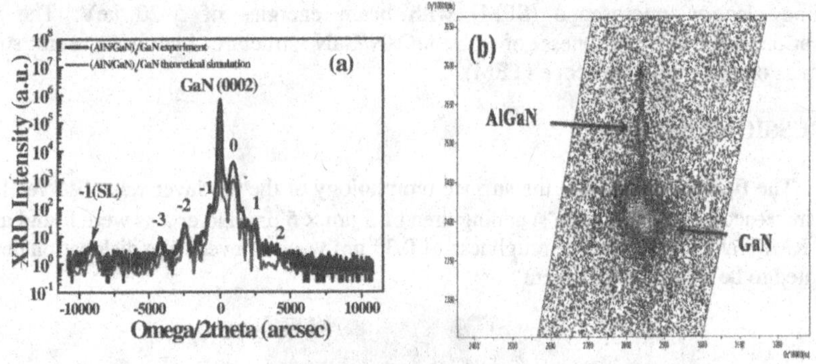

Fig. 3 (a) high-resolution x-ray diffraction pattern at (0002) plane for AlGaN/GaN MQWs using ALD AlGaN barrier and (b) the RSM of AlGaN/GaN MQWs obtained from $(10\bar{1}5)$ diffraction.

A large strain existed in the MQWs, because the RSM result demonstrated that the AlGaN epilayer is fully strained. In addition, the previous report indicated that the strain in the epi-layer can exerted a net force on the dislocation, and the dislocation could be bent and terminated at an epilayer edge, without threading through the epilayer to the top surface of the epi-layer [15]. Therefore, as can been seen, Fig. 4 shows a typical threading dislocation bent and suppressed at the boundary of MQWs without extending into the MQWs. This result suggests that the ALD grown AlGaN barrier with AlN/GaN SLs can reduce the defects in MQW and improve the surface morphology. Additionally, the inset shows the enlarged image for the MQWs and the scale bar is 10 nm. It can be clearly observed that the QWs and SLs exhibited sharp interfaces with good periodicity, indicating that the high quality SLs and MQWs were successful formed by the ALD technique, and we also obtained that the AlGaN barrier consists of six pairs AlN/GaN SLs with AlN thickness of 0.43 nm and GaN thickness of 0.77 nm, respectively, to form a AlGaN barrier with thickness of 7.2 nm, and the GaN well has a thickness of 3 nm. Hence, the thickness of SLs and MQWs is in good agreement with the result estimated from HRXRD in Fig. 3 (a).

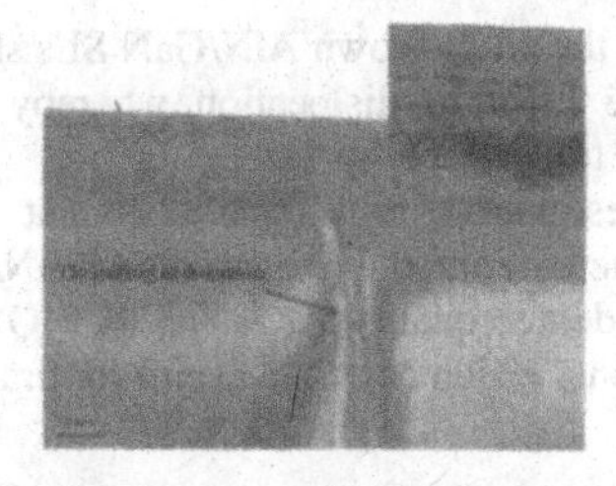

Fig. 4 Cross-sectional TEM micrograph of AlGaN/GaN MQWs with the ALD grown AlGaN barrier. Inset shows the enlarged image of MQWs.

Finally, Fig. 5 shows the PL spectra of the grown AlGaN/GaN MQWs at 13K. As can be seen, an emission peak at 3.71 eV (334 nm) was observed. In particular, the emission line has a narrow linewidth of about 45 meV at 13K. This linewidth is lower than that of the halide vapor phase epitaxy grown AlGaN/GaN MQW data by a percentage of 30-40 [16]. In addition, the CL image showed in the inset of Fig. 5 has a relatively uniform emission intensity without dark spots, suggesting again that the ALD grown AlGaN/GaN sample has a relatively superior crystalline quality.

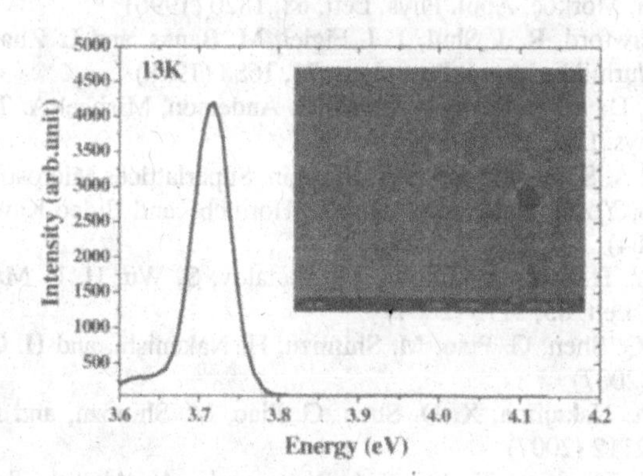

Fig. 5 The 13K PL spectra of AlGaN/GaN MQWs by using the ALD AlGaN barrier composed of AlN/GaN SLs. Inset shows CL image taken at E= 3.71 eV.

CONCLUSIONS

We have successfully grown high quality AlGaN/GaN MQWs on sapphire substrate by using the ALD grown AlGaN barrier consisted of AlN/GaN SLs. The SEM and AFM data show smooth surface morphology with a low surface roughness RMS value of about 0.35 nm and no surface crack found. The cathodoluminescence measurements indicate uniform luminescence pattern at room temperature. The TEM and HRXRD measurements show that the grown structure has sharp interfaces between SLs layers and QWs with good periodicity. In addition, the TEM images revealed that the ALD grown AlN/GaN SLs structure could be used to improve the by effectively suppress the threading dislocation, whereby the pits density was estimated to be as low as 2×10^8 cm^{-2} on the top surface.

According to our investigation, it is concluded that a structure of AlN/GaN SLs as AlGaN barrier was used for heteroepitaxial growth of AlGaN/GaN MQWs by ALD technique makes it possible to obtain a device-quality AlGaN/GaN MQWs structure with atomically flat interfaces and low-defect regions, which could be useful for practical applications.

ACKNOWLEDGMENTS

This work was supported by the MOE ATU program and in part by the National Science Council of Taiwan under contract Nos. NSC 96-2221-E009-067 and US Air Force Research Laboratory.

REFERENCES

[1] Norio Iizuka, Kei Kaneko, Nobuo Suzuki, Takashi Asano, Susumu Noda, and Osamu Wada, Appl. Phys. Lett. **77,** 648 (2000)
[2] Ümit Özgür, Henry O. Everitt, Lei He, and Hadis Morkoç, Appl. Phys. Lett. **82,** 4080 (2003)
[3] T. J. Schmidt, X. H. Yang, W. Shan, J. J. Song, A. Salvador, W. Kim, Ö. Aktas, A. Botchkarev, and H. Morkoç, Appl. Phys. Lett. **68,** 1820 (1996)
[4] J. Han, M. H. Crawford, R. J. Shul, J. J. Figiel, M. Banas, and L. Zhang, Y. K. Song, H. Zhou, and A. V. Nurmikko, Appl. Phys. Lett. **73,** 1688 (1998)
[5] Joan M. Redwing, David A. S. Loeber, Neal G. Anderson, Michael A. Tischler, and Jeffrey S. Flynn, Appl. Phys. Lett. **69,** 1 (1996)
[6] Greg Sun, Richard A. Soref, and Jacob B. Khurgin, Superlattices Microstruct. **37,** 107 (2005)
[7] Takayoshi Takano, Yoshinobu Narita, Akihiko Horiuchi, and Hideo Kawanishi, Appl. Phys. Lett. **84,** 3567 (2004)
[8] V. Adivarahan, W. H. Sun, A. Chitnis, M. Shatalov, S. Wu, H. P. Maruska and M. Asif Khan, Appl. Phys. Lett. **85,** 2175 (2004)
[9] Y. Kawakami, X.Q. Shen, G. Piao, M. Shimizu, H. Nakanishi, and H. Okumura, J. Crystal Growth **300,** 168 (2007)
[10] Y. Kawakami, A. Nakajima, X. Q. Shen, G. Piao, M. Shimizu, and H. Okumura, Appl. Phys. Lett. **90,** 242112 (2007)
[11] H. Tokunaga, A. Ubukata, Y. Yano, A. Yamaguchi, N. Akutsu, T. Yamasaki, and K. Matsumoto, J. Cryst. Growth. **272,** 348 (2004)
[12] Y S Park, C M Park, S J Lee, Hyunsik Im, T W Kang, Jae-Eung Oh, Chang Soo Kim, and Sam Kyu Noh, Semicond. Sci. Technol. **20,** 775 (2005)
[13]G. S. Huang, T. C. Lu, H. H. Yao, H. C. Kuo, S. C. Wang, Chih-Wei Lin, and Li Chang, Appl. Phys. Lett. **88,** 061904 (2006)
[14] A. Torabi, W. E. Hoke, J. J. Mosca, J. J. Siddiqui, R. B. Hallock, and T. D. Kennedy, J. Vac. Sci. Technol. **B 23,** 1194 (2005)
[15] J. Knall, L. T. Romano, D. K. Biegelsen, R. D. Bringans, H. C. Chui, J. S. Harris, Jr., D. W. Treat, and D. P. Bour, J. Appl. Phys. **76,** 2697 (1994)
[16] M. Smith, J. Y. Lin, H. X. Jiang, A. Salvador, A. Botchkarev, W. Kim, and H. Morkoc, Appl. Phys. Lett. **69,** 2453 (1996)

Mater. Res. Soc. Symp. Proc. Vol. 1068 © 2008 Materials Research Society 1068-C06-08

High Optical Quality GaN Nanopillars Grown on (111) Si Using Molecular Beam Epitaxy

Agam Prakash Vajpeyi, G. Tsiakatouras, A. Adikimenakis, K. Tsagaraki, M. Androulidaki, and Alexandros Georgakilas
Microelectronics Research Center, Institute of Electronic Structure and Laser (IESL), FORTH, and Department of Physics, University of Crete, Herakilon, 71110, Greece

ABSTRACT

The spontaneous growth of GaN nanopillars on (111) Si by plasma assisted molecular beam epitaxy has been investigated. The growth of GaN nanopillars on Si is driven by the lattice mismatch strain energy on Si and the high surface energy of the nitrogen stabilized (0001) GaN surface. A higher growth rate of nanopillars compared to a compact GaN film suggests the diffusion of Ga atoms from the uncovered substrate areas to the nucleated GaN nanopillars. The GaN nanopillars were characterized by field-emission scanning electron microscopy (FE-SEM), photoluminescence, and micro Raman spectroscopy. SEM image revealed that average diameter of GaN nanopillars was in the range of 70-100nm and an average height of 600nm. The photoluminescence (PL) spectra indicate the good emission property of the nanopillars. The low temperate PL spectrum exhibited an emission peak at 3.428eV besides a sharp excitonic peak. PL and Raman spectra indicate that GaN nanopillars are fully relaxed from lattice and thermal strain.

INTRODUCTION

Hetroeptaxial growth of III-V semiconductors on silicon has been very interesting for the monolithic integration of III-V optoelectronic and microwave device on Si.[1,2] In particular, the III-nitride semiconductors offer unique capabilities for application in optoelectronic and high power microwave devices.[3,4] One of the major problems with III-nitride materials hetroepitaxial growth is related with high density of threading dislocations, of the order of 10^8–10^{10} cm^{-2}, which adversely affect the device performance. Nanodimensional nitrides such as nanowires (NWs) / nanopillars (NPs) are attractive materials for the reduction of the dislocation density. In addition to the fact that the NPs geometry prevents the propagation of threading dislocations, it also allows the device dimensions to be scaled down. The NPs could be free of threading dislocations as their small lateral length could allow initially formed threading dislocations to move out of the crystal. In addition, the free surface at the sidewalls permits elastic relaxation of the strain.[5] Different techniques such as metal-organic vapor phase epitaxy, [6, 7] molecular beam epitaxy,[8,9] chemical beam epitaxy, [10] and laser ablation [11, 12] have been used to grow a variety of semiconductor NWs with high crystalline quality. Recently, Calarco *et al*[3] reported the nucleation density of GaN NWs and their evolution in terms of growth time for a fixed set of growth parameters. It has been also reported that the NW morphology is affected by the V/III flux ratio and the growth temperature.[14-16] In this paper, we report on the growth and optoelectronic properties of GaN NPs spontaneously formed on n-type (111) silicon, using nitrogen radio-frequency plasma source MBE (RFMBE). Photoluminescence spectra at 20K show the strong excitonic emission line at 3.47eV from the NPs sample. The near-band-edge excitonic transitions in the NPs are studied by temperature-dependent photoluminescence spectroscopy and an activation energy of bound exciton with neutral donor ((D^oX) was found to be 26meV, which is in agreement with previously reported value by Tiginyanu et al.[25] PL and Raman spectra showed that the GaN NPs are completely stress-free compared to GaN film grown on Si.

EXPERIMENTAL DETAILS

The GaN NPs were grown without any external catalyst on n-type (111) silicon substrates at a V/III flux ratio of 5 and the substrate temperature of 750°C, using RFMBE.[17] The Si wafers were chemically cleaned[18] and then loaded into the MBE system. The Si surface oxide was removed by heating in the growth chamber at 800°C for 10 min. Then, the plasma was turned on at 300 Watt with a 1.35 standard cubic centimeter per minute (sccm) nitrogen flow rate. The growth experiments were monitored by *in situ* reflection high-energy electron diffraction (RHEED). The sample surface morphology was characterized by field-emission high resolution scanning electron microscopy (JEOL 6700FESEM). The optical properties of the GaN NPs were investigated by photoluminescence spectroscopy at 20K (20K-PL), and micro-Raman spectroscopy. PL spectra were recorded using 325 nm line excitation wavelength from a He-Cd laser. Micro-Raman measurements was carried out using 514.5 nm line excitation, where the scattered light was dispersed through the JY-T64000 triple monochromatic system attached to a liquid nitrogen cooled charge coupled device detector. The spatial and spectral resolution of the visible Raman setup is about 1.0 μm and 0.2 cm^{-1}, respectively.

DISCUSSION

During the initial stage of the growth, the RHEED shows a rings pattern probably indicating the formation of polycrystalline GaN and silicon nitride layer on the major part of the wafer's surface. The formation of amorphous silicon nitride layer has been previously reported by Kim et al and Grandal et.al.[9, 20] After MBE growth of 5-15 mins, the ring like RHEED pattern gradually changed into a spotty one, indicative for the formation of (0001) oriented GaN NPs . The transition from ring like RHEED pattern to spotty one depends on the growth temperature and V/III flux ratio. In general, growth at higher temperature needed longer time to show the spotty RHEED pattern, because of reduced coverage of the substrate surface by the GaN NPs at higher temperature.

GaN NPs could be formed only by using a growth temperature higher than 700°C and nitrogen rich growth condition (N/Ga flux ratio larger than 1). The driving force for the three dimensional GaN growth mode on Si is the reduction of the lattice mismatch strain energy and the high surface energy of a nitrogen stabilized (0001) GaN surface. Figure 1a and 1b show the plan and cross-sectional SEM images of GaN NPs. SEM images revealed that the average diameter of the NPs varies from 70-100 nm while the average height is 600 nm. The NPs grown on Si (111) are not perfectly vertically aligned relative to the Si substrate but slightly tilted. The tilting of NPs might be due to the inclined incidence of the nitrogen beam on the substrate surface. We have also observed that GaN NPs show a tapering effect[14]. The tapering of the NPs can be evaluated by the diameter difference between the top and bottom of the individual NP. The average diameter of GaN NPs changes from 50 to 75 nm when measured from bottom to top part of the NP. Such a tapering effect for GaN NPs grown on silicon has been previously reported by Meijers et al.[14]. The tapering effect is controlled using optimized growth conditions such as growth temperature and V/III flux ratio. The growth rate of NPs is found to be 20%

higher than the nominal growth rate of compact film formation, which would correspond to a NPs height of 498nm. A higher growth rate of NPs compared to the compact film suggests the diffusion of Ga adatoms from the uncovered substrate areas to the nucleated GaN NPs.

FIG. 1: SEM images of GaN NPs grown on (111) Si with V/III flux ratio of 5 at substrate temperature of 750°C (a) plane view (b) cross-sectional view.

Figure 2 shows typical PL spectra recorded at 20K, for the GaN NPs and a reference compact GaN / Si film where emission peaks from GaN NPs are observed at 3.470, 3.452, and 3.428eV, while the GaN film shows emission peaks at 3.466 and 3.264 eV. The PL emission at 3.470 eV is associated with exciton bound with neutral donor (D^oX). The PL peak at 3.452eV in the NPs may be related to defect mode as in the case of GaN film, the emission peak at 3.452eV (Y_1 line) was tentatively assigned to the inversion domain interface[18]. The PL emission peak at 3.428eV was observed in GaN NPs samples. The origin of this PL band has been attributed to an exciton bound to structural defects at the GaN/Si NP interfaces[21, 22]. In Fig. 2, the PL data also show an intensity enhancement of the exciton bound with neutral donor (D^oX) emission from the NPs compared to the reference GaN film. The PL intensity enhancement in the NPs sample should be related to improved structural quality. Wang et al previously reported fifty times PL intensity enhancement in GaN NPs compared to the GaN film. [23] A blue shift of about 4.0 ± 0.5 meV is observed for the (D^oX) transition from the GaN NPs when compared to the GaN film and such a blue-shifted (D^oX) emission can be associated with relaxation of tensile stress in the NP samples.

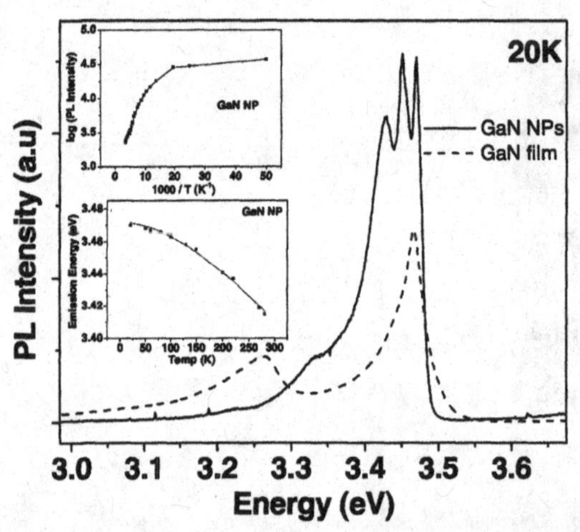

FIG. 2: 20K PL spectra of GaN NPs. The spectrum recorded from the GaN film is also shown for a comparison. The inset shows temperature dependence of D^oX PL intensity and peak energy.

We have also studied the evolution of integrated D^oX PL intensity as a function of temperature for the GaN NPs. The experimental data points were fitted according to the following equation, [24]

$$I(T) = \frac{I(0)}{1 + C\exp(-E_A / kT)} \quad (1)$$

where $I(T)$ is the integrated intensity at temperature T, $I(0)$ is the integrated intensity at 0 K and E_A is the activation energy. For the GaN NPs, activation energy of 26 meV was estimated from the best-fitted curve for temperature range of 20 K – 300 K. Such observation agrees well with the reported value of activation energy of GaN nanocoloumns prepared by photoelectrochemical etching[25] and represents bulk-like excitonic properties.

The temperature dependence of the D^oX peak energy for the GaN NPs are well fitted by Varshni's equation[26]

$$E_g(T) = E_g(0) - \alpha \frac{T^2}{\beta + T} \quad (2)$$

where $E_g(0)$ is the D^oX energy at 0 K and α, and β are the Varshni's thermal coefficients. The solid line in this figure is obtained by using least-square fitting. For GaN NPs, the best fitted values were given by $E(0) = 3.470$ eV, $\alpha = 5.3 \times 10^{-4}$ and $\beta = 494$K. These values are in close agreement to the reported value for D^oX transition.[27]

Figure 3 presents typical Raman spectra of the GaN NPs along with a reference GaN film grown on Si (111). The spectra were recorded in the $z(xx)\bar{z}$ scattering geometry. The inset in Fig. 3 shows an expansion of the spectra near the vicinity of the E_2 (high) phonon peak and revealed the shifting of the peak from the GaN NPs in comparison to the GaN film.

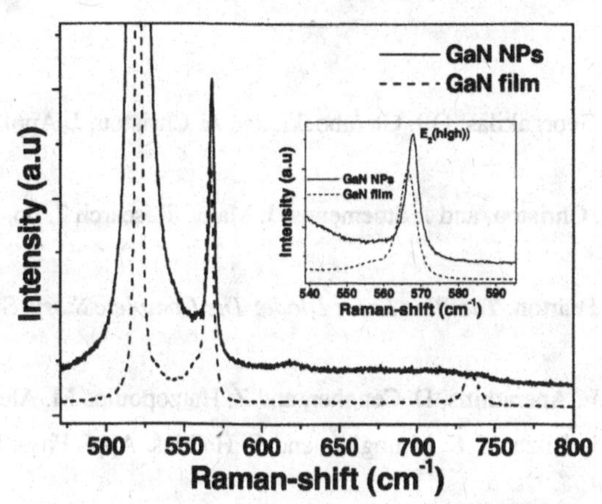

FIG. 3: Raman spectra of GaN NPs along with GaN film grown on (111) Si. The inset shows an expansion of the spectrum in the vicinity of the E_2 (high) phonon peak.

The Raman spectra are dominated by the strong E_2 (high) optical phonon peaks besides the strong silicon peaks at 520.5 cm^{-1}. The Raman spectrum of the GaN film shows strong E_2 (high) and A_1(LO) modes at 566.4 and 734.0 cm^{-1}, respectively, which are in agreement with Raman selection rules for wurtzite GaN. The E_2 (high) phonon line from GaN film shows that GaN film grown on Si(111) possesses tensile strain of 0.25 ± 0.05 GPa when compared to strain-free 400 μm thick freestanding GaN grown by HVPE. The amount of strain was evaluated using the proportionality factor of 4.3cm^{-1} GPa^{-1} for hexagonal GaN.[28] The E_2(high) phonon line from the GaN NPs is centered at 567.5 cm^{-1} which is the value reported for the freestanding GaN film. In contrast to GaN film, a very weak and broad A_1(LO) mode was observed in NPs which is the characteristics feature exhibited by nanomaterials.[29] The weak and asymmetrical broad A_1(LO) mode in nanodimensional structures was attributed to the relaxation in $q = 0$ selection rule owing to interruption of lattice periodicity in a nanocrystalline material.

CONCLUSION

In conclusion, SEM observations revealed that the average size of spontaneously grown GaN nanopillars on (111) Si varies from 70 to100nm for average height of 600nm. GaN NPs are grown on Si substrate due to lattice mismatch strain energy and the high surface energy of nitrogen stabilized (0001) GaN surface. A higher growth rate of nanopillars compared to the compact film suggests the diffusion of Ga atoms from the uncovered substrate areas to the nucleated GaN pillars. Low temperature photoluminescence spectra at 20K show strong excitonic emission peaks and indicate the good structural quality of GaN nanopillars. PL and Raman spectra show that GaN nanopillars are fully relaxed from lattice and thermal strain.

ACKNOWLEDGEMENT

The authors would like to acknowledge the support provided by European Commission through the "PARSEM" Marie Curie Project (MRTN-CT-2004-005583).

REFERENCES

1. A. Dimoulas, P. Tzanetakis, A. Georgakilas, O.J. Glembocki, and A. Christou, J. Appl. Phys. 67, pp. 4389-4392 (1990)
2. A. Georgakilas, A. Dimoulas, A. Christou, and J. Stoemenos, J. Mater. Research 7, pp. 2194-2204 (1992)
3. S. Nakamura, G. Fasol, and S.J. Pearton, *The Blue Laser Diode: The Complete Story*, Springer Verlag, 2nd Edition (2000)
4. A. Georgakilas, G. Deligeorgis, E. Aperathitis, D. Cengher, and Z. Hatzopoulos M. Alexe, V. Dragoi, U. Go¨sele E. D. Kyriakis-Bitzaros, K. Minoglou, and G. Halkias, Appl. Phys. Lett 81, 5099 (2002)
5. F. Glas, Phys. Rev. B 74, 121302(R) (2006).
6. F. Qian, Y. Li, S. Gradecak, D. L. Wang, C. J. Barrelet, and C. M. Lieber, Nano Lett. 4, 1975 (2004).
7. J. Su *et al* Appl. Phys. Lett. 86, 13105 (2005).
8. R. Calarco, M. Marso, T. Richter, A. I. Aykanat, R. Meijers, A. V. Hart, T. Stoica, and H. Luth, Nano Lett. 5, 981 (2005).
9. Y. H. Kim, J. Y. Lee, S. H. Lee, J. E. Oh, and H. S. Lee, Appl. Phys. A 80, 1635 (2005).
10. L. E. Jensen, M. T. Björk, S. Jeppesen, A. I. Persson, B. J. Ohlsson, and L. Samuelson, Nano Lett. 4, 1961 (2004).
11. A. M. Morales and C. M. Lieber, Science 279, 208 (1998).
12. X. F. Duan and C. M. Lieber, J. Am. Chem. Soc. 122, 188 (2000).
13. R. Calarco, R. J. Meijers, R. K. Debnath, T. Stoica, E. Sutter, and H. Luth, Nano Lett. 7, 2248 (2007).

14. R. Meijers, T. Richter, R. Calarco,T. Stoica, H. P. Bochem, H. Marso, and H Luth, J. Cryst. Growth 289, 381 (2006).
15. K. A. Bertness, A. Roshko, N.A. Sanford, J. M. Barker, and A. V. Davydov, J. Cryst. Growth 287, 522 (2006).
16. Y. S. Park, S. H. Lee, J.E. Ob, C. M. Park, and T. W. Kang, J. Cryst. Growth 282 313 (2005).
17. Iliopoulos E., Adikimenakis A., Dimakis E., Tsagaraki K., Konstantinidis G., Georgakilas A., J. Cryst. Growth 278 426 (2005)
18. Kayambaki M., Callec R., Constantinidis G., Papavassiliou Ch., Loechtermann E., Krasny H., Papadakis N., Panayotatos P., Georgakilas A., J. Cryst. Growth 157 300 (1995)
19. J. Grandal, M. A. Sánchez-García, E. Calleja, E. Luna and A. Trampert, Appl. Phys. Lett. 91, 021902 (2007).
20. E. Calleja, M.A. Sanchez-Garcıa, F. Calle, F.B. Naranjo, E. Munoz, U. Jahn, K. Ploog, J. Sanchez, J.M. Calleja, K. Saarinen, P. Hautojarvi, Mat. Sci. Eng. B 82 (2001)
21. Hung-Ying Chen, Hon-Way Lin, Chang-Hong Shen, and Shangjr Gwo, Appl. Phys. Lett. 89, 243105, (2006)
22. E. Calleja, M. A. Sánchez-García, F. J. Sánchez, F. Calle, F. B. Naranjo, E. Muñoz, U. Jahn, and K. Ploog, Phys. Rev. B 62, 16826 (2000)
23. Wang, X., Sun, X., Fairchild, M., Hersee, S.D., Appl. Phys. Lett. 89, 233115 (2006)
24. Wei-Tse Hsu, Kuo-Feng Lin, and Wen-Feng Hsieh, Appl. Phys. Lett. 91, 181913, (2007)
25. I.M. Tiginyanu, V. V. Ursaki, V. V. Zalamai, S. Langa, S. Hubbard, D. Pavlidis, and H. Foll, *Appl. Phys. Lett*. 83, 1551 (2003).
26. Y. P. Varshni, *Physica* 34, 149 (1967).
27. C. M. Park, Y. S. Park, Hyunsik Im and T. W. Kang, Nanotechnology 17, 952, (2006)
28. L. S. Wang, K. Y. Zang, S. Tripathy, and S. J. Chua, Appl. Phys. Lett. 85, 5881 (2004)
29. A. K. Arora, M. Rajalakshmi, T. R. Ravindran and V. Sivasubramanian, J. Raman Spectrosc, 38: 604-617 (2007)

GaN and Related Alloys on Silicon Growth and Integration Techniques

Mater. Res. Soc. Symp. Proc. Vol. 1068 © 2008 Materials Research Society 1068-C03-07

Effect of Graded AlxGa1-xN Interlayer Buffer on the Strain of GaN Grown on Si (111) Using MOCVD Method

KungLiang Lin[1], Edward-Yi Chang[1], Tingkai Li[2], Wei-Ching Huang[1], Yu-Lin Hsiao[1], Douglas Tweet[2], Jer-shen Maa[2], and Sheng-Teng Hsu[2]

[1]Department of Materials Science and Engineering, National Chiao Tung University, 1001 University Road, Hsinchu, 300, Taiwan

[2]Sharp Laboratories of America, Inc., Camas, WA, 5700

ABSTRACT

GaN film grown on Si substrate with AlN/$Al_xGa_{1-x}N$ buffer is studied by low pressure metal organic chemical vapor deposition (MOCVD) method. The $Al_xGa_{1-x}N$ film with Al composition varying from 0~ 0.66 was used. The correlation of the Al composition in the $Al_xGa_{1-x}N$ film with the stress of the GaN film grown was studied using high resolution X-ray diffraction including symmetrical and asymmetrical ω/2θscans and reciprocal space maps. It is found that with proper design of the Al composition in the $Al_xGa_{1-x}N$ buffer layer, crack-free GaN films can be successfully grown on Si (111) substrates using AlN and $Al_xGa_{1-x}N$ buffer layers.

INTRODUCTION

Growth of GaN on Si (111) is of particular interest to the industry because of it large area in comparison with other substrates and the possibility of integrating conventional Si-based devices with group III-nitride devices on a single wafer. Due to the considerable differences in lattice parameters and thermal expansion coefficients (CTE) between GaN and Si substrates, the growth of high quality crack-free GaN films on Si substrate poses serious difficulties. Thus, strain is an important issue of group-III-nitride growth on Si to achieve devices as GaN-based light emitting diodes and field effect transistor on Si substrates. Therefore, reduction of the stress and crack density in the GaN film was required for the growth of GaN on the Si (111) substrate. To obtain high quality GaN film on Si substrates, the design of the interlayer structure between the GaN and the Si substrate is crucial. Using a low temperature AlN (LT-AlN) interlayer contributes to a reduction in the growth stress.[1-6] However, the AlN film grown at low temperature is of inferior quality to the AlN grown at high temperature, the LT-AlN layer cannot stop the edge threading dislocations and may introduce new dislocations, which influences the quality of the GaN film grown. Due to a significant lattice mismatch between AlN and Si, cracks and higher defect density were formed on the high temperature (1100ºC) grown AlN epi-layer. However, compressive stress (about 9 GPa) were generated on GaN epitaxially grown on high quality AlN due to the lattice mismatch between AlN and GaN which compensates the tensile stress caused by the CTE mismatch.[7] $Al_xGa_{1-x}N$ intermediate buffer was used to reduced the lattice mismatch between GaN and AlN to relax the tensile stress induced by the thermal mismatch between GaN an Si.[8] In this letter, we report the properties of the GaN layer grown on a multilayer buffer including $Al_xGa_{1-x}N$ interlayer with fixed Al composition (x was set at a fixed value between 0.5 to 0.25), a graded $Al_xGa_{1-x}N$ layer (x varies from 1 to 0.66) and along with multilayer AlN on the Si (111) substrate using MOCVD system. It is found that AlN multilayer buffer combined with graded $Al_xGa_{1-x}N$ layers and an $Al_{0.42}Ga_{0.58}N$ interlayer can greatly reduce the crack density and surface roughness of the GaN layers grown and thus improve the structural and optical properties of the GaN layers.

EXPERIMENT

The multilayer AlN, the $Al_xGa_{1-x}N$ layers and the GaN films were grown by EMCORE D-180 MOCVD reactor on 6" Si (111) wafers. Trimethylgallium (TMGa), Trimethylaluminum (TMAl), and ammonia (NH_3) were the source gases for Ga, Al, and N, respectively. The Si (111) substrates were chemically cleaned before epitaxial growth to obtain a hydrogen-free surface.[9] Sample A as listed in Table 1 is a reference sample that contains a GaN film grown on Si (111) substrate with an multilayer AlN buffer layer. The first multilayer AlN buffer consisted of a 30nm high temperature AlN (HT-AlN) layer, a 50nm LT-AlN layer and a 120nm HT-AlN layer and was grown at 50 torr. The composition and thickness of the graded $Al_xGa_{1-x}N$ ($0 \leq x \leq 1$) layer and the $Al_xGa_{1-x}N$ with fixed Al composition are listed in Table 1.

Table 1. Properties and Compositions of the different types of buffers used in this study

Sample	GaN Thickness (μm)	*AlN Thickness (nm)	Graded $Al_xGa_{1-x}N$ Thickness (nm) x=0~0.66	Fixed $Al_xGa_{1-x}N$ *x* fraction	AlN strain (ε_a)	GaN strain (ε_a)
A	1	200	0	0	0.378	0.146
B	1	200	500	0.25	0.335	0.137
C	1	200	500	0.50	0.312	0.132
D	1	200	500	0.42	0.3	0.12

* Multilayer AlN(High Temperature/Low Temperature/ High Temperature AlN)

For the samples grown, the Al composition for the fixed $Al_xGa_{1-x}N$ layer was set at 0.25, 0.5 and 0.42 for sample B, C, D respectively. The thickness for the graded $Al_xGa_{1-x}N$ and fixed composition $Al_xGa_{1-x}N$ were set at 500nm and 200nm respectively. The growth temperature was 800ºC for the 50nm LT-AlN layer and 1035ºC for other layers of the buffer. The GaN epi-layers were grown at 1035ºC. The structures of the samples grown are listed in Table 1 and the growth details are as described in reference10.[10] High-resolution X-ray diffractometry (HRXRD), X-ray energy dispersive spectroscopy (EDS) measurement, scanning electron microscopy (SEM) and Optical microscope (OM) were used to investigate the composition, structure, cross-section image, and the surface morphologies of the films. The stress states in the films were studied by Raman spectrum.

DISCUSSION

Since the Al composition in the $Al_xGa_{1-x}N$ layers could be accurately controlled by the MOCVD growth, a graded $Al_xGa_{1-x}N$ interlayer with a total thickness of about 500 nm was grown on the AlN composite buffers layer. The Al composition (x) of the fixed $Al_xGa_{1-x}N$ interlayer was set at different TMGa and TMAl with NH_3 source ratio along the growth direction. The AlN multilayer and the $Al_xGa_{1-x}N$ films were used as the buffer for the GaN film grown on Si. The Al composition in the $Al_xGa_{1-x}N$ layers were determined by HRXRD measurement of the lattice constant, assuming that Vegard's law was obeyed by the $Al_xGa_{1-x}N$ films. Figure 1 shows the 2θ scan XRD patterns of the 1μm GaN grown on different types of buffers with different combinations of graded $Al_xGa_{1-x}N$ and fixed $Al_xGa_{1-x}N$ layers in samples A to D as listed in Table 1.

Figure 2 (a) and (b) reveal the θ- rocking curve of the mosaic GaN (004) and AlN (004), the full width at half maximum (FWHM) data indicate that the GaN grown on the multilayer buffer with a graded $Al_xGa_{1-x}N$ interlayer has a single crystalline structure. For sample D, the values of mosaic GaN (004) and AlN (004) FWHMs are smaller than those of samples A to C. It means that sample D has the best crystal qualities for AlN and GaN. As the Al mole fraction x of the fixed $Al_xGa_{1-x}N$ changed from 0.25 to 0.42, the crystal quality of GaN also varied due to the lattice mismatch between GaN and AlN.

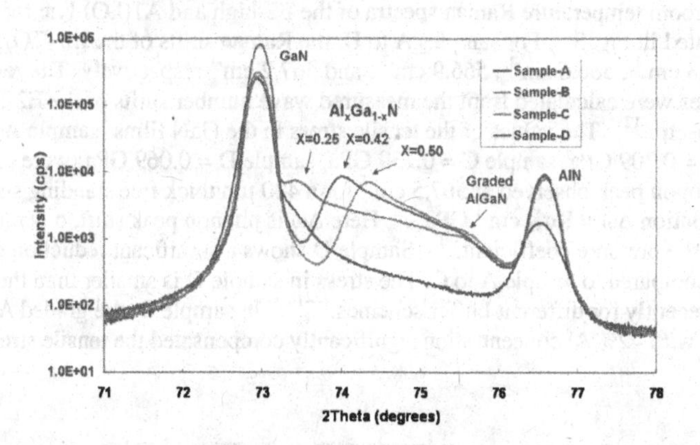

Fig.1 XRD ω-2θ scans of the GaN on different types of composite buffers

However, when the Al mole fraction increased to 0.50, the crystallinity of GaN became worse. For the four samples studied, the composite buffer with $Al_{0.42}Ga_{0.58}N$ layer resulted in the lowest strain in GaN and AlN films. It means that the Al mole fraction x = 0.42 in the fixed $Al_xGa_{1-x}N$, the lattice mismatch between $Al_xGa_{1-x}N$ at other films can compensate the strains induced in the GaN and AlN films. In other words, as the Al composition in the interlayer was graded, the thermal expansion coefficient of the graded $Al_xGa_{1-x}N$ interlayer was gradually changed from that of AlN to GaN. Therefore, the thermal stress may be accommodated within the graded $Al_xGa_{1-x}N$ interlayer. It is expected that a complete elimination of cracks can be achieved through a further optimization of the composition and thickness of the graded $Al_xGa_{1-x}N$ interlayer.

The thermal stress is calculated from the average of thermal expansion coefficient of Si and GaN. The thermal stress ε can be calculated using the following relation,[11] $\varepsilon = (\alpha_{GaN} - \alpha_{Si})(T_{growth} - T_{RT})$, where α_{GaN}, α_{Si} are the thermal expansion coefficients of GaN ($5.59 \times 10^{-6} K^{-1}$) and Si ($3.59 \times 10^{-6} K^{-1}$), respectively. T_{growth} and T_{RT} are the growth temperature (1035ºC) and room temperature (25ºC), respectively. The tensile strain of GaN grown on Si after cooling down was ε = 0.202%. Table 1shows that the strain in the GaN film is smaller than the theoretical nature tensile stress, it means that AlN and AlGaN interlayer can effectively reduce the thermal stress induced during the growth.

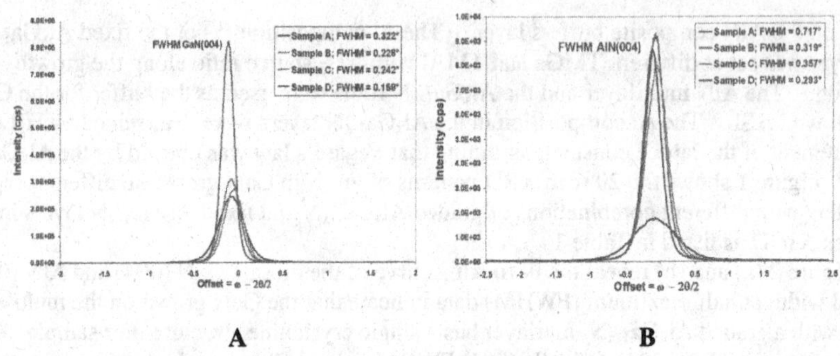

Fig.2 (A) GaN (004) XRD data for GaN grown on different types of buffers(B) AlN (004) XRD data for AlN grown on different types of buffers

The room temperature Raman spectra of the E2-high and A1(LO) line for the samples are presented in Fig. 3. For samples A to D, the Raman shifts of the E2 (TO) phonon peak were 566.8 cm^{-1}, 566.6 cm^{-1}, 566.9 cm^{-1}, and 567.2 cm^{-1} respectively. The residual stress in the samples were calculated from the measured wave number shifts of the E2-high mode in Raman spectra.[12] The values of the tensile stress in the GaN films (sample A = 0.162 GPa, sample B = 0.209 GPa, sample C = 0.139 GPa, sample D = 0.069 GPa) were calculated using the E2 phonon peak observed at 567.5 cm^{-1} for a 400 μm thick free standing strain free GaN and the relation $\Delta\omega = K\sigma_{xx}$ cm^{-1} GPa^{-1}. Here $\Delta\omega$ is phonon peak shift, σ is biaxial stress and K= 4.3 is the pressure coefficient.[13] Sample D shows a significant reduction of the in-plane stress as compared to sample A to C. The stress in sample D is smaller than the stress values reported recently for different buffer schemes.[14~16] In sample D, the graded $Al_xGa_{1-x}N$ interlayer with 42% Al concentration significantly compensated the tensile stress in the AlN film.

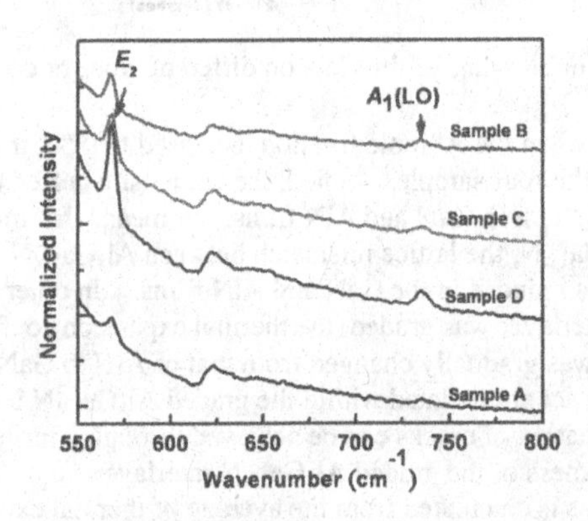

Fig.3 Raman spectrum for GaN films grown on different types of buffers

Figure 4 shows the XRD reciprocal space mappings (RSMs) of sample D which used $Al_{0.42}Ga_{0.58}N$ film/graded $Al_xGa_{1-x}N$/multilayer AlN composite buffer. In this figure, the AlN reciprocal-lattice points are distributed on the top of the solid line box region, which shows the AlN multilayer is under tensile strain (ε_a) about 0.3%. The GaN reciprocal-lattice points distributed in the bottom of the solid line box region, which indicates the GaN film is under small tensile strain (ε_a) about 0.12%. The thin graded $Al_{0.66}Ga_{0.34}N$ and $Al_{0.42}Ga_{0.58}N$ reciprocal-lattice points distributed between AlN and GaN positions were under compressive strain of about -0.5% and -0.92% respectively, which compensated the tensile stress formed in the AlN film and resulted in the crack free GaN film on Si.

The cross-sectional scanning electron microscopy (SEM) image of the sample D is shown in figure 5.

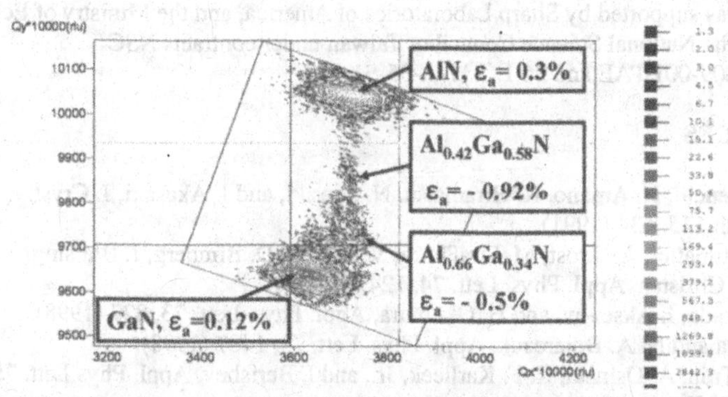

Fig. 4. Reciprocal space mappings (RSMs) for GaN with multilayer AlN and graded AlGaN buffer on Si (111) substrate of sample D.

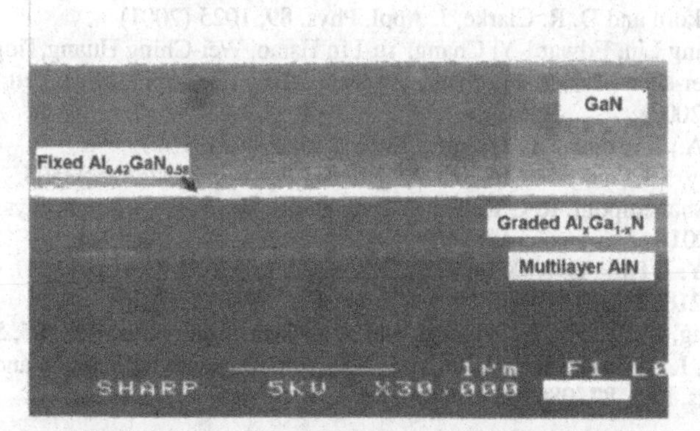

Fig 5. Cross-sectional SEM image of the sample D

CONCLUSION

A composite buffer using multilayer HT-AlN/LT-AlN/HT-AlN films structure combined with $Al_xGa_{1-x}N$ film was proposed to reduce the tensile stress in the AlN film and increase the compressive stress of the GaN film grown. Two types of $Al_xGa_{1-x}N$ layers one with graded Al concentration, one with fixed Al concentrations were incorporated in the composite buffer. The correlation of the Al composition in the $Al_xGa_{1-x}N$ layer with the stress in the GaN film grown was studied. It is found that a composite buffer layer consists of a graded $Al_xGa_{1-x}N$ layer (*x* varies from 0~0.66), and a fixed $Al_{0.42}Ga0_{.58}N$ interlayer incorporated and a multilayer HT-AlN/LT-AlN/HT-AlN films, the tensile stress in the AlN film can be significantly reduced to 0.069 GPa as observed from the Raman shift, and a 1 μm crack-free GaN film was successful growth on the 6" Si (111) substrate using this proposed composite buffer layers.

ACKNOWLEDGMENTS

This work was supported by Sharp Laboratories of America, and the Ministry of Economic Affairs and the National Science Council of Taiwan under contracts NSC 96-2752-E-009-001-PAE and 96-EC-17-A-05-S1-020.

REFERENCES

1. T. Takeuchi, H. Amano, K. Hiramatsu, N. Sawaki, and I. Akasaki, J. Cryst. Growth **115**, 634 (1991).
2. A. Strittmatter, A. Krost, M. Straßburg, V. Tu¨rck, D. Bimberg, J. Bla¨sing, and J. Christen, Appl. Phys. Lett. **74**, 1242 (1999).
3. Y. Nakada, I. Aksenov, and H. Okumura, Appl. Phys. Lett. **73**, 827 (1998).
4. S. Guha and N. A. Bojarczuk, Appl. Phys. Lett. **73**, 1487 (1998).
5. C. A. Tran, A. Osinski, R. F. Karlicek, Jr., and I. Berishev, Appl. Phys.Lett. **75**, 1494(1999).
6. J. W. Yang, A. Lunev, G. Simin, A. Chitnis, M. Shatalov, M. Asif Khan, J. E. Van Nostrand, and R. Gaska, Appl. Phys. Lett. **76**, 273 (2000).
7. S. Zamir, B. Meyler, and J. Salzman, Appl. Phys. Lett. **78**, 288(2001).
8. A. Dadgar, J. Bläsing, A. Alam, M. Heuken, A. Krost, Jpn. J. Appl. Phys.39 (2002) L1183
9. E. V. Etzkorn and D. R. Clarke, J. Appl. Phys. **89**, 1025 (2001)
10. Kung-Liang Lin, Edward-Yi Chang, Yu-Lin Hsiao, Wei-Ching Huang,Tingkai Li, Doug Tweet, Jer-shen Maa, Sheng-Teng Hsu Ching-Ting Lee, Appl. Phys. Lett. 91, 222111(2007)
11. Slack, G.A.; Bartram, S.F. *J. Appl. Phys.*, v. 46, p. 89, 1975.
12. S.Tripathy, S.J.Chua, P.Chen, and Z.L.Miao, J.Appl.Phys.92(2002)3503.
13. Guha, F.Shahedipour, R. C.Keller, V. Yang and B. W. Wessels, Appl. Phys. Lett. 78,58(2001)
14. K. Koh, Y. J. Park, E. K. Kim, C. S. Park, S. H. Lee, J. H. Lee, and S. H. Choh, J. Cryst. Growth **218**, 214 (2000).
15. L. S. Wang, K. Y. Zang, S. Tripathy, and S. J. Chua. Appl. Phys. Lett. **85**, 5881 (2004).
16. M. Jamil, J. R. Grandusky, V. Jindal, F. Shahedipour-Sandvik, S. Guha and M. Arif. Appl. Phys. Lett. **87**, 082103 (2005).

Mater. Res. Soc. Symp. Proc. Vol. 1068 © 2008 Materials Research Society 1068-C03-08

Growth optimization for high quality GaN films grown by metal-organic chemical vapor deposition

Jung Hun Jang, A M Herrero, Seungyoung Son, B Gila, C Abernathy, and V Craciun
Materials Science and Engineering, University of Florida, Gainesville, FL, 32611

ABSTRACT

GaN layers were grown on c-plane sapphire substrates by using a conventional two step growth method via metal organic chemical vapor deposition (MOCVD). The effect of different growth conditions used in the deposition of the low temperature nucleation layer and high temperature islands growth on the crystalline quality of the GaN layers was investigated by high resolution X-ray diffraction (HRXRD) and transmission electron microscopy (TEM). The polar (tilt) and azimuthal (twist) spread were estimated from the full width at half maximum (FWHM) values of the omega rocking curves (ω-RCs) recorded from planes parallel and perpendicular to the sample surface, respectively. It was found from the XRD and TEM study that the edge and mixed type threading dislocations were dominant defects so that the relevant figure of merit (FOM) for the crystalline quality should be considered only by the FWHM value of ω-RC of the surface perpendicular plane. The results showed that the mixed- and edge-types dislocation densities were strongly influenced by the growth conditions used in the deposition of the nucleation layer and high temperature islands growth.

INTRODUCTION

High quality GaN film has been obtained by a two-step growth method in which about 20 nm thick nucleation layer (NL) was deposited at low temperature (LT), followed by high temperature (HT) island growth [1,2]. The growth conditions, such as process temperature, TMGa flow rate, V/III ratio and chamber pressure used in each growth step, have a strong influence on the microstructure of the final GaN films. Figge et al. reported that the number of the nucleation sites during the nucleation layer deposition could be controlled by using V/III ratio [3]. Wood et al. showed the effect of nucleation layer annealing temperature and buffer layer growth temperature on their microstructure and morphology [4]. This work on the growth optimization has been carried out in order to find where the threading dislocations are generated from and how they can be reduced.

EXPERIMENT

The GaN films were grown on c-plane sapphire substrates using a Veeco P75 vertical MOCVD reactor via a conventional two-step growth method [1,2]. Trimethylgallium (TMGa), ammonia (NH_3), and hydrogen (H_2) were used as the Ga and N precursors, and carrier gas, respectively. The growth process was monitored by the reflectance transient method [5]. A 20 nm thick GaN nucleation layer (NL) was grown initially at 532~550 ℃. Then, the temperature was ramped up to 1028-1096 ℃ for high temperature three-dimensional (3D) islands growth. The growth conditions for the high temperature islands and low temperature nucleation layer were controlled by changing TMGa flow rate with the same V/III ratio and growth temperature, which was shown in table I. Finally, about 2 μm thick main GaN layers were deposited with a growth rate of 1.8 μm/h in 100 Torr. A PANanalytical MRD X'Pert system was used for ω-RCs and ϕ-scans in skew and grazing incidence geometries. For the cross sectional TEM images of the GaN layers, a JEOL TEM 200CX was used and the sample preparation was carried out by using the focused ion beam (FIB) technique.

Table I. The growth conditions used during the deposition of nucleation layer and 3D islands.

Growth conditions		Samples A	B	C	D	E
Nucleation layer	Temperature (°C)	550	550	550	532	542
	V/III	19880	19880	19880	19880	19880
3D island growth	Temperature (°C)	1094	1095	1028	1085	1096
	TMGa (sccm)	10	8	10	9	9
	V/III	5964	6088	5964	6012	6012
	Time (min)	10.0	16.7	18.5	16.6	19.3

DISCUSSION

The microstructure of GaN films with high dislocation density could be described by the mosaic spread, which consists of the polar (tilt) and azimuthal (twist) spreads [6]. Measurement of the tilt is very straightforward and directly obtained by in-plane high resolution X-ray diffraction setting. However, the twist angle measurement is not simple because it requires the special geometry, such as transmission or edge geometry [7,8]. Here, the grazing incidence X-ray diffraction (GIXD) technique was used in order to directly obtain the twist parameter because it is fast and accurate [9]. Fig. 1(a) represents a 360° ϕ-scan of the $\{10\bar{1}0\}$ plane recorded by

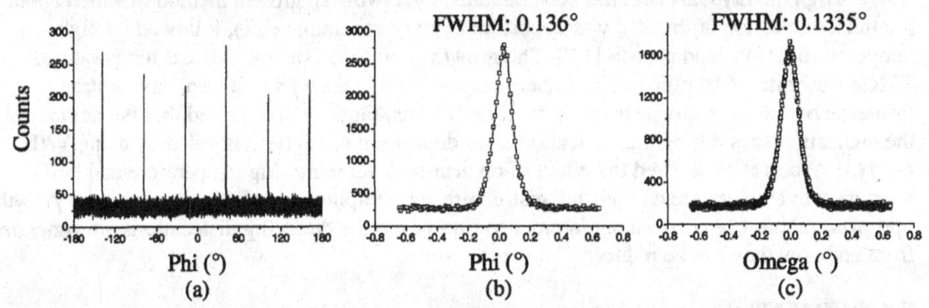

Figure 1. (a) Grazing incidence $\{10\bar{1}0\}$ 360° ϕ-scan (b) $(10\bar{1}0)$ ϕ-scan (c) $(10\bar{1}0)$ ω-RC of sample 'B'. The whole phi scan represents a six-fold symmetry of wurtzite structure of GaN. The FWHMs measured from ω-RC and ϕ-scan are almost identical.

GIXD technique. The six diffraction peaks are well defined, which represent the six-fold symmetry of the basal plane in the wurtzite crystal structure. The diffraction profiles of the $(10\bar{1}0)$ ϕ-scan and ω-RCs are also shown in Fig. 1(b) and (c). When the inclination angle (χ) is

90°, the scan directions of ω-RC and ϕ-scan of the surface normal plane are exactly the same and their full width at half maximum (FWHM) values should also be the same, as shown in Fig. 1(b) and (c). However, ϕ-scans, except when the ϕ-axis is co-axial with the ω-axis, always have a component of out-of plane resolution, which is typically two orders of magnitude worse than the in-plane resolution on a high resolution diffractometer. Hence, a grazing incidence ω-RC of surface perpendicular plane (χ=90°) was used in order to obtain the twist angle.

Now, the FWHM values of the in-plane, asymmetric plane and out of-plane ω-RCs of the GaN films used in this study are shown in table II. In this study, the instrumental broadening was neglected because it contributed to only a few arc sec. The tilt and twist angles of the high defective GaN layers are associated with the screw and edge type dislocations ($l = [0001]$, $\vec{b}_{screw} = [0001]$, $\vec{b}_{edge} = \frac{1}{3}[11\bar{2}0]$), respectively [10,11]. Heying et. al. reported that FWHM of $(10\bar{1}2)$ ω-RC could be a relevant figure of merit (FOM) for the crystal quality because this can be broadened by both type dislocations [12]. As shown in table II, the FWHM value of the ω-RC of the surface normal plane is much larger than that of the surface parallel plane, which is indicative

Table II. FWHM values (arc sec) of in-, asymmetric and out of-planes ω-RCs. The FWHM value of the $(10\bar{1}0)$ ω-RC is much higher than that of the (0002) ω-RC.

Planes	X	Samples					Si reference sample	
		A	B	C	D	E	X	FWHM
(0002)	0.0°	195	199	204	216	194	0.0°	11
$(10\bar{1}2)$	42.9°	419	346	330	312	332	35.6°	11
$(30\bar{3}2)$	70.2°	537	444	399	387	414	72.1°	8
$(10\bar{1}0)$	90.0°	593	490	477	443	423	90.0°	13

that the edge type threading dislocations are dominant. Therefore, the crystalline quality of GaN layer should be determined only by the edge dislocation density. Typical bright field images from sample 'A' and 'E' are shown in Fig. 2. Under $\vec{g} = 0002$, the type of the threading dislocations that can be observed are pure screw or mixed ($l = [0001]$, $\vec{b}_{mixed} = \frac{1}{3}[11\bar{2}3]$), while the edge or mixed dislocations are present under $\vec{g} = 11\bar{2}0$. From Fig. 2, it was found that the edge and mixed type dislocations were dominant defects in the GaN films, which is consistent with XRD data mentioned above. By Heinke. et. al., the FWHM of $(30\bar{3}2)$ ω-RC could be a figure of merit for the edge dislocation density because this plane has a high inclination angle (~71°) with respect to the surface plane. They reported that the twist angle was 1.14±0.04 times the FWHM value of the $(30\bar{3}2)$ ω-RC [13]. In our case, the sample 'D' shows the minimum FWHM values of the $(30\bar{3}2)$ ω-RC, but the sample 'E' shows the minimum FWHM values of the $(10\bar{1}0)$ ω-RC and the best crystalline quality of all GaN films. This discrepancy might be due to the

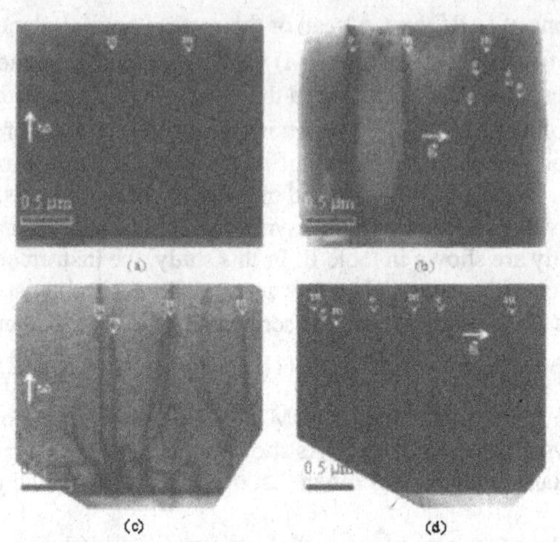

Figure 2. Cross sectional bright field images of sample 'A' (a, b) and 'E' (c, d). The defects observed under (a) and (c) $\vec{g}=0002$ are screw or mixed type dislocations and ones under (b) and (d) $\vec{g}=11\bar{2}0$ are edge type dislocations.

dependent distribution of the tilt and twist angles. Srikant et al. used a mathematical formulation to obtain a relationship between the tilt and twist distributions [14]. Therefore, a FOM for the crystalline quality should be determined by the FWHM value of the ω-RC of the surface perpendicular plane. In this aspect, the grazing incidence X-ray diffraction is a more suitable technique than above mentioned theoretical method for the study of the growth optimization since it requires only one omega rocking curve and the twist angle can be directly measured without any theoretical computation.

Sample A, B and C were grown on the same nucleation layer, but under different high temperature island growth conditions in order to find out the effect of the 3D growth mode time, which can be defined as the transition time from 3D to 2D growth mode. It increased with decreasing the growth temperature and TMGa flow rate, as shown table I. The FWHM values of their (0002) ω-RCs are almost identical around 200 arc sec, but those of $(10\bar{1}0)$ ω-RCs, which is caused by the pure edge or mixed type dislocations, decrease with increasing 3D growth mode time. Therefore, it was found that the defects generated during the high temperature 3D island growth were edge type threading dislocation. Many reports showed that the edge type dislocations were generated from the junctions between the misoriented high temperature islands [15,16,17,18]. The 3D growth mode time is correlated to the average grain size; a longer t_{3D} gives a larger grain size, resulting in the reduction of the edge dislocations. This study showed that the edge threading dislocations can be reduced by optimizing 3D growth mode time. Next, the nucleation layer would be thought as another source of defects generation because it acts as a template for high temperature main GaN films. As shown above, since the mixed threading dislocation density was not changed even though using different growth conditions used during the high temperature island growth, we believe that the mixed dislocations could be controlled by changing the growth condition for the nucleation layer. In addition, the initial GaN layers grown at low temperature were found to have a zinc blend crystal structure with high density of stacking faults in one set of {111} planes, and then they were transformed into the wurtzite crystal structure by introducing Shockley partial dislocations during the high temperature ramping [15,19]. In this time, the edge type dislocation can be also generated from the reaction between the sub-boundary dislocations and these Shockley partial dislocations. Since this large stacking disorder of nucleation layers was caused by a low growth temperature, it has been thought that the growth temperature must be an important parameter to give a great influence on the crystalline quality of the main GaN layers. Sample 'D' and 'E' were deposited on the nucleation layer grown at relatively lower temperature than sample 'A', 'B', and 'C'. Sample 'E' grown at 542 °C for the nucleation layer and with 3D growth mode time of 19.3 min showed the highest crystalline quality of our samples.

CONCLUSIONS

The growth optimization study was performed by using the values of ω-RC measured by grazing incidence X-ray diffraction. The threading dislocation density has been controlled by using different growth conditions during the deposition of the nucleation layer and high temperature islands growth. The results from XRD and TEM investigations showed that the pure edge and mixed type dislocations were dominant defects in our GaN films. Therefore, the relevant figure of merit (FOM) for the crystalline quality should be determined by FWHM values of ω-RC of the surface perpendicular planes. In our study, GaN film grown at 542 °C for the nucleation layer and with 3D growth mode time of 19.3 min showed the best crystalline quality so far.

ACKNOWLEDGMENTS

The authors would like to thank Major Analytical Instrumentation Center (MAIC) in University of Florida for help with XRD and TEM measurements. The authors gratefully acknowledge the support from AFOSR Contract No. FA8650-04-2-1619.

REFERENCES

1. I. Akasaki and H. Amano, Tech. Dig. Int. Electron Devices Meet **96**, 231 (1996)
2. K. Lorenz, M. Gonsalves, W. Kim, W. Narayanan, and S. Mahajan, Appl. Phys. Lett. **77**, 3391 (2000)
3. S. Figge, T. Bottcher, S. Einfeldt, and D. Hommel, J. Crys. Growth **221**, 262 (2000)
4. D. A. Wood, P. J. Parbrook, R. J. Lynch, M. Lada, and A. G. Cullis, Phys. Stat. Sol. (a) **188**, 641 (2001)
5. S. M. Hubbard, G. Zhao, D. Pavlidis, W. Sutton, and E. Cho, J. Crys. Growth **284**, 297 (2005)
6. R. Chierchia, T. Bottcher, H. Heinke, S. Einfeldt, S. Figge, and D. Hommel, J. Appl. Phys. **93**, 8918 (2003)
7. V. V. Ratnikov, R. Kjutt, and T. Shubina, J. Appl. Phys. **88**, 6252 (1998)
8. H. Amano, T. Takeuchi, H. Sakai, S. Yamaguchi, C. Wetzel, and I. Akasaki, Mater. Sci. Forum **264-268**, 1115 (1998)
9. J. H. Jang, A. M. Herrero, B. Gila, C. Abernathy, and V. Craciun, J. Appl. Phys. **103**, 063514 (2008)

10. T. Metzger, R. Hopler, E. Born, O. Ambacher, M. Stutzmann, R. Stommer, M. Schuster, H. Gobel, S. Christiansen, M. Albrecht, and H. P. Strunk, Philosophical Magazine A 77, 1013 (1998)
11. D. Kapolnek, X. H. Wu, B. Heying, S. Keller, B. P. Keller, U. K. Mishra, S. P. DenBaars, and J. S. Speck, Appl. Phys. Lett. **67**, 1541 (1995)
12. B. Heying, X. H. Wu, S. Keller, Y. Li, D. Kapolnek, B. P. Keller, S. P. DenBaars, and J. S. Speck, Appl. Phys. Lett. **68**, 643 (1996)
13. H. Heinke, V. Kirchner, S. Einfeldt, D. Hommel, Appl. Phys. Lett. **77**, 2145 (2000)
14. V. Srikant, J. S. Speck, and D. R. Clarke, J. Appl. Phys. **82**, 4286 (1997)
15. X. H. Wu, P. Fini, E. J. Tarsa, B. Heying, S. Keller, U. K. Mishra, S. P. DenBaars, and J. S. Speck, J. Cryst. Growth **189/190**, 231 (1998)
16. V. Narayanan, K. Lorenz, Wook Kim, and S. Mahajan, Appl. Phys. Lett. **78**, 1544 (2001)
17. B. Moran, F. Wu, A. E. Romanov, U. K. Mishra, S. P. Denbaars, and J. S. Speck, J. Cryst. Growth **273**, 38 (2004)
18. L. Lu, B. Shen, F. J. Xu, J. Xu, B. Gao, Z. J. Yang, G. Y. Zhang, X. P. Zhang, J. Xu, and D. P. Yu, J. Appl. Phys. **102**, 033510 (2007)
19. J. Narayan, Punam Pant, A. Chugh, H. Choi, and J. C. C. Fan, J. Appl. Phys. **99**, 054313 (2006)

Mater. Res. Soc. Symp. Proc. Vol. 1068 © 2008 Materials Research Society 1068-C01-08

Investigation of Blistering Phenomena in Hydrogen-Implanted GaN and AlN for Thin Film Layer Transfer Applications

R. Singh[1], R. Scholz[2], S. H. Christiansen[2], and U. Goesele[2]
[1]Physics, Indian Institute of Technolgy Delhi, Hauz Khas, New Delhi, 110016, India
[2]Experimental II, Max Planck Institute of Microstructure Physics, Weinberg 2, Halle, 06120, Germany

ABSTRACT

High fluence hydrogen implantation-induced blistering phenomena in GaN and AlN have been investigated for potential thin film layer transfer applications. GaN and AlN were implanted with 100 keV H_2^+ ions with various ion fluences in the range of 5×10^{16} to 2.5×10^{17} cm^{-2}. After implantation the samples were annealed at higher temperatures up to 800°C in order to observe the formation of surface blisters. In the case of GaN only those samples that were implanted with a fluence of 1.3×10^{17} cm^{-2} or higher showed surface blistering after post-implantation annealing. For AlN the samples those were implanted with a fluence of 1.0×10^{17} or 1.5×10^{17} cm^{-2} displayed surface blistering after post-implantation annealing. Cross-sectional transmission electron microscopy was utilized to observe the microscopic defects that eventually caused surface blistering. Large area microcracks, as revealed in the XTEM images, were clearly observed in the case of both GaN and AlN after post-implantation annealing. A comparison of the hydrogen implantation-induced blistering in GaN and AlN has also been presented.

INTRODUCTION

Wide bandgap nitride semiconductors such as GaN and AlN have a wide range of applications in the area of optoelectronics as well as high frequency, high power electronic devices [1-3]. These nitrides are mostly grown epitaxially on lattice and thermal mismatched substrates like sapphire, SiC or even on Si due to the fact that free-standing bulk substrates of GaN and AlN are very expensive and are mostly available in small sizes [4-6]. The heteroepitaxial growth of nitride epitaxial layers on foreign substrates leads to the formation of growth-related defects like dislocations, stacking faults, twins etc. that occur to relax the strain. The high density of dislocations in the nitride epitaxial layers grown on hetero-substrates has deleterious effects on the performance and reliability of the devices fabricated utilising these layers. One of the methods to fabricate low-cost and high structural quality substrates, comparable to free-standing substrates of GaN and AlN, for the epitaxial growth of group-III nitrides would be direct wafer bonding and layer transfer of thin GaN films via a high fluence hydrogen implantation and layer splitting upon annealing [7, 8]. The free-standing GaN and AlN substrates can be utilised to transfer multiple layers on other substrates. This process is based upon the agglomeration of hydrogen implantation-induced platelets upon annealing and the subsequent formation of over-pressurized microcracks. For the case of the implanted wafer bonded to a handle wafer, splitting of a thin slice of material parallel to the bonding interface occurs [7, 8]. For this process to occur a narrow parameter window of implantation fluence, annealing temperature and time has to be defined since the layer splitting is a strongly material dependent process. The physical mechanisms leading to the process of layer splitting can be conveniently investigated by studying the development of surface blisters in hydrogen implanted and annealed but unbonded wafers [9]. In the present report, we present our results related to high fluence hydrogen implantation of free-standing GaN and epitaxial AlN layers after post-implantation annealing.

EXPERIMENT

The free-standing GaN substrates used in the present investigation were hydride vapor phase epitaxy (HVPE) grown 2-inch wafers having thickness of about 350 μm. These GaN wafers were implanted at room temperature with 100 keV H_2^+ ions with a fluence of 1.3×10^{17} cm^{-2}. The AlN epitaxial layers used for hydrogen implantation were about 2 μm in thickness. These layers were grown on 2-inch (0001) c-plane sapphire substrates using hydride vapor phase epitaxy (HVPE). These AlN/sapphire wafers were purchased commercially from TDI, Inc. USA. The free-standing GaN wafers and AlN epitaxial layers were implanted at room temperature with 100 keV H_2^+ ions with various fluences in the range of 5×10^{16} to 2.5×10^{17} cm^{-2}. During implantation the sample surface normal was inclined at ~7° relative to the incident ion beam in order to avoid channeling effects. The hydrogen implantation was performed at Ion Beam Services (IBS), France. After implantation the wafers were cut into small pieces (~3×3 mm^2) and annealed at different temperatures in the range of 300-800°C in order to observe the formation of surface blisters. The annealing was carried out in an air ambient box type furnace. The formation of optically detectable surface blisters on FS-GaN and AlN epitaxial layers was observed using a Nomarski optical microscope. The microstructural characterization of the implantation-induced damage in GaN and AlN was carried out using cross-sectional transmission electron microscopy (XTEM). The XTEM measurements were performed using a Philips CM20T machine operated at 200 kV.

RESULTS AND DISCUSSION

Free-standing GaN samples were hydrogen-implanted with a fluence of 1.3×10^{17} cm^{-2}. In our previous work involving hydrogen implantation of GaN/sapphire epitaxial layers, it was found that the minimum hydrogen fluence required for blistering in GaN after post-implantation annealing was 1.3×10^{17} cm^{-2} [9, 10]. Below this fluence blistering did not occur in GaN even after post-implantation annealing at higher temperatures of 800°C. Hence for FS-GaN we selected the minimum hydrogen fluence of 1.3×10^{17} cm^{-2}. The surface of this sample after post-implantation annealing is shown in figure 1. The surface blisters could be clearly observed on the GaN surface. The cross-sectional TEM image of the implanted and annealed FS-GaN is shown in figure 2. Large area microcracks could be clearly seen in the damage band. This damage band extends between 130 – 420 nm from the surface of GaN. The hydrogen filled microcracks lead to the formation of surface blisters upon high temperature annealing when annealed for time longer than the blistering time at that particular temperature [9]. In our earlier work, it was shown that the damage band was decorated with hydrogen filled nanovoids in the as-implanted GaN samples [11]. These nanovoids served as precursors for the formation of large area microcracks.

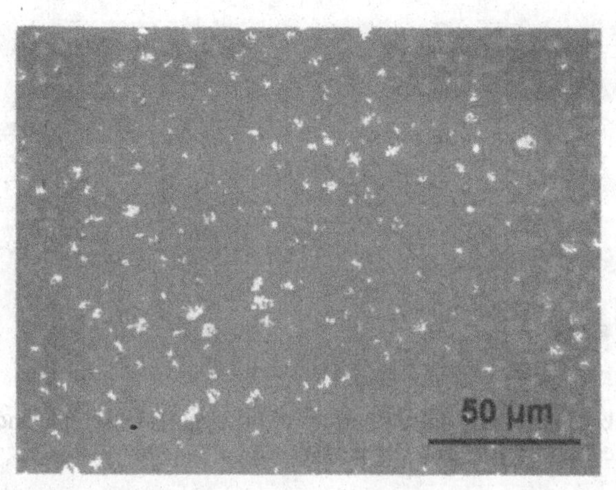

Figure 1. Nomarski optical image of the surface of FS-GaN implanted by 100 keV H_2^+ ions and annealed at 500°C for 30min.

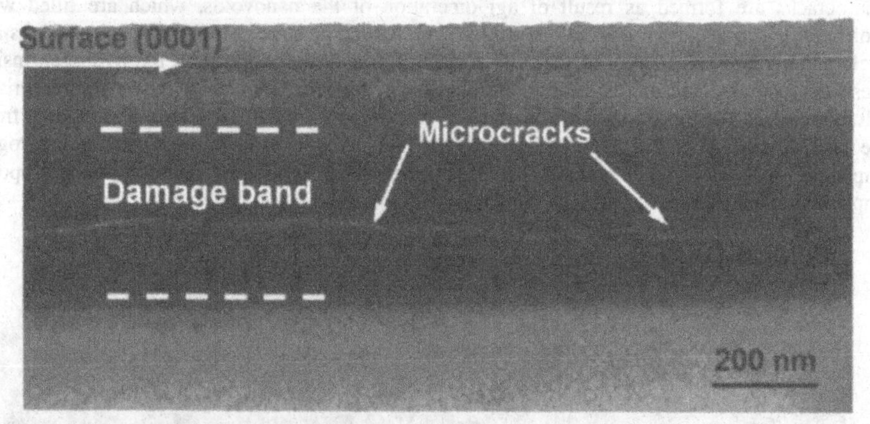

Figure 2. Cross-sectional transmission electron image of FS-GaN sample implanted by 100 keV H_2^+ ions and annealed at 500°C for 30min.

In the case of AlN epitaxial layers, the surfaces of the virgin AlN samples were found to be quite rough, as evident from the Nomarski optical image of figure 3a. The surfaces were found to have a large number of hillocks having the lateral size of a few microns. In the case of hydrogen-implanted AlN, the wafers implanted with a fluence of 2.0×10^{17} cm^{-2} or higher exhibited surface blisters in the as-implanted state. In contrast, the AlN wafers that were implanted with a fluence of 1.0×10^{17} or 1.5×10^{17} cm^{-2} showed surface blisters only after post-implantation annealing at higher temperatures. The sample implanted with a fluence of 5.0×10^{16} cm^{-2} did not show any blistering even after post-implantation annealing up to 800°C for 2 hrs [12]. A typical Nomarski optical image of the AlN surface implanted with a fluence of 1.0×10^{17} cm^{-2} and subsequently annealed at 700°C for 5min is shown in figure 3b. The surface blisters have diameters in the range of 5–15 µm.

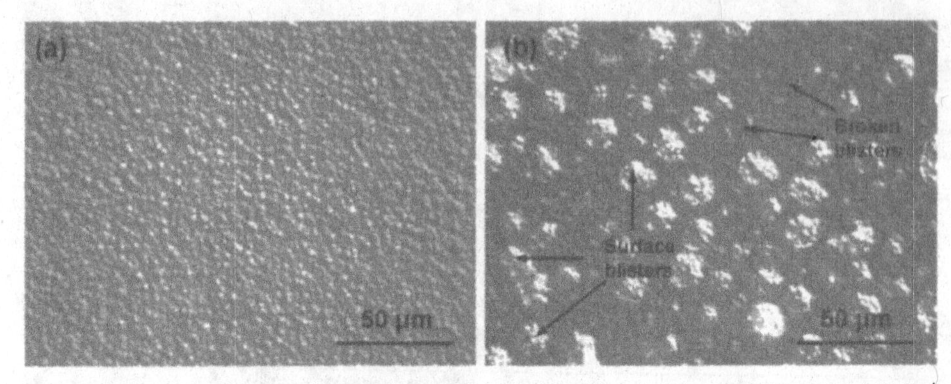

Figure 3. Nomarski optical image of the surface of (a) virgin AlN sample and (b) 100 keV H_2^+ implanted and annealed (500°C/1 hr) AlN sample.

The thickness of the AlN layers is about 2 µm. When the hydrogen implanted AlN layers are annealed at higher temperatures, large area microcracks are formed, as shown in figure 4. These microcracks are formed as result of agglomeration of the nanovoids, which are filled with molecular hydrogen, upon thermal annealing [13]. As can be seen in Fig. 4, the microcracks are preferentially formed parallel to the AlN surface. Upon further annealing, the hydrogen inside these microcracks pressurizes them to become larger in size. Ultimately these overpressurized microcracks lead to the formation of surface blisters as seen in Fig. 3b. Hence it is evident from the present study that the nanovoids formed in the damage band of high fluence hydrogen implanted AlN serve as precursors for the formation of large area microcracks after post-implantation annealing.

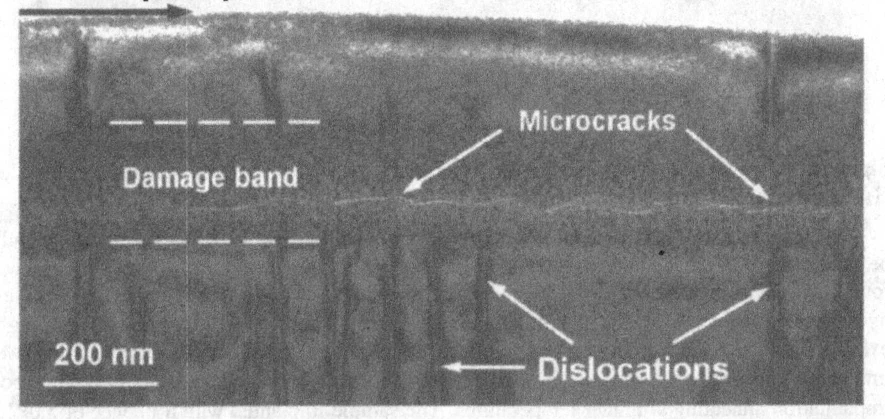

Figure 4. Cross-sectional transmission electron image of the hydrogen implanted AlN epilayer after post-implantation annealing at 500°C/10 min.

In the earlier reports concerned with high fluence hydrogen implantation of semiconductors such as Si, Ge and SiC [7, 14, 15], it has been observed that the implantation leads to the formation of penny-shaped nanoplatelets in the damage band, which serve as precursors for the formation of microcracks upon high temperature annealing. However, in the case of AlN hydrogen implantation results in the development of nanovoids inside the damage band [13]. Similar observations were also made in the case of hydrogen implanted GaN(0001) [11, 16] where the damage band was observed to be decorated with nanovoids. These nanovoids led to the formation of microcracks after post-implantation annealing. Both AlN and GaN have wurtzite crystal structure and this could perhaps be the reason for nanovoids formation in the hydrogen-implanted AlN and GaN.

CONCLUSIONS

High fluence hydrogen implantation-induced blistering process has been investigated for GaN and AlN. The minimum fluences required for the blistering to occur after post-implantation annealing were 1.3×10^{17} and 1.0×10^{17} cm^{-2} for GaN and AlN, respectively. It was observed using XTEM measurements that the implantation-induced damage bands in both GaN and AlN were decorated with molecular hydrogen filled nanovoids. These nanovoids led to the formation of large area microcracks in the damage band upon high temperature annealing of the implanted samples. Hence on both the semiconductors the nanovoids served as precursors for the formation of large area microcracks that eventually caused the formation of surface blisters upon annealing at higher temperatures for longer times.

ACKNOWLEDGMENTS

The authors are thankful to Mrs. S. Hopfe for the XTEM specimen preparation. The work is supported financially by the Max Planck Society (MPG), Germany under the cooperation scheme of Max Planck Partner Group at Indian Partner Institutions and co-funded by the Department of Science and Technology (DST), India. The authors also thank the German Federal Ministry of Education and Technology (BMBF) for partial support in the framework of the CrysGaN-project.

REFERENCES

1. S. J. Pearton, J. C. Zolper, R. J. Shul and F. Ren **86**, 1 (1999).
2. Y. Taniyasu, M. Kasu, and T. Makimoto, *Nature* **441**, 325 (2006).
3. Z. Ren, Q. Sun, N. Mizuhara, S.-Y. Swon, J. Han, K. Davitt, Y.K. Song, A. V. Nurmikko, H.-K. Cho, W. Liu, J. A. Smart and L. J. Schowalter, *Appl. Phys. Lett.* **91**, 051116 (2007).
4. P. Gibbart, *Rep. Prog. Phys.* **67**, 667 (2004).
5. See the web-site: www.crystal-is.com.
6. N. Mizuhara, M. Miyananga, S. Fujiwara, H. Nakahata, and T. Kawase, D. Contributorname, *Phys. Stat. Sol. (c)* **4**, 2244 (2007).
7. M. Bruel, *Electron. Lett.* **31**, 1201 (1995).
8. S. H. Christiansen, R. Singh and U. Goesele, *Proc. IEEE* **94**, 2060 (2006).
9. R. Singh, I. Radu, U. Gösele and S. H. Christiansen, *Phys. Stat. Sol. (c)* **3**, 1754 (2006).

10. R. Singh, I. Radu, G. Bruederl, C. Eichler, V. Haerle, U. Goesele and S. H. Christiansen, *Semicond. Sci. Technol.* **22**, 418 (2007).
11. I. Radu, R. Singh, R. Scholz, U. Goesele, S. Christiansen, G. Bruederl, C. Eichler and V. Haerle, *Appl. Phys. Lett.* 89, 031912 (2006).
12. R. Singh, R. Scholz, S. H. Christiansen and U. Gösele, *Semicond. Sci. Technol.* **23**, 045007 (2008).
13. R. Singh, R. Scholz, S. H. Christiansen and U. Gösele (communicated).
14. L. D. Cioccio, Y. Letiec, F. Letertre, C. Jaussaud and M. Bruel, *Electron. Lett.* 32, 1144 (1996).
15. J. M. Zahler, A. Fontcuberta I Morral, M. J. Griggs, H. A. Atwater and Y. J. Chabal, *Phys. Rev. B* **75**, 035309 (2007).
16. A. Tauzin, T. Akatsu, M. Rabarot, J. Dechamp, M. Zussy, H. Moriceau, J.F. Michaud, A.M. Charvet, L.Di Cioccio, F. Fournel, J. Garrione, B. Faure, F. Letertre, and N. Kernevez, *Electron. Lett.* **41**, 668 (2005).

Mater. Res. Soc. Symp. Proc. Vol. 1068 © 2008 Materials Research Society 1068-C03-13

Dependence of GaN Defect Structure on the Growth Temperature of the AlN Buffer Layer

Yuen-Yee Wong[1], Edward Yi Chang[1], Tsung-Hsi Yang[2,3], Jet-Rung Chang[2], Yi-Cheng Chen[2], and Jui-Tai Ku[4]

[1]Material Science and Engineering, National Chiao Tung University, 1001, Ta-Hsueh Rd., Hsinchu, Taiwan

[2]Electronics Engineering, National Chiao Tung University, 1001, Ta Hsueh Rd., Hsinchu, Taiwan

[3]Microelectronic and Information System Research Center, National Chiao Tung University, 1001, Ta Hsueh Rd., Hsinchu, Taiwan

[4]Electrophysics, National Chiao Tung University, 1001, Ta Hsueh Rd., Hsinchu, Taiwan

ABSTRACT

The defect structure of the GaN film grown on sapphire by plasma-assisted molecular beam epitaxy (PAMBE) technique was found to be dependent on the AlN buffer layer growth temperature. This buffer growth temperature controlled the defect density in GaN film but had shown contrary effects on the density of screw threading dislocation (TD) and edge TD. The density of screw TD was high on lower temperature buffer but low on the higher temperature buffer. Meanwhile the density of edge TD had shown the opposite. Further examinations have suggested that the defect structure was closely related to the stress in the GaN film, which can be controlled by the growth temperature of the AlN buffer. Using the 525°C AlN buffer, optimum quality GaN film with relatively low screw and edge TDs were achieved.

INTRODUCTION

The wide bandgap GaN material has generated huge interest in the recent years due to its potential to fabricate high power and high frequency high electron mobility transistors (HEMTs). Metalorganic chemical vapor deposition (MOCVD) and molecular beam epitaxy (MBE) are among the most commonly used methods to achieve high quality GaN material. Currently, GaN materials are normally produced using MOCVD especially for the light emitting diodes (LEDs) application due to its high production capability. But for the high performance electronic devices, MBE may be the method of choice. Researchers have shown that GaN with best electrical properties were grown by MBE [1-3]. The advantages of MBE include the in-site diagnostic technique using RHEED for real-time crystal growth monitoring, a carbon-free and hydrogen-free growth environment, smooth surface, sharp interfaces and low point defect density. However, to achieve high quality GaN material using MBE has it own difficulties. Unlike the availability of high quality GaAs substrate for the growth of GaAs devices using MBE, the lack of large-size native substrate for GaN has resulted in growing GaN on a foreign substrate such as sapphire, silicon carbide and silicon. Due to the large mismatches in crystal lattice and thermal expansion coefficient, the GaN films grown on these substrates are associated with large density of dislocation. Furthermore, the relatively lower growth temperature of MBE as compared to that of MOCVD also causes the MBE grown GaN to have lesser quality. The typical growth temperature of MBE (700-800°C) is only about one third of the GaN melting point T_m, which is significantly lower than the temperature required ($0.5T_m$) for effective dislocation gliding and annihilating. Therefore, the growth of GaN using MBE includes a thin AlN layer as the buffer layer. AlN has the intermediate lattice constant between the GaN and thesapphire or SiC. Another reason to use AlN is to achieve the Ga-face GaN when grown by MBE on sapphire [4-6]. Without the AlN, the GaN can be N-face when grown by MBE. A Ga-face GaN is needed to obtain the high quality crystal material.

AlN as buffer layer for GaN grown on sapphire has been used as early as in the beginning of 1980s. [7-8]. With AlN buffer, the quality of GaN grown on sapphire has improved significantly. Throughout the years, many researches led to investigate the role played by the AlN and its effect on the quality of the GaN film. For instance, Koblmueller et. al., [9] using MBE, have grown the AlN with different III/V ratios and growth temperatures. They have observed that the AlN film morphology was rough under N-rich condition. On the other hand, smooth surface was achieved when the Al/N ratio was slightly larger than 1. However, metal droplet would form on the smooth surface if excessive Al was used. The effect of different Al/N ratio buffers on GaN microstructure and electron mobility were further studied by Storm and co-workers [10-11]. They have shown that when the Al/N ratio of the AlN buffer layer increased from below to above unity, the electron mobility in the AlGaN/GaN heterostructures and also the buffer leakage current increased. Higher threading dislocation density GaN film was grown on N-rich AlN buffer. But this film has lower electron mobility. The dislocation density was reduced when high Al/N ratios were used but the buffer leakage was also increased as a result of the formation of cubic phase GaN at the GaN/AlN interface because of Al metal droplets. The crystal quality of GaN was also studied by Shim et. al. [12] with GaN grown on the different growth temperature AlN buffers. The growth temperature of the AlN buffer was varied from 750 to 1020°C, and with single and 2-step (double layer AlN with different growth temperatures) methods. Among these buffer layers, lowest dislocation density GaN was achieved when grown on the 2-step AlN, which could act as an effective dislocation filter layer.

The above examples have shown that the AlN buffer layer has great influence on the properties of the GaN film. In this paper, we have studied the effect of the AlN growth temperature on the GaN defect structure. The crystal properties of the GaN film were characterized using x-ray diffraction method.

EXPERIMENT

MBE system from the ULVAC Inc. equipped with radio frequency plasma for N source was used to grow the GaN and AlN materials throughout this study. Reflection high-energy electron deflection (RHEED) was used to monitor the growth process. Epi-ready sapphire substrates were thermally cleaned inside the MBE chamber before the materials deposition. Then, nitridation was carried out in order to form an initial layer for the growth of AlN buffer. A change of RHEED pattern could be used to observe the transformation of the topmost layers of sapphire into AlN. This step is necessary to achieve good quality GaN film. The AlN was then deposited and followed by a 2 um thick GaN. In order to investigate the dependence of GaN crystal quality on the growth parameters of AlN buffer, the growth condition for the GaN was kept constant throughout the study. On the other hand, the AlN growth temperature was changed from 450 to 840°C. From the RHEED, streaky patterns observed during the growth of all the above layers indicating that atomically flat surfaces were achieved for each of them. RHEED was also used to monitor the GaN surface change during the cooling process after the growth was completed. A change from 1x1 to 2x2 surface reconstruction was used to confirm that the GaN is of Ga-face. The crystal quality was checked ex-situ using the high resolution x-ray diffraction (HRXRD, Bede D1 system). In the XRD analysis, defects present in the GaN were determined by performing the ⊦-scans (rocking curve) on the symmetric plane (0002) and the asymmetric plane (10-12). The broadening of the rocking curve corresponded to the defect density in the GaN film. Mosaicity in the (0002) and (10-12) planes were a result of the screw and edge type dislocations, respectively. Heying et. al. [13] have demonstrated the reliability of using XRD to determine the defect structure by comparing the XRD result with transmission electron microscopy analysis.

DISCUSSION

The XRD results are shown in figure 1. Full width at half-maximum (FWHM) of the 0002 and 10-12 plane rocking curves are plotted for the GaN samples grown on different temperature AlN buffers. The AlN growth temperature had contrary effects on the densities of these two kinds of dislocations. From figure 1(a) the screw TD in GaN film (indicated by the value of FWHM of the 0002 plane) was inversely proportional to the AlN growth temperature. In our study, FWHM value as low as 1 arcmin was achieved when GaN was grown on the AlN buffer at 840°C. On the other hand, the edge TD in the GaN film (value of FWHM of the (10-12) plane in figure 1 (b)) was suppressed with the used of lower buffer temperature.

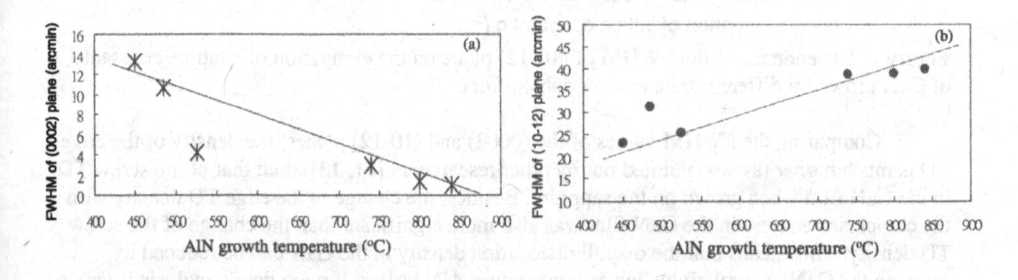

Figure 1. FWHM of (0002) plane (a) and (10-12) plane (b) rocking curves for GaN films grown on different AlN buffer layer temperatures.

In order to investigate the origin of the dislocations in the GaN films grown on different temperature AlN buffers, we have also checked the stresses present in the GaN films. Using the same XRD system, we have determined the C lattice constant of the GaN films. In comparison to the standard value of the GaN lattice constant (C_o=5.185 angstroms, [14]), the elongations of the lattice constant ($C-C_o$) of the GaN films were determined. The results are shown in figure 2. Also shown in figure 2 are the FWHM of the (10-12) plane. The result has explained that the GaN films grown on the sapphire are all compressively stressed ($C-C_o>0$). This is because large mismatches in lattice and thermal expansion coefficient between the GaN and sapphire have induced compressive stress in the GaN film. Furthermore, the result shown in figure 2 also reveals that the edge TD in the GaN film is increased proportionally with the stress present in the film. On the other hand, as suggested by the result in figure 1, the screw TD decreased accordingly. The origin of the lower stress for the GaN grown on lower temperature AlN buffer is unknown. But from the relatively high screw TD density generated in this GaN, it has been suggested that the somewhat inferior and less perfect atomic arrangement of the AlN buffer growth at lower temperature and that the GaN film may have aided the stress relief mechanism in the GaN film. The more relaxed GaN film thus has less edge TD. In contrast, when higher temperature was used to grow the AlN buffer, better atomic arrangement was achieved. But the stress in the GaN film was unable to be relieved. Higher edge TD density was thus generated, particularly by the large thermal stress created upon wafer cooling.

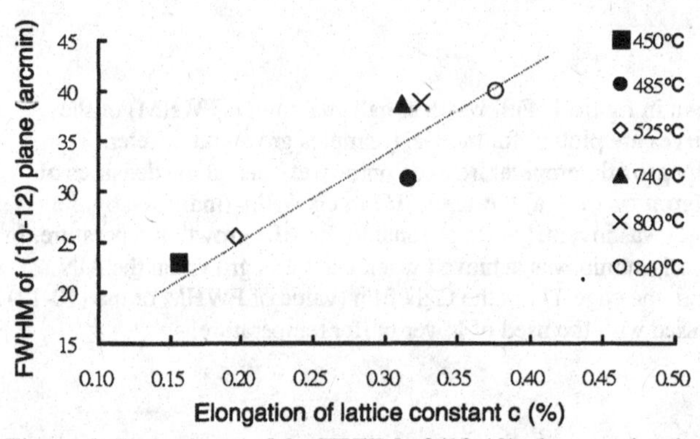

Figure 2. Dependence of the FWHM of (10-12) plane on the elongation of c lattice constant of GaN grown on different temperature AlN buffers.

Comparing the FWHM values of the (0002) and (10-12) planes, the density of the edge TD is much higher (as also pointed out by other researchers [**11, 13**]) than that of the screw TD in the GaN film when grown on the sapphire. Besides, the change in the edge TD density with the compressive stress in the GaN film was also more significant than the change of the screw TD density. This means that the overall dislocation density in the GaN can be reduced by growing the GaN on a relatively lower temperature AlN buffer. If more detail analysis is carried out on figure 1, optimum growth temperature for the AlN buffer layer can be determined to be about 525°C. On this buffer, the FWHM values of both the (0002) and (10-12) planes (4 and 25 arcmin, respectively) are relatively smaller as compared to others.

CONCLUSIONS

In summary, we have studied the dependence of the GaN defect structure on the AlN buffer growth temperature for GaN grown on sapphire by plasma-assisted MBE. Screw and edge dislocation densities in the GaN were affected by the growth temperature of the AlN buffer. The study has shown that the screw TD density was reduced using higher growth temperature buffer while the edge TD density was minimized using lower growth temperature buffer. We have attributed this phenomenon to the inherent stress in the GaN films. The lower temperature buffer seemed to provide a better way to relieve the stress in the GaN film. Therefore, lower edge TD density was achieved. Even though the screw TD was increased in the low temperature regime, the overall dislocation density was reduced due to the more significant reduction of the edge TD density on low temperature buffer. In our study, we have found that GaN film grown on the 525°C buffer had relatively low dislocation density for both the screw and edge types, and this provide the best quality film for further GaN device structure growth.

ACKNOWLEDGMENTS

This work was financially supported by the Ministry of Economic Affairs and the National Science Council of Taiwan under the research grants of 95-EC-17-A-05-S1-020, and NSC 95-2752-E-009-001-PAE and NSC 96-2221-E-009-236. The authors would like to thank ULVAC Taiwan Inc., Micheal Chen, Chien-Ying Chen and Stanley Wu for the MBE maintenance support.

REFERENCES

1. A. Corrion, C. Poblenz, P. Waltereit, T. Palacios, S. Rajan, U. K. Mishra and J.S. Speck, *IEICE Trans. Electron.* E89-C, 906 (2006).
2. M. J. Manfra, N. G. Weimann, J. W. P. Hsu, L. N. Pfeiffer, K. W. West and S. N. G. Chuet, *Appl. Phys. Lett.* 81, 1456 (2002).
3. L. K. Li, J. Alperin, W. I. Wang, D. C. Look and D. C. Reynolds, *Appl. Phys. Lett.* 76, 742 (2000).
4. O. Ambacher, J. Smart, J. R. Shealy, N. G. Weimann, K. Chu, M. Murphy, W. J. Schaff, L. F. Eastman, R. Dimitrov, L. Wittmer, M. Stutzmann, W. Rieger and J. Hilsenbeck, *J. Appl. Phys.* 85, 3222 (1998).
5. M. Stutzmann, O. Ambacher, U. Karrer, A. L. Pimenta, R. Neuberger, J. Schalwig, R. Dimitrov, P. J. Schuck and R. D. Grober, *Phys. Stat. Sol. (b)* 228, 505 (2001).
6. M. Sumiya and S. Fuke, *MRS Internet. J. Nitride Semocond.* Res. 9, 1 (2004).
7. S. Yoshida, S. Misawa and S. Gonda, *Appl. Phys. Lett.* 42, 427 (1982).
8. H. Amano, N. Sawaki, I. Akasaki and Y. Toyoda, *Appl. Phys. Lett* 48, 353 (1985).
9. G. Koblmueller, R. Averbeck, L. Geelhaar, H. Riecher, W. Hösler, and P. Pongratz, *J. Appl. Phys.* 93, 9591 (2003).
10. D. F. Storm, D. S. Katzer, S. C. Binari, B. V. Shanabrook, L. Zhou, and D. J. Smith, *Appl. Phys. Lett.* 85, 3786 (2004).
11. L. Zhou, D. J. Smith, D. F. Storm, D. S. Katzer, S. C. Binari, and B. V. Shanabrook, *Appl. Phys. Lett.* 88, 011916 (2006).
12. B. Shim, H. Okita, K. Jeganathan, M. Shimizi and H. Okumura, *Jpn. J. APpl. Phys.* 42, 2265 (2003).
13. B, Heying, X. H. Wu, S. Keller, Y. Li, D. Kapolnek, B. P. Keller, S. P. DenBaars, and J. S. Speck, *Appl. Phys. Lett.* 68, 643 (1995).
14. A. Trampert, O. Brandt and K. H. Ploog, .Crystal Structure of Group III Nitrides., *Gallium Nitride (GaN) I*, ed. J. I. Pankove and T. D. Moustakas (Academic Press, 1999) pp.167-192.

Mater. Res. Soc. Symp. Proc. Vol. 1068 © 2008 Materials Research Society 1068-C03-05

Molecular Beam Epitaxy of AlN Layers on Si (111)

Jean-Christophe Moreno, Eric Frayssinet, Fabrice Semond, and Jean Massies
CRHEA-CNRS, rue Bernard Gregory, Valbonne, 06560, France

ABSTRACT

In this work, aluminum nitride (AlN) thin films are epitaxially grown by Molecular Beam Epitaxy (MBE) on silicon substrates in order to fabricate thin film bulk acoustic resonators (TFBARs). Using atomic force microscopy, scanning electron microscopy and X-ray diffraction, we study the quality of AlN films as a function of different silicon surface preparation techniques. Finally, acoustic picosecond measurements are presented.

INTRODUCTION

In recent years, AlN thin films have attracted much attention for optoelectronic and acoustic applications [1-4]. AlN is potentially a very interesting material because it is a high-band gap piezoelectric semiconductor material having a high cohesive energy and furthermore it is compatible with the silicon technology. Present AlN-based FBAR filters are fabricated with AlN films that are deposited by sputtering [5, 6]. Although sputtered AlN films are polycrystalline, films having a thickness above 1 μm are sufficiently well oriented that the piezoelectric coupling (Kt^2) is close to the value expected for bulk single crystals. Nowadays the wireless telecommunications development needs to handle higher frequencies where submicron AlN films would be required. However, the degradation of the quality of submicron AlN films deposited by sputtering as well as the related increases of insertion losses is reported and seems to be a limiting factor to further increases operating frequencies. On the other hand epitaxial AlN thin films are expected to behave like bulk single crystals and their quality is not expected to depend a lot with the film thickness.

During the last 10 years, an efficient growth process has been developed at CRHEA about the epitaxy of group III-nitride on silicon substrates [7]. Although mainly dedicated to the fabrication of GaN-based transistor devices, this has resulted in the ability to grow smooth epitaxial AlN films on silicon substrates. So, we propose to use epitaxial AlN thin films grown on silicon substrates to realize advanced high frequency bulk acoustic wave devices for filtering and time reference (oscillators) applications.

EXPERIMENT

AlN epitaxial films are grown on 2" silicon (111) wafers. Obviously in the silicon technology, the (111) surface orientation is much less used than the (100) but it turns out that the hexagonal surface symmetry of (111) orientation is more appropriate to grow the AlN wurtzite phase. Also, using such surface orientation only one AlN polarity is grown, actually only the Al-polarity can be grown, and no inversion domains are usually observed. Three different silicon surface preparation techniques are assessed in this work. The first one consists in an *ex-situ* standard chemical cleaning followed by an H-passivation using a diluted HF solution with a subsequent de-ionized water rinsing [8, 9]. Then wafers are loaded in the MBE growth chamber. The second surface preparation involves high temperature (1200°C) hydrogen annealings in a chemical vapor deposition reactor prior loading wafers in the MBE growth chamber. This surface preparation, like the first one, results in an oxide-free and H-terminated Si(111) surface. Before growth, the H-passivation is removed *in-situ* by a thermal

annealing at about 600°C for about ten minutes and then the well known 7x7 surface reconstruction appears as observed by reflection of high energy electron diffraction (RHEED). Concerning the third surface preparation called Ultra High Vaccuum (UHV) thermal annealing, "as received" substrates covered by native oxide are directly taken from the box and loaded in the outgassing chamber where wafers are heated up to about 500°C during several hours. After outgassing, samples are transferred inside the growth reactor and the native oxide is removed in-situ using several thermal flashes up to 900°C. Upon cooling, the 1x1-7x7 surface reconstruction transition appears around 830°C. In all cases the 7x7 surface reconstruction is the starting point of the AlN epitaxy. Ammonia is used as the N-precursor. The silicon substrate temperature is monitored using a calibrated pyrometer and it is maintained at 900°C during the AlN growth. 250 nm thick AlN films are grown and the surface morphology of the growth front as well as the crystalline quality is followed *in-situ* by RHEED during the growth. Also the growth rate as well as the thickness of AlN films is monitored using a home-made reflectometry set-up.
More details about the growth parameters are reported elsewhere [7].

DISCUSSION

Before to assess the AlN epilayer quality, we carried out an AFM study of silicon substrate surfaces depending on the surface preparation technique used. Images displayed in Fig.1 show how looks like the silicon surface morphology just before to initiate the growth. Actually, for those images silicon substrates are taken out of the growth reactor once a well-developed 7x7 surface reconstruction is observed by RHEED. Using a wet H-passivation (Fig.1-A), although the surface is very smooth, terraces and step edges are hardly seen. Some features likely related to mechanical polishing are observed suggesting that almost no silicon has been etched away during the wet chemical treatment. On the other hand, the high temperature hydrogen annealing surface preparation method leads to a very smooth surface with clearly resolved terraces and straight monolayer height step edges. Polishing features are removed suggesting that some material has been etched away during the H_2 high temperature process. Concerning the last surface preparation method involving ultra-high vacuum (UHV) thermal annealings of "as received" silicon substrates, we do see many small islands having typically a density of 10^9 cm^{-2}. Also terraces and step edges are clearly seen but they do not form a straight regular network. Actually, it turns out that step edges are pinned by the small islands leading to the formation of terraces having an irregular shape. In agreement with the interpretation of features seen on the RHEED pattern, it is believed that islands are SiC crystallites but so far we were not able to confirm that by cross-section TEM images. It is suggested that carbon impurities are trapped in the native oxide and once this last one is thermally removed *in-situ* inside the growth reactor, carbon species being less volatile they do form Si-C bonds at the silicon surface.

AFM and SEM images of 250 nm thick AlN layers are shown in Fig.2. Using « as received » Si substrates followed by UHV thermal annealings (Fig. 2-C) result in a poor AlN surface morphology as shown by the presence of many islands. These islands are likely AlN grains nucleated on top of the biggest SiC islands seen on the silicon surface prior to the growth. Some islands are perfectly aligned along the growth axis and an almost perfect hexagonal shape emerge from the surface (as shown in the inset) but most of them are tilted and inclined top facets are emerging from the surface. On the other hand, growth on H-terminated surfaces, prepared either by wet etching or either by hydrogen annealings, results in smooth AlN surface morphologies (Fig. 2-A and 2-B).

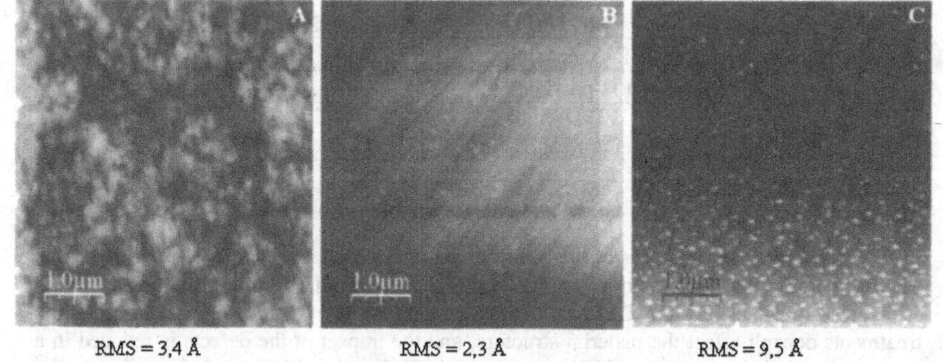

Fig.1: AFM pictures showing (111) oriented Si surfaces after different surface treatment A) after a wet H-passivation, B) after a CVD H_2 high temperature annealing, C) after an UHV annealing

Due to the huge thermal expansion coefficient mismatch between AlN and silicon, AlN epilayers usually exhibit a high tensile strain. By measuring the curvature of 2" wafers before and after the growth, strain calculations are carried out using the Stoney formula [10]. Results are summarized in Tab.1. All layers are in tensile strain but surprisingly the amount of strain depends strongly on the substrate surface preparation method used.

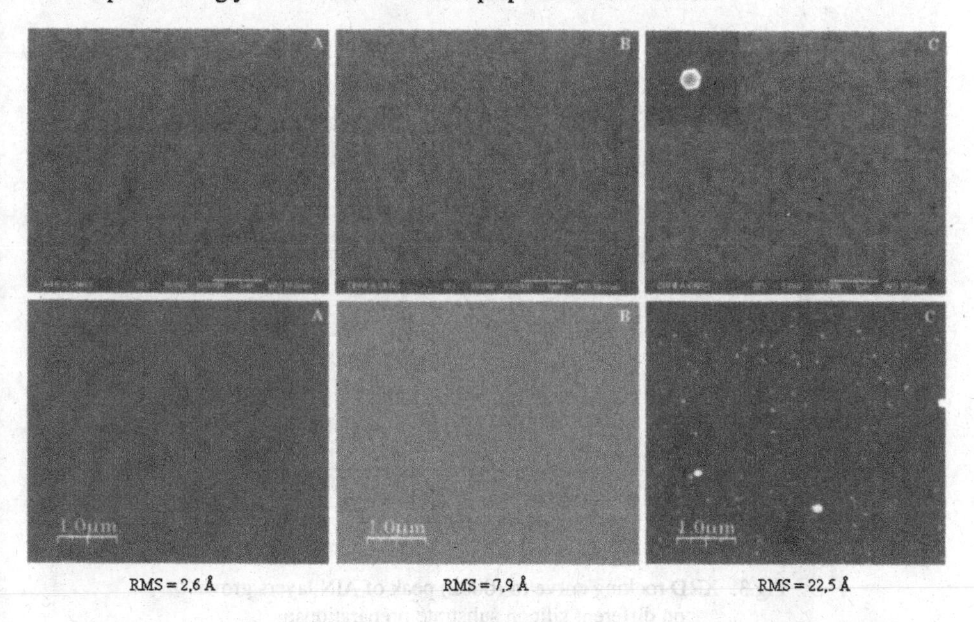

Fig.2: SEM (up) and AFM (down) images of AlN surfaces epitaxially grown by MBE on silicon. A) after a wet H-passivation, B) after a CVD H_2 high temperature annealing, C) after an UHV annealing

Si(111)	Strain (MPa)
UHV annealing	610
wet H-passivation	998
CVD H_2 high temperature annealing	1342

Tab.1: Strain in AlN layers.

The structural properties of these three layers have been assessed by X-ray diffraction. The full width at half maximum of the (002) peak is very similar for all the samples, around 1200 arcseconds (see Fig. 3). This result suggests that the morphological improvement of the treatments doesn't affect the material structure, and the impact of the defects is reduced in a very local scale in comparison with the matrix. The (302) peak, which give information on the edge-dislocations density, can't be measured owing to its low intensity (4% of (002) peak intensity). However, enhancements of the piezoelectric coefficient d_{33} and of electromechanical coupling factor Kt^2 are expected with such low values compared to films obtained by sputtering [11, 12]. The acoustic wave speed in the [001] direction (along the growth axis) is measured by acoustic picoseconds. There are not many differences between these three samples. The acoustic wave speed is around 11700 m/s, which is the best value ever reported to our knowledge. On sputtered AlN, this value range in between 9000 and 11000 m/s [13].

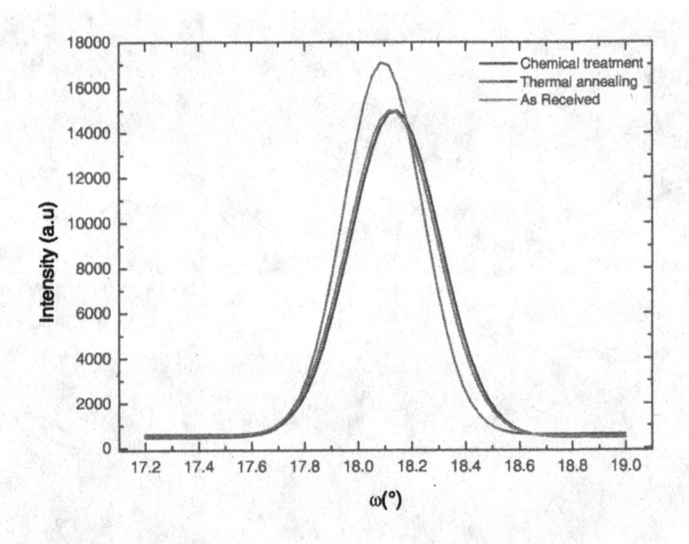

Fig.3: XRD rocking curve of (0002) peak of AlN layers grown on different silicon substrate preparations

CONCLUSIONS

250 nm thick AlN films grown by MBE on silicon (111) substrates using different surface preparations are studied. MBE allows to grow much better crystalline quality AlN layers than those grown by sputtering and depending on the surface substrate preparation very flat defect free AlN surfaces could be obtained. On the other hand a huge tensile strain is usually measured in those AlN thin films grown by MBE and this strain is strongly dependent to the silicon surface preparation. Moreover, structural properties and acoustic wave speed values suggest that high operating frequency BAW devices could be fabricated using AlN thin films grown by MBE.

ACKNOWLEDGMENTS

The authors thank P. Emery (ISEN, Lille, France) for acoustic picosecond measurements and want to acknowledge STMicroelectronics which support this work within the project Nano2013.

REFERENCES

[1] Y. Taniyasu, M. Kasu, T. Makimoto, Nature 441 (2006) 325
[2] C. Caliendo, Appl. Phys. Lett. 92, 033505 (2008)
[3] J. Olivares, E. Iborra, M. Clement, L. Vergara, J. Sangrador and A. Sanz-Hervás, Sensors and actuators A 123-124 590 (2005)
[4] V. Cimalla, J Pezoldt and O. Ambacher, J. Phys. D: Appl. Phys. 40 (2007) 6386-6434
[5] M.A. Dubois, P. Muralt, Appl. Phys. Lett. 74, 3032 (1999)
[6] K.H. Chiu, J.H. Chen, H.R. Chen, R.S. Huang, Thin Solid Films 515 (2007) 4819-4825
[7] F. Semond, Y. Cordier, N. Grandjean, F. Natali, B. Damilano, S. Vézian, J. Massies, Phys. Stat. Sol. (a) 188, (2001) 501.
[8] V. A. Burrows et al., Appl. Phys. Lett. 53, 998 (1988)
[9] J. Wang et al., Microelectronic Engineering 56 (2001) 221-225
[10] P. H. Townsend, D. M. Barnett, T. A. Brunner, J. Appl. Phys., 62 (1987) 4438-4444
[11] F. Martin et al., J. Vac. Sci. Technol. A 2004/22(2)/361
[12] H.P. Loebl et al., Materials Chemistry and Physics 79 (2003) 143-146
[13] Private communication

Mater. Res. Soc. Symp. Proc. Vol. 1068 © 2008 Materials Research Society 1068-C06-03

AlGaN Transition Layers on Si (111) Substrates - Observations of Microstructure and Impact on Material Quality

John C. Roberts, James W. Cook, Jr., Pradeep Rajagopal, Edwin L. Piner, and Kevin J. Linthicum
Nitronex Corporation, 2305 Presidential Drive, Durham, NC, 27703

ABSTRACT

AlGaN based transition layers are one proven solution used for the growth of device quality GaN layers on Si(111) substrates. Examination of the microstructure of GaN on silicon epitaxy grown with these AlGaN transition layers, by transmission electron microscopy (TEM) and other methods, reveals some interesting properties that can help explain how high quality III-N epitaxy can be performed in a system with significant thermal and lattice mismatch. The existence of a thin amorphous SiN_x layer at the silicon substrate/transition layer interface has been documented. This amorphous SiN_x layer accommodates stress at this interface as well as affecting a reduction of the formation of misfit dislocations. Relatively low screw dislocation densities (less than ~10^7 cm^{-2}) are also observed in GaN on silicon films grown with AlGaN based transition layers. These microstructural characteristics contribute to the overall high quality of the AlGaN/GaN HEMT epitaxial structures that are grown on Si(111) substrates utilizing AlGaN transition layers.

INTRODUCTION

AlGaN/GaN based high electron mobility transistors (HEMTs) have emerged from the realm of research and development and are being evaluated and designed into systems in the commercial and military sectors that require high power and high efficiency amplification with excellent linearity and/or bandwidth. To date, the vast majority of devices being considered for such systems are based on some type of heteroepitaxial growth, due to the lack of a commercially available GaN substrate that is cost effective and readily available in suitable quantities. The most common substrates used for high power AlGaN/GaN based devices are silicon carbide, silicon, and sapphire, each of which has specific advantages and disadvantages.

Silicon has numerous advantages as a substrate choice for III-N heteroepitaxy. It is an extremely mature substrate technology, where wafers 100 mm in diameter and larger are readily available from a multiplicity of vendors for a few tens of dollars per wafer. Due to the maturity of the silicon wafer industry, substrate quality is extremely high, and wafer to wafer consistency is superb. The availability of very large diameter silicon substrates currently enables the GaN on silicon approach as the only platform with an immediate roadmap to wafer sizes 150 mm in diameter and beyond. From a manufacturing standpoint, choosing silicon as the substrate also leverages the capability to use existing high volume silicon process services and assembly houses (e.g., wafer thinning, via technology, dicing, and plastic over-mold packaging).

There are also challenges associated with using silicon as the substrate for III-N materials. Growth of high quality GaN on Si(111) can be achieved only by addressing the significant levels of lattice misfit and thermal expansion coefficient (TEC) mismatch between the silicon substrate and III-N epitaxial layer. For example, based on the difference in lattice constant between wurtzitic AlN and Si(111), an estimated maximum of ~3 x 10^{12} cm^{-2} misfit dislocations near the Si/AlN interface would be expected. Additionally, the ~56% difference between thermal expansion coefficients between GaN and Si is expected to result in significant film stress that will manifest itself during cooldown after epitaxial growth. This stress must be adequately accommodated by the film-substrate combination to result in crack free GaN films. There are a variety of ways of implementing a transition layer to mitigate the effect of the lattice and TEC mismatch between Si(111) and GaN, including the use of super lattices, AlN interlayers, and lateral epitaxial over-growth [1, 2, 3, 4]. At Nitronex we have adopted a multi-step AlN/AlGaN transition layer that results in the growth of high quality crack-free GaN suitable for the manufacturing of state of the art AlGaN/GaN based HEMT structures for high-power and high frequency applications [5,6]. This paper will discuss some of the details of this transition layer and the effect it has on overall material quality.

EXPERIMENT

The AlN, AlGaN and GaN epitaxial layers discussed in this paper were grown on 100 mm diameter Si(111) substrates in a custom-built, cold wall, rotating disc MOCVD reactor at nominally 1020 °C. Trimethylgallium (TMG) and trimethylaluminum (TMA) precursors were carried by Pd-diffused hydrogen and ammonia (NH_3) was used as the N precursor. A typical full epitaxial stack consists of an AlN nucleation layer, followed by two AlGaN layers, which completes the transition layer portion of the stack. A thicker GaN buffer layer is grown on top of this transition layer, which would normally be followed by an AlGaN device layer to complete the HEMT structure. Scanning transmission electron microscopy (STEM), TEM, and atomic force microscopy (AFM) characterization, which were outsourced, were used to study the microstructure of the epitaxial films. White light reflectance thickness mapping was used to characterize layer thicknesses. Characterization of the two-dimensional electron gas (2DEG) channel was determined by van der Pauw Hall effect measurement.

DISCUSSION

Growth of crack-free GaN

Growth of crack-free GaN on Si(111) was achieved by using a multi-stack transition layer, all of which is deposited at high temperature (1020 °C). AlN is used as the first component of this transition layer, thus avoiding the potential of a Ga-Si eutectic reaction which would compromise the epitaxial template of the silicon substrate. The second and third components of the transition layer consist of AlGaN, each with a successively lower AlN content. The AlN/AlGaN transition layer stack described here has been successfully used over a total thickness range of ~600-1100nm. A number of combinations of component layer compositions and thickness have been effectively implemented for the growth of crack free GaN. A continuous AlGaN grade has also been used between the AlN and the GaN buffer layer, but a step-wise transition layer has advantages with manufacturability and allows precise monitoring of the composition and thickness of each layer by x-ray metrology and other techniques. Using a high temperature AlN layer alone was less effective at eliminating cracking in the GaN buffer layer that is ultimately deposited on the transition layer stack. Mosaic cracking of the GaN buffer across a majority of the 100 mm diameter wafer was observed when high temperature AlN was the only transition layer used. The multi-step transition layer, using high temperature

AlN and AlGaN layers, is better able to accommodate the cool-down stress from the high growth temperature and results in an uncracked GaN buffer layer over the entire 100 mm diameter wafer. A total epitaxial thickness of more than 2 μm can routinely be achieved with this technique with total wafer warp of less than 30 μm for the 100 mm diameter wafer.

Reduced misfit dislocation density

Close inspection of TEM images at the silicon substrate/AlN nucleation layer boundary reveals the presence of an amorphous silicon nitride based layer (α-SiN_x) that resides between the diamond (Si) and hexagonal (AlN) crystalline lattices. The timing of NH_3 introduction into the growth chamber prior to AlN growth initiation is an important parameter that can influence the formation of this α-SiN_x layer. The TEM micrograph of Figure 1a shows that this layer is approximately 15 – 20 Å thick and is continuous across the field of view of the TEM specimen. The presence of this α-SiN_x layer has been confirmed in multiple samples analyzed by TEM or STEM techniques.

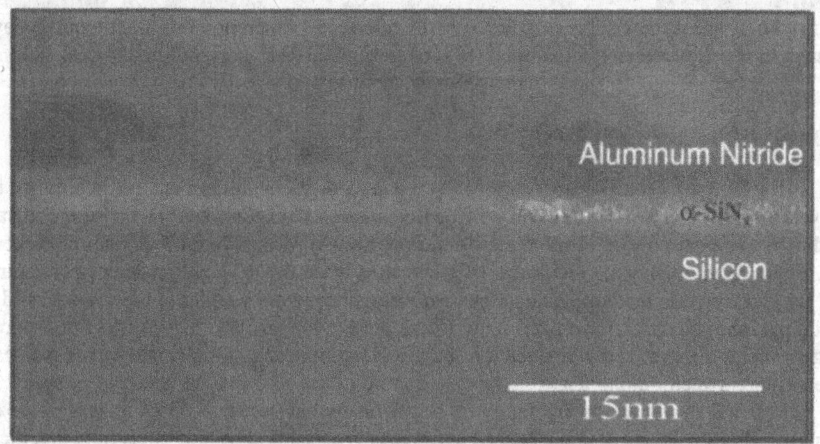

Figure 1a. TEM micrograph of the diamond Si(111)/hexagonal AlN interface, showing the existence of an amorphous SiN_x layer.

Examination of a higher resolution TEM image of this region, shown in Figure 1b, reveals that misfit dislocations are absent. The amorphous SiN_x layer accommodates the lattice mismatch and presumably a portion of the stress at the silicon/AlN interface in lieu of the creation of misfit dislocations. The high resolution TEM image shows continuity in the atomic arrangement of the silicon substrate and the AlN nucleation layer on either side and up to the α-SiN_x region. The micrograph of Figure 1a is representative of many TEM samples from multiple GaN on silicon epiwafers. The SiN_x is always present and, therefore, because the silicon and AlN are not in direct contact, misfit dislocations are not created. Thus, the misfit dislocation density is well below ~3 x 10^{12} cm^{-2} one would expect due to ideal lattice mismatch accommodation and, more precisely, below ~1 x 10^{10} cm^{-2}, based on TEM observations.

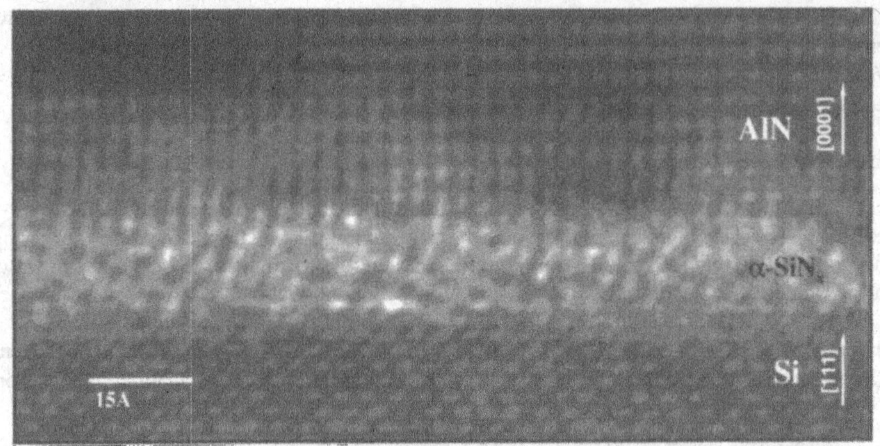

Figure 1b. Lattice image showing the Si / AlN interface. An amorphous silicon nitride layer is shown to reside between the diamond (Si) and hexagonal (AlN) crystalline lattices. No misfit dislocations are observed.

Reduced screw dislocation density

The Burgers vector of dislocations in the GaN on silicon system can be determined by analysis of TEM images. By applying diffraction contrast, images taken in two-beam diffraction conditions can be used to determine the Burgers vector (**b**) by applying the invisibility criteria (i.e., **g • b** = 0 where g is the reciprocal lattice vector) when the dislocations are out of contrast [7]. By applying this method, dislocations with Burgers vectors parallel to the **c**, **c + a**, and **a** crystallographic direction have been shown to be present in GaN [8]. Edge dislocations have a Burgers vector along **a**, screw dislocations have a Burgers vector along **c**, and mixed dislocations have a Burgers vector along **a + c**.

Figure 2 shows two-beam condition bright field cross-sectional TEM images of the same sample. In Figure 2a, **g** = (0001) and in Figure 2b, **g** = (1-100). For reference, note that the "*" symbol indicates identical spatial positions on the images. The dotted lines on Figure 2a indicate where dislocation contrast is visible for the g = (0001) condition in the GaN buffer and upper AlGaN transition layer. The dislocations seen in the g = (0001) condition are also seen for the g = (1-100) condition (Figure 2b) and are also highlighted by dotted lines. The one-to-one correlation of dislocations present in the **g** = (0001) and **g** = (1-100) condition indicate dislocations with mixed (edge + screw) character. Note that in the **g** = (0001) condition, there are no dislocations independent of those that appear in the **g** = (1-100) condition. From **g • b** analysis, this signifies the absence of screw dislocations in this particular TEM sample. On the other hand, there are a number of dislocations present in the **g** = (1-100) condition that are not present in the **g** = (0001) condition. In this circumstance, from **g • b** analysis, this signifies the presence of edge dislocations in the TEM sample. (The lower AlGaN transition layer and the AlN nucleation layer indicate similar dislocation defect character, but are not individually noted for sake of clarity.)

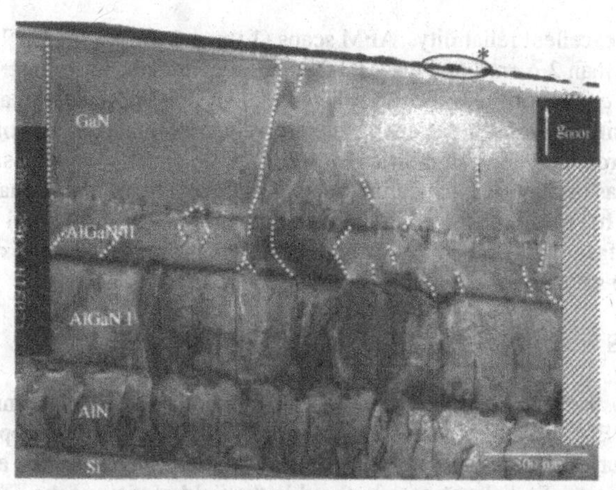

Figure 2a. TEM image of an AlGaN/GaN HEMT epi structure on an AlN/AlGaN-based transition layer. The g-vector is parallel to the c-axis (i.e., g_{0001}). The dashed lines overlay and are intended to highlight dislocations in the upper AlGaN transition layer and GaN buffer layer.

Figure 2b. TEM image of the same sample of Figure 2a, but with the g-vector oriented parallel to the a-axis (i.e., $g_{1\text{-}100}$), two beam condition. The dashed lines, all of which overlay dislocations in this image, are in the same positions as those that were drawn in for Figure 2a.

The apparent absence of screw dislocations as determined by the g • b technique has been confirmed on multiple occasions for TEM specimens prepared from GaN on silicon films using multi-stack AlN/AlGaN based transition layers. Based on the spatial extent of the TEM specimens, it can be stated that we observe an apparent screw dislocation density of less than ~1 x 10^7 cm^{-2}. The density of edge and mixed dislocations, based on this and other TEM analyses, is estimated to in the 1 to 5 x 10^9 cm^{-2} range.

The use of multi-stack AlN/AlGaN transition layers enables the deposition of state of the art AlGaN/GaN HEMT epitaxial structures that, upon fabrication into devices, exhibit high performance and excellent reliability. AFM scans (1 μm x 1 μm) of the GaN buffer layer yield a roughness of less than 2 Å rms, and the AlGaN device layer/GaN buffer interface is smooth and abrupt. The high quality of the GaN buffer layer and AlGaN device layer made possible by the AlN/AlGaN transition layer scheme enable electron channel characteristics of $Al_{0.26}Ga_{0.74}N$ /GaN HEMTs to routinely have mobilities exceeding 1500 cm^2/Vs (no AlN sub-barrier). Shubnikov de Haas oscillations, indicating the existence of a two-dimensional electron gas of excellent quality, have been observed in these structures [9]. The GaN on Si material platform enables the manufacture of state of the art high power, high frequency transistors with excellent reliability that are suitable for the wireless infrastructure marketplace.

CONCLUSIONS

The microstructure of GaN on silicon films utilizing AlN/AlGaN transition layers has been studied by TEM and other techniques, revealing several interesting properties. A thin, amorphous SiN_x layer is observed sandwiched between the silicon substrate and AlN nucleation layer. There are no misfit dislocations observed in the field of view of the TEM image over which the amorphous layer is seen. This result implies that misfit dislocation densities on the order of or less than ~1 x 10^{10} cm^{-2} may be expected at the substrate/AlN interface, which is significantly less than would be expected based on the difference in lattice constant between the two materials. The amorphous layer accommodates a portion of the stress that results from both the lattice and thermal mismatch between the silicon substrate and AlN layer. No screw dislocation are observed in the TEM images presented, implying a screw dislocation density no greater than ~10^7 cm^{-2} in the GaN buffer grown on top of the AlN/AlGaN transition layer stack. These microstructural characteristics help explain how high quality AlGaN/GaN HEMTs epitaxial structures can be grown on Si(111) substrates.

REFERENCES

1. Min-Ho Kim, Young-Gu Do, Hyon Chol Kang, Do Young Noh and Seong-Ju Park, Appl. Phys. Lett. **79**, 2713 (2001).
2. Eric Feltin, B. Beaumont, M. Laugt, P. de Mierry, P. Vennéguès, H. Lahrechè, M. Leroux, and P. Gibart, App. Phys. Lett. **79**, 3230 (2001).
3. A. Dadgar, M. Poschenrieder, J. Bläsing, K. Fehse, A. Diez, and A. Krost, Appl. Phys. Lett. **80**, 3670 (2002)
4. Yankun Fu, and Daniel A. Gulino, J. Vac. Sci. Technol. A **18**(3), 965 (2000).
5. Johnson, J.W., Piner, E.L., Vescan, A., Therrien, R., Rajagopal, P., Roberts., J.C., Brown, J.D., Singhal, S., & Linthicum, K.L., "12W/mm AlGaN-GaN HFETs on silicon substrates." IEEE Electron Device Letters, 25(7), 459-461. (2004).
6. Therrien, B., Singhal, S., Johnson, J.W., Nagy, W., Borges, R., Chaudhari, A., Hanson, A.W., Edwards, A., Marquart, J., Rajagopal, P., Park, C., Kizilyalli, I.C., & Linthicum, K.J., "A 36 mm GaN-on-Si HFET Producing 368W at 60V with 70% Drain Efficiency". 2005 IEEE International Electron Devices Meeting (IEDM), Washington, DC. (2005).
7. P.B. Hirsh, et al., Electron Microscopy of Thin Crystals (Krieger, New York, 1977).
8. F. Ponce, et al., APL 69(6) (1996) 770
9. S. Elhamri, R. Berney, W. C. Mitchel, W. D. Mitchell, and J. C. Roberts, P. Rajagopal, T. Gehrke, E. L. Piner, and K. J. Linthicum,"An Electrical Characterization of a Two Dimensional Electron Gas in GaN/AlGaN on Silicon Substrates.", J. Appl. Phys. 95, 7982 (2004).

Mater. Res. Soc. Symp. Proc. Vol. 1068 © 2008 Materials Research Society 1068-C06-04

Effects of Stress-Relieving AlN Interlayers in GaN-on-Si Grown by Plasma-Assisted Molecular Beam Epitaxy

Adam Adikimenakis[1,2], Suman-Lata Sahonta[3], George Dimitrakopulos[3], Jaroslav Domagala[4], Philomela Komninou[3], and Alexander Georgakilas[1,2]
[1]MRG - IESL, FORTH, Vassilika Vouton, Heraklion, 71110, Greece
[2]Physics Department, University of Crete, Heraklion, 71203, Greece
[3]Physics Department, University of Thessaloniki, Thessaloniki, 54124, Greece
[4]Institute of Physics, Polish Academy of Science, Warsaw, Poland

ABSTRACT

The insertion of an AlN interlayer for tensile strain relief in GaN thin films grown on Si (111) on-axis and vicinal substrates by nitrogen rf plasma source molecular beam epitaxy has been investigated. The 15 nm AlN interlayer was inserted between the bottom 0.5 micron GaN layer and the top 1.0 micron GaN layer. The interlayer was very effective to reduce the tensile stress in the overall 1.5 micron GaN/Si film to the level required for complete avoidance of microcracks, which were present in high densities in GaN/Si heterostructures grown without an AlN interlayer. The strain of the AlN interlayer, as well as the strain in all the layers of the entire GaN/Si heterostructure was analyzed by x-ray diffraction (XRD) and transmission electron microscopy (TEM) measurements. Reciprocal space map in XRD indicated that the 15 nm AlN interlayer was coherently strained with the GaN films. However TEM observations revealed that the AlN interlayer was partially relaxed in local regions. The AlN interlayer was also observed to interfere with the GaN growth process. In particular, above morphological features such as V-defects, GaN was overgrown with a large density of threading dislocations and inversion domain boundaries.

INTRODUCTION

The heteroepitaxial growth of III-V semiconductors on Si has attracted significant research efforts for more than two decades [1-11]. The most ambitious target of this work concerns the monolithic integration of III-V optoelectronic and high frequency electronic devices with Si digital electronic circuits. However, the replacement of conventional substrates (used for III-V growth) with the low cost, large diameter and high quality Si substrates is also a strong motivation. For these reasons, significant research efforts have been recently devoted to the GaN growth on (111) Si [2-10] and much attention has been paid on the realization of good quality, crack-free GaN films on Si (111) layers.

The epitaxial orientation relationship between GaN and Si is $(0001)_{GaN}/(111)_{Si}$, $<\bar{1}\bar{1}20>_{GaN}/<1\bar{1}0>_{Si}$, for which the large lattice mismatch (16.9% with respect to GaN) induces strain, leading to misfit and threading dislocations in the epitaxial films. Moreover, the difference in the thermal expansion coefficients between the two materials leads to thermally induced tensile stress, causing microcracks in the GaN films [4]. Microcracks can be avoided by utilising one or more intermediate AlN layers [2,3] in order to introduce compressive residual strain in the overgrown GaN layers, due to the lattice mismatch between GaN and AlN. The remaining tensile stress is then reduced below the threshold value for microcrack formation [4].

High quality GaN films have been grown on Si (111) by MOVPE [2,3] and NH_3-source MBE [5,6] in the last few years. The nitrogen RF plasma source MBE (RFMBE) method has also shown promising results [7,8]. In particular, RFMBE offers the possibility of lowering the growth temperature, which could reduce the thermally induced tensile stress in the GaN/Si layers. However, up to present, HEMT and LED devices have been realized on (111) Si substrates only with MOVPE and NH_3-MBE grown materials [9, 10].

In this paper, we studied the structural properties of the intermediate AlN layer used to suppress crack formation in the RFMBE grown GaN films on Si (111). High resolution transmission electron microscopy (HRTEM) and high resolution X-ray diffraction (HR-XRD) measurements have been employed in order to visualize the properties of such a thin layer. Its influence on the structural quality of the overgrown GaN layers has also been investigated.

EXPERIMENT

The GaN layers were grown by RFMBE using a RIBER 32P MBE system equipped with an O.A.R. HD25 nitrogen RF plasma source [11]. All substrates were etched, ex-situ, in H_2SO_4:H_2O_2 (4:1), HF:H_2O (1:10) and HCl:H_2O_2:H_2O (1:1:6) [12] and heated, in-situ, up to 800 °C in high vacuum in order to remove the surface oxide. A 30 nm AlN seed layer was deposited at 800 °C, followed by 1.5 μm of GaN at 700 °C, with an intermediate AlN layer of 15 nm, grown after the first 0.5 μm of GaN. Finally, a 30 nm thick $Al_{0.3}Ga_{0.7}N$ layer was grown. A large series of samples was grown on Si (111) 3-inch wafers. Three samples with a total GaN thickness of 2 μm were, also, grown on Si (111), two using an AlN interlayer and one without AlN interlayer. Another sample of 1 μm GaN thickness was grown without AlN interlayer.

Growth process was monitored by in-situ RHEED, operating at 12 KV. Surface morphology was studied with optical microscope and Atomic Force Microscopy, AFM (D.I. Nanoscope IIIa), while High Resolution X-Ray Diffraction (HRXRD) measurements were carried out in a BEDE D1 and a Philips X'Pert MRD system. Specimens for cross section transmission electron microscopy (XTEM) observations were prepared by mechanical thinning, followed by Ar^+ ion milling. Both conventional TEM and high resolution TEM (HRTEM) were performed on a 200 kV Jeol 2011 microscope (Cs = 0.5 mm, point resolution = 0.20 nm).

RESULTS AND DISCUSSION

The 15 nm AlN interlayer, inserted between the bottom 0.5 μm GaN and the top 1 μm GaN layers, proved very effective to reduce the tensile stress arising from the difference of the thermal expansion coefficients between GaN and Si. Complete avoidance of microcracks was achievable for GaN thicknesses up to 2 μm in total, as was evidenced from observations with optical microscope and AFM. However, the 2 μm GaN thin films grown without the AlN interlayer exhibited cracking. Thus, it is evident that the AlN interlayer is an effective method to reduce the tensile strain in the GaN films grown on (111) Si and avoid crack formation.

The crystalline structure and quality of the samples were evaluated by HRXRD using the (0002) symmetric and the $(20\bar{2}1)$ asymmetric reflections of GaN. All samples were of rather good crystalline quality with (0002) rocking curve FWHM ranging from 670 to 720 arcsec and $(20\bar{2}1)$ rocking curve FWHM ranging between 1260 and 1730 arcsec. Edge threading dislocation densities (calculated from the twist angle of the $(20\bar{2}1)$ reflection [13]) ranged from $8X10^9$ to $1.6X10^{10}$ cm^{-2}.

High resolution XRD reciprocal space maps (RSMs) have been recorded around the $(10\bar{1}5)$ reflection of GaN. Due to the very low intensity of the peaks, a double axis configuration with a slit in front of the detector was used. Such RSM is presented in Figure 1, along with the in plane lattice parameters for both AlN layers used (namely the seed and the interlayer). The AlN peaks in the RSM were identified by their relative intensities in the map. The AlN seed layer was twice as thick as the AlN interlayer, thus corresponding to the peak with the higher intensity. Similar RSMs have been recorded for all the optimized GaN/Si samples that were chosen for detailed HRXRD analysis. Due to the fact that no analyzer was used for the RSM, the peaks appear more elongated than they should normally be. The 15 nm AlN interlayer appears to be almost pseudomorphically grown to GaN, while the AlN seed layer appears relaxed. Also, the 30 nm $Al_{0.3}Ga_{0.7}N$ layer is strained to GaN, as it was expected. These results are in agreement with the observations of Blasing et Al [4], for AlN interlayers grown at high temperature by MOCVD.

The a-lattice constants of the two AlN layers were extrapolated from the RSM (using the 2 theta diffraction angle) and they were found to be 3.10 Å for the seed layer and 3.17 Å for the AlN interlayer. Therefore, the AlN seed layer appears to be completely relaxed, within the accuracy of this method. This result was also confirmed by quantitative HRTEM using geometrical phase analysis (GPA) [14] to obtain maps of lattice constant variation within the sample. HRTEM also showed that this relaxation is accomplished by a regular array of misfit dislocations at the AlN/Si interface.

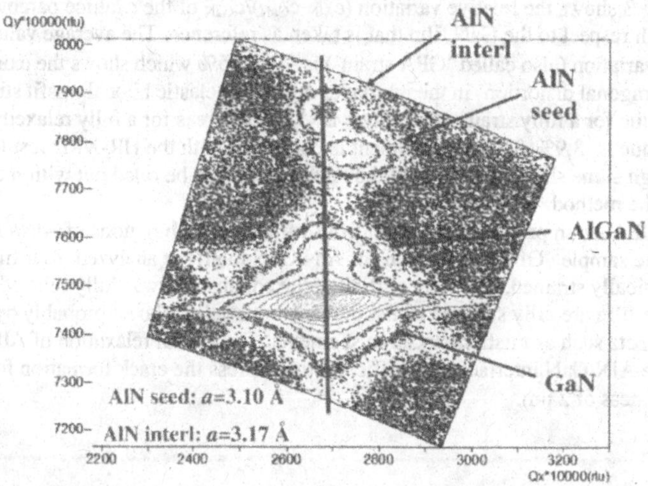

Figure 1: Reciprocal space map around the $(10\bar{1}5)$ reflection of GaN showing all the layers grown on Si (111) substrate. The 15 nm AlN interlayer and the 30 nm AlN seed (or nucleation) layer grown on Si (111) are denoted on the figure. The extracted AlN in plane lattice constants *a*, also are shown in the inset.

HRTEM images of the AlN interlayer have been recorded along a $[1\bar{1}00]$ zone axis to reveal only (0002) basal plane periodic lattice fringes. The image negatives were digitized at 4000 dpi. Square "box" sections of image of size 4096 x 4096 pixels (and in some cases 2048 x 2048 pixels) were taken and the (0002) spots in their Fourier transforms were masked to show the phase and strain state of the AlN film. Lattice parameter profiles of the 15 nm AlN interlayer were taken along lines of type AB (as shown in Figure 2) in different areas of the specimen, and integrated over 1000 pixels to improve averaging. All masks were Gaussian of size g/2. Strain measurements were taken only from areas with well-defined interfaces (i.e. the top part of the strain map in Figure 2) and no extended defects present, in order to measure accurately the strain in the AlN film.

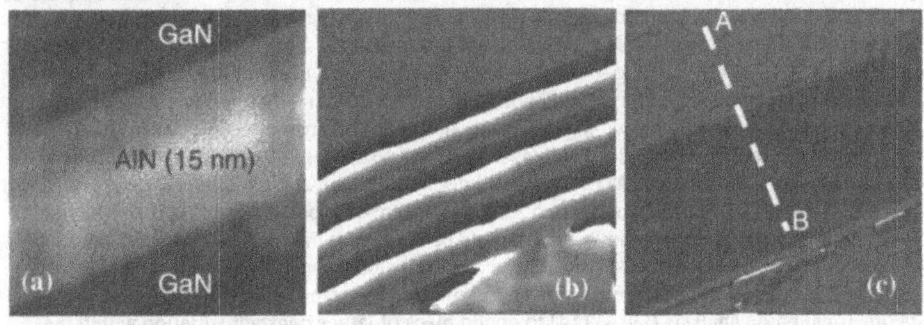

Figure 2: (a): HRTEM image along $[1\bar{1}00]$ showing the AlN interlayer, (b): 0002 phase map, (c): 0002 *d*-spacing profile.

Figure 3 shows the relative variation $(c_{AlN}-c_{GaN})/c_{GaN}$ of the *c* lattice parameter in the AlN interlayer with respect to the GaN film that is taken as reference. The average value obtained for this relative variation (also called 'GPA strain') is -4.8±0.5% which shows the existence of the so-called 'tetragonal distortion' in the interlayer due to the elastic biaxial misfit strain. The calculated value for a fully strained interlayer is -5.3% whereas for a fully relaxed film the calculated value is -3.9%. This measurement is consistent with the HR-XRD results discussed above although some slight relaxation of the AlN film cannot be ruled out within the attained accuracy of the method.

Ten GPA strain profiles were taken from various smooth regions of AlN/GaN interface throughout the sample. Of the ten regions of AlN/GaN interface analyzed, four films were pseudomorphically strained, five were partially relaxed, and one was fully relaxed. Hence some regions of the film are fully strained whilst others are partially relaxed probably by the presence of nearby defects such as misfit dislocation segments. The partial relaxation of AlN at local regions of the AlN/GaN interface was sufficient to suppress the crack formation for a total GaN epilayer thickness of 2 μm.

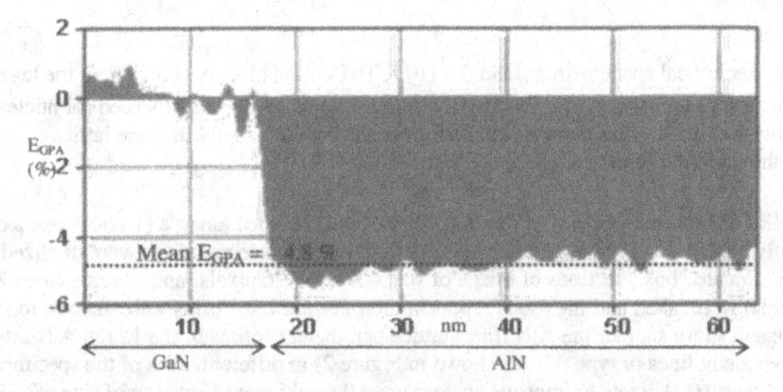

Figure 3: Profile of the relative variation of the *c* lattice parameter across the AlN interlayer/GaN interface averaged over 1000 pixels

An XTEM image of the whole GaN/Si structure can be seen in Figure 4. Emanation of threading dislocations (TDs) and narrow inversion domains was evident, particularly after the 15 nm AlN interlayer. Defects impinging on the lower interface of the 15 nm interlayer are observed to introduce V-defects (arrowed in the image) in the overlying GaN layers. Above the V-defects, GaN was overgrown with large defect density as a result either of growth on exposed facets or the coalescence of laterally overgrown wings (HRTEM image in the inset of Figure 4). Additional TDs and inversion domain boundaries (IDBs) are generated in the upper GaN due to surface roughness induced by the V-defects.

It is thus evident that the thin AlN interlayers can degrade the crystalline quality of the GaN films grown on Si (111), by introducing additional defects in the film. However, the presence of such layers is necessary to avoid cracking due to tensile thermal stress. It is suggested that the steps and facets of V-defects and other areas of film roughness may pin the misfit dislocations so that they may not glide on the basal plane to relieve stress uniformly throughout the film, or undergo other rearrangements to minimize their relative energies. As a result, they are not distributed regularly throughout the film and may accumulate locally to relax isolated patches of the interlayer leaving the rest elastically strained.

Figure 4: Two-beam BF image along [11-20] with g = 0004 of the whole heterostructure grown on Si (111). V-defects in the AlN interlayer are shown in the inset.

CONCLUSIONS

The 15 nm AlN interlayer inserted in RFMBE grown GaN/Si heterostructures was found to be an effective way to suppress crack formation due to tensile thermal stress. HR-XRD RSM indicated that the AlN interlayer is coherently strained to the GaN matrix, while the AlN seed layer that was directly grown on Si, was completely relaxed. Strain profiles recorded with HRTEM revealed that the AlN interlayer was partially relaxed at local areas. Irregular distribution of misfit dislocations due to V-defects, and areas of film roughness may be the cause of this phenomenon. This was sufficient to suppress the crack formation for a total GaN thickness of 2 μm. Finally, V-shape surface defects of the AlN interlayer were found to introduce a significant number of defects in the overlying GaN layers.

ACKNOWLEDGEMENTS

Support by the European Community's Marie Curie Research Training Network PARSEM (MRTN-CT-2004-005583) is gratefully acknowledged. Authors of FORTH also acknowledge support from IST-FET project "ULTRAGAN", contract N^{0} 006903.

REFERENCES

1 A. Georgakilas, J. Stoemenos, K. Tsagaraki, Ph. Komninou, N. Flevaris, P. Panayotatos, and A. Christou, J. Mater. Res. **8**, pp. 1908-1921 (1993)
2 A. Reiher, J. Blasing, A. Dadgar, A. Diez, and A. Krost, J. of Crystal Growth **248** (2003) , 563-567
3 A. Dadgar, M. Poschenrieder, J. Blasing, K. Fehse, A. Diez and A. Krost, Appl. Phys. Lett. **80** (2002) , 3670-3672
4 J. Blasing, A. Reiher, A. Dadgar, A. Diez, and A. Krost, Appl. Phys. Lett. 81, 15 (2002), 2722-2724
5 F. Semond, B. Damilano, S. Vezian, N. Grandjean, M. Leroux and J. Massies, Appl. Phys. Lett. **75** (1999), 82-84
6 S.A. Nikishin, N.N. Faleev, V.G. Antipov, S. Francoeur, L. Grave de Peralta, G.A. Seryogin, H. Temkin, T.I. Prokofyeva, M. Holtz and S.N.G. Chu, Appl. Phys. Lett. **75** (1999), 2073
7 E. Calleja, M.A. Sanchez-Garcia, F.J. Sanchez, F. Calle, F.B. Naranjo, E. Munoz, S.I. Molina, A.M. Sanchez, F.J. Pacheco and R. Garcia, J. Of Crystal Growth **201/202** (1999), 296
8 M. Androulidaki, A. Georgakilas, F. Peiro, K. Amimer, M. Zervos, K. Tsagaraki, M. Dimakis and A. Cornet, Phys. Stat. Sol. (a) **188** (2001), 515-518
9 A. Dadgar, C. Hums, A. Diez, J. Blasing, A. Krost, Journal of Crystal Growth **297** (2006) 279–282
10 Y. Cordier et Al., Journal of Crystal Growth **251** (2003) 811–815
11 E. Iliopoulos, A. Adikimenakis, E. Dimakis, K. Tsagaraki, G. Konstantinidis, A. Georgakilas, J. of Crystal Growth **278**, 426–430 (2005)
12 M. Kayambaki, R. Callec, G. Constantinidis, Ch. Papavassiliou, E. Lochtermann, H. Krasny, N. Papadakis, P. Panayotatos and A. Georgakilas, J. of Crystal Growth **157** (1995) pp. 300-303
13 T. Metzger, R. Hopler, E. Born, O. Ambacher, M. Stutzmann, R. Stommer, M. Schuster, H. Gobel, S. Christiansen, M. Albrecht, and H. P. Strunk, Phil. Mag. A **77**, 1013 (1998)
14 M. J. Hÿtch, E. Snoeck E, and R. Kilaas, Ultramicroscopy **74** (1998) 131-146

Mater. Res. Soc. Symp. Proc. Vol. 1068 © 2008 Materials Research Society 1068-C06-02

InGaN Thin Films Grown by ENABLE and MBE Techniques on Silicon Substrates

Lothar A. Reichertz[1,2], Kin Man Yu[1], Yi Cui[1], Michael E. Hawkridge[1], Jeffrey W. Beeman[1], Zuzanna Liliental-Weber[1], Joel W. Ager III[1], Wladyslaw Walukiewicz[1], William J. Schaff[3], Todd L. Williamson[4], and Mark A. Hoffbauer[4]

[1]Materials Sciences Division, Lawrence Berkeley National Laboratory, Berkeley, CA, 94720

[2]University of California at Berkeley, Berkeley, CA, 94720

[3]School of Electrical and Computer Engineering, Cornell University, Ithaca, NY, 14853

[4]Chemistry Division, Los Alamos National Laboratory, Los Alamos, NM, 87545

ABSTRACT

The prospect of developing electronic and optoelectronic devices, including solar cells, that utilize the wide range of energy gaps of InGaN has led to a considerable research interest in the electronic and optical properties of InN and In-rich nitride alloys. Recently, significant progress has been achieved in the growth and doping of InGaN over the entire composition range. In this paper we present structural, optical, and electrical characterization results from InGaN films grown on Si (111) wafers. The films were grown over a large composition range by both molecular beam epitaxy (MBE) and the newly developed "energetic neutral atomic-beam lithography & epitaxy" (ENABLE) techniques. ENABLE utilizes a collimated beam of ~2 eV nitrogen atoms as the active species which are reacted with thermally evaporated Ga and In metals. The technique provides a larger N atom flux compared to MBE and reduces the need for high substrate temperatures, making isothermal growth over the entire InGaN alloy composition range possible. Electrical characteristics of the junctions between n- and p-type InGaN films and n- and p-type Si substrates were measured and compared with theoretical predictions based on the band edge alignment between those two materials. The predicted existence of a low resistance tunnel junction between p-type Si and n-type InGaN was experimentally confirmed.

INTRODUCTION

The band gap of InGaN can be tuned from 0.65 eV to 3.4 eV [1]. This makes it a very attractive material for optoelectronic devices and, in particular, for solar cells, as this energy range is a perfect match to the part of the solar spectrum that reaches the surface of the Earth. InGaN is therefore a material of choice for a "full-spectrum-photovoltaic" development and there is a considerable research interest in the electronic and optical properties of InN and In-rich nitride alloys [2].

Growing InGaN on Si is interesting for various reasons. First of all, InGaN structures could be imbedded into the mature Si technology, and an obvious application in PV would be to improve the efficiency of a standard single junction Si solar cell by adding a matched InGaN top cell. The thermodynamic efficiency limit [3] of 29% for a Si solar cell under 1 sun standard illumination (AM1.5G) will increase to 42.5% already for only one additional cell with an optimum band gap of 1.8 eV [4]. To obtain this band gap, an InGaN film with a composition of 45% In is necessary.

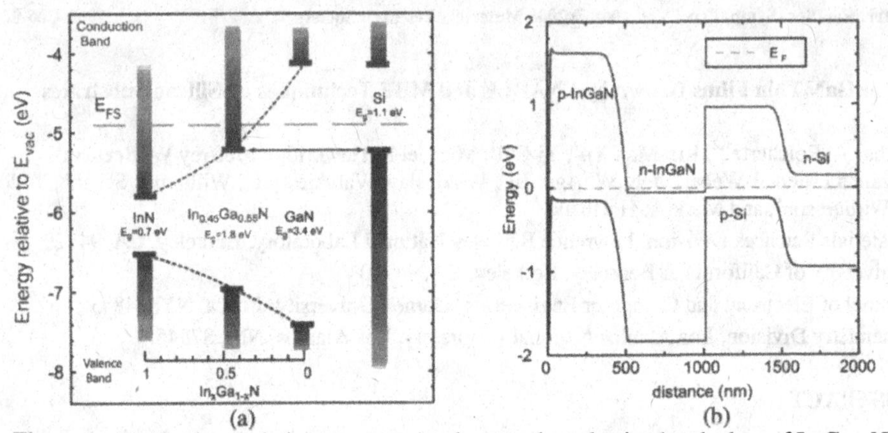

Figure 1. (a) Absolute (relative to vacuum) valence and conduction band edges of $In_xGa_{1-x}N$ and Si. E_{FS} marks the Fermi stabilization energy at -4.9 eV (from [5]). (b) Band diagrams for an $In_{0.45}Ga_{0.55}N$/Si tandem solar cell, assuming $2x10^{17}cm^{-3}$(p,n) in InGaN and $2x10^{16}cm^{-3}$(p,n) in Si.

Series-connected multijunction solar cells require an efficient tunnel junction to provide electron hole recombination between the sub cells. Such tunnel junctions are usually formed by adding additional layers of highly doped material. The InGaN/Si hybrid cell offers a unique advantage here as a consequence of the band alignment between InGaN and Si. Figure 1(a) shows the absolute positions of the band edges for InGaN and Si. It can be seen that the conduction band of InGaN aligns with the valence band of Si for a 45% In composition. Figure 1(b) shows the calculated band diagram of a pn- $In_{0.45}Ga_{0.55}N$ /pn-Si combination [5]. Doping and composition can be chosen such that there is no band bending between n-type InGaN and p-type Si, and thus an ohmic contact is predicted and no heavily doped additional layer is required. Growth of high quality InN and GaN epitaxial films on Si (111) substrates using molecular beam epitaxy (MBE) technique has been demonstrated [6,7].

EXPERIMENT

For this study we investigated InGaN on Si structures that were grown by the newly developed "energetic neutral atomic-beam lithography & epitaxy" (ENABLE) [8] and by gas-source molecular beam epitaxy (MBE) [9]. ENABLE utilizes a collimated beam of ~2 eV nitrogen atoms as the active species which are reacted with thermally evaporated Ga and In metals. The technique provides a larger N atom flux compared to MBE and reduces the need for high substrate temperatures making isothermal growth over the entire InGaN alloy composition range possible. For growth on Si (111) typical substrate temperatures were ~600 °C and growth rates were approximately 2 to 3 μm/hr. InN buffer layers were used to facilitate InGaN lattice matching. MBE growth of $In_xGa_{1-x}N$ films with x = 0 to x = 0.5 on Si (111) was initiated by removal of residual oxide. This was achieved by a sequence of heating and Ga exposure steps. The cleaning procedure was followed by growth of a 30 to 50 nm AlN buffer layer. The InGaN layers were deposited at 630 °C.

Structural characterization of each film was performed using Rutherford backscattering spectrometry (RBS), ion channeling, x-ray diffraction (XRD), and transmission electron microscopy (TEM). Optical characterization was done with photoluminescence (PL) measurements, and electrical properties were studied with Hall effect. Surface morphology and metallic cluster defects were studied by atomic force microscopy (AFM) and optical microscopy. Electrical characteristics of the junctions between InGaN films and n- or p-type Si substrates were determined by I-V measurements in the dark and with simulated AM1.5 illumination.

RESULTS AND DISCUSSION

Figure 2 shows a TEM image of an InN film which illustrates the capabilities of the new ENABLE method. This micrograph shows a columnar growth of the InN film. Column diameters are in the range of 500 nm to 1000 nm. Some voids are observed at the InN/GaN interface, which help in stress relaxation and result in much better structural quality of the top layer compared to the area close to the interface. The sample surface show slight corrugation, not exceeding 50 nm. RBS channeling analysis (with channeling minimum yield < 4%) confirms a good structural perfection of this film compared to standard MBE grown InN grown on the same substrate. Residual donor concentrations below $10^{19}/cm^3$ were measured and electron mobilities higher than 1000 cm^2/Vs were achieved. It was noted that the mobility values were generally higher than in MBE grown films with comparable electron concentrations.

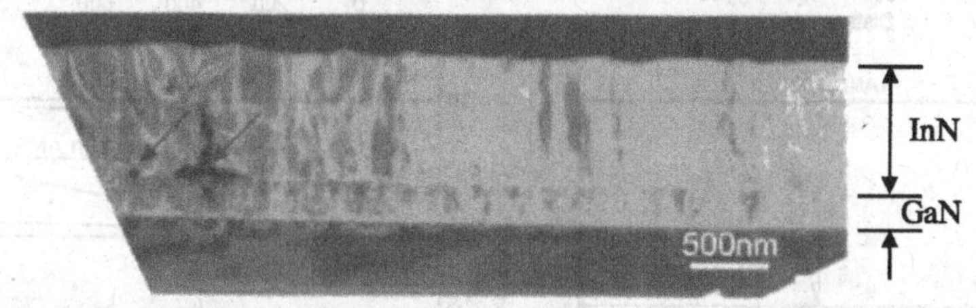

Figure 2. TEM image of an ENABLE grown InN film grown on a sapphire substrate with a GaN buffer layer. Arrows mark the formation of voids at the GaN/InN interface.

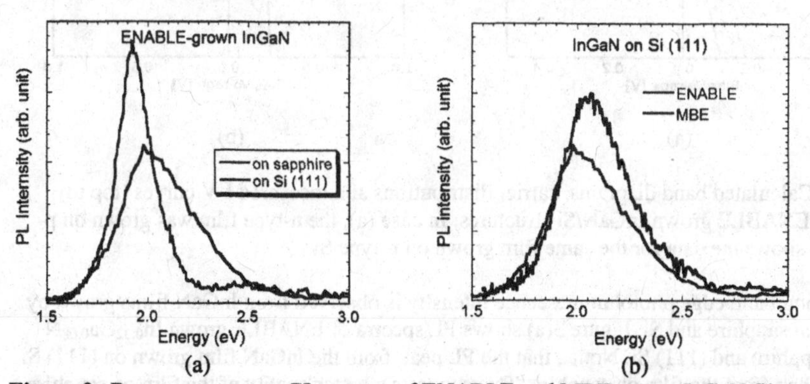

Figure 3. Room temperature PL spectra of ENABLE and MBE grown InGaN films.

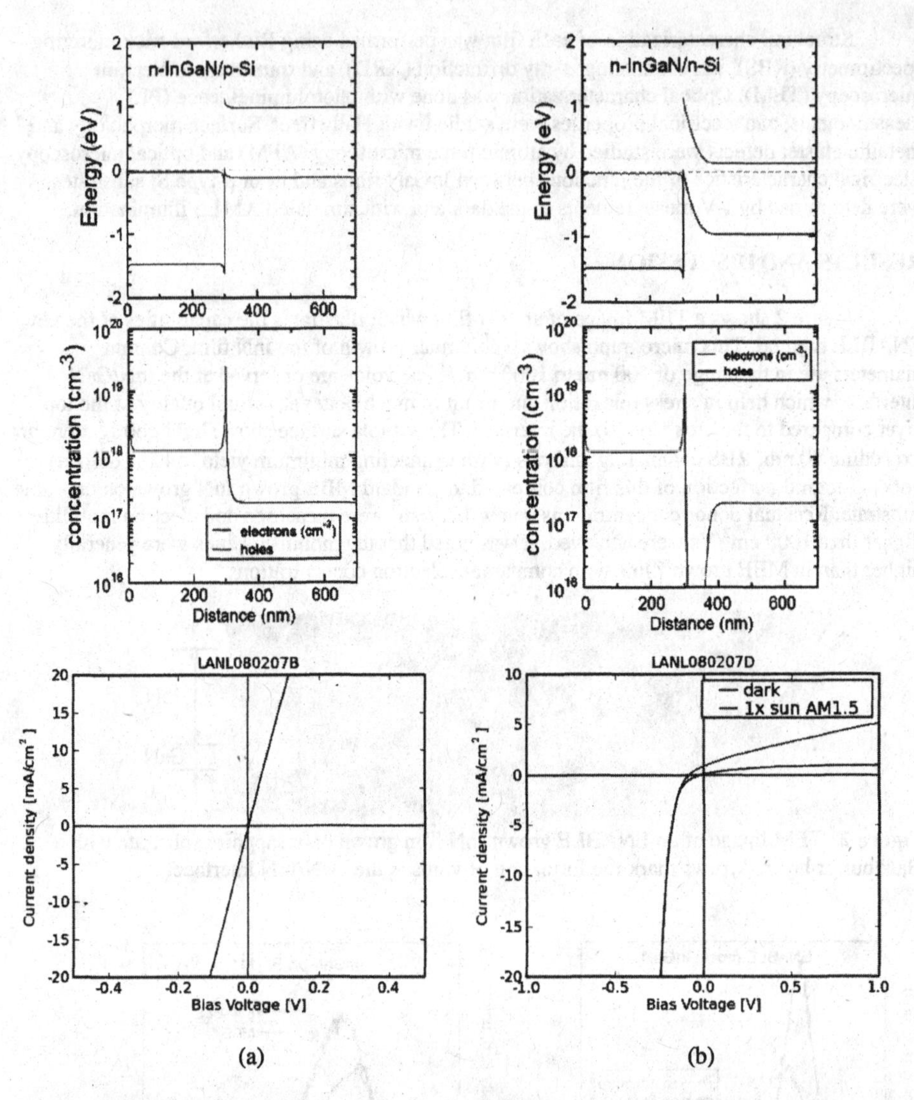

Figure 4. Calculated band diagrams, carrier distributions and measured I-V curves (top to bottom) of ENABLE grown InGaN/Si structures. In case (a), the n-type film was grown on p-type Si, (b) shows the data for the same film grown on n-type Si.

Strong band edge photoluminescence intensity is observed from InGaN films grown by ENABLE on sapphire and Si. Figure 3(a) shows PL spectra of ENABLE grown $In_{0.32}Ga_{0.68}N$ films on sapphire and (111) Si. Notice that the PL peak from the InGaN film grown on (111) Si is broader than from the film on sapphire. This suggests a better quality of the film on sapphire.

Figure 3(b) shows a comparison of the PL from InGaN films with similar composition grown by ENABLE and MBE on (111) Si. A very similar PL peak can be observed in both films.

Channeling-RBS and TEM investigations of the ENABLE-grown InGaN film reveal that the film is polycrystalline. This can be attributed to the direct growth of the film on Si without a AlN buffer layer. XRD measurement showed a single phase film with a (0001) preferred orientation. The polycrystalline nature of this film also results in a weaker and broader PL signal as well as a higher residual donor concentration of $6x10^{19}cm^{-}3$. The reduced film quality is also noticeable in the relatively large AFM surface roughness of 30 nm RMS. In case of the MBE samples, an AlN nucleation layer was first grown on the Si substrate and therefore an epitaxial InGaN film was obtained. MBE films typically showed an AFM roughness of less than 10 nm RMS.

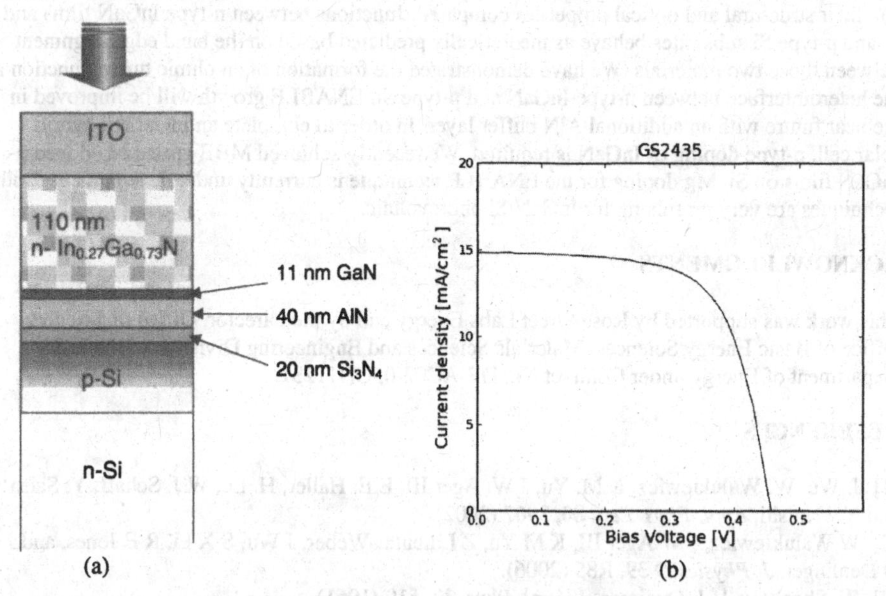

Figure 5. (a) MBE grown InGaN film on an n-type Si wafer with a diffused p-type top layer and (b) corresponding I-V curve under 1x sun (AM1.5) illumination.

In order to test the predicted electrical transport between InGaN and Si, undoped n-type InGaN of 1 μm thickness was simultaneously grown on p-type and n-type Si substrates. The results of the current-voltage tests are shown in figure 4. The InGaN/Si heterojunctions behave electrically as predicted by the calculated band diagrams in figure 4. On p-type Si, a linear I-V curve is measured (figure 4(a)), demonstrating the perfectly ohmic character of the tunnel junction. Figure 4(b) shows for comparison the same ENABLE film grown on n-type Si. The InGaN acts like a metal in this case because of the very low position of the conduction band and depletes the n-type Si. A Schottky type junction is predicted and a rectifying behavior is indeed observed. The light sensitivity to 1x sun illumination is rather small in this case. We attribute this to light absorption mostly outside the rather small and buried active region and to imperfect crystallinity.

Finally, as a first stage towards the realization of an InGaN/Si double junction tandem solar cell, we fabricated a n-InGaN/p-Si/n-Si test structure as shown in the schematics in figure 5(a). A 110 nm thick n-type InGaN layer was grown on a pn-Si-structure by MBE. This example demonstrates the effectiveness of the tunnel junction. Figure 5(b) shows the I-V characteristics of this structure under 1x sun AM1.5 illumination. Photocurrent is generated at the pn-junction in the Si and transported to the ITO top contact. PV properties of this structure are comparable to a reference pn-Si sample.

CONCLUSIONS AND FUTURE WORK

InGaN films with a wide range of In contents were grown on Si by ENABLE and MBE and their structural and optical properties compared. Junctions between n-type InGaN films and n- and p-type Si substrates behave as theoretically predicted based on the band edge alignment between those two materials. We have demonstrated the formation of an ohmic tunnel junction at the heterointerface between n-type InGaN and p-type Si. ENABLE growth will be improved in the near future with an additional AlN buffer layer. In order to complete an InGaN/Si hybrid solar cell, p-type doping of InGaN is required. We recently achieved MBE grown Mg doped p-InGaN films on Si. Mg doping for the ENABLE technique is currently under development. Both techniques are very promising for InGaN/Si photovoltaic.

ACKNOWLEDGMENTS

This work was supported by Rose Street Labs Energy and by the Director, Office of Science, Office of Basic Energy Sciences, Materials Sciences and Engineering Division, of the U.S. Department of Energy under Contract No. DE-AC02-05CH11231.

REFERENCES

[1] J. Wu, W. Walukiewicz, K.M. Yu, J.W. Ager III, E.E. Haller, H. Lu, W.J. Schaff, Y. Saito and Y. Nanishi, *Appl. Phys. Lett.* **80,** 3967 (2002).
[2] W Walukiewicz, J W Ager III, K M Yu, Z Liliental-Weber, J Wu, S X Li, R E Jones, and J D Denlinger, *J. Physics D* **39**, R85 (2006).
[3] W. Shockley, H.J. Queisser, *J. Appl. Phys.* **32**, 510 (1961).
[4] M.A. Green, *Third Generation Photovoltaics: Advanced Solar Energy Conversion*, (1st ed. 2003. 2nd printing Berlin: Springer, 2006.) p. 59.
[5] J.W. Ager III, L.A. Reichertz, D. Yamaguchi, L. Hsu, R.E. Jones, K.M. Yu, W. Walukiewicz, and W.J. Schaff, Group III-nitride alloys for multijunction solar cells, *Proc. 22nd European Photovoltaic Solar Energy Conference and Exhibition,* Milan, Italy (2007).
[6] Chung-Lin Wu, Jhih-Chun Wang, Meng-Hu Chan, Tom T. Chen, and Shangir Gwo, *Appl. Phys. Lett.* **83,** 4530 (2003).
[7] C. L. Wu, C. H. Shen, H. Y. Chen, S. J. Tsai, H. W. Lin, H. M. Lee, S. Gwo, T. F. Chuang, H. S. Chang, T. M. Hsu, *J. Crystal Growth* **288,** 247 (2006).
[8] A.H. Mueller, E.A. Akhadov, and M.A. Hoffbauer, *Appl. Phys. Lett.* **88,** 041907 (2006).
[9] H. Lu, W. J. Schaff, J. Hwang, H. Wu, G. Koley, and L.E. Eastman, *Appl. Phys. Lett.* **79,** 1489 (2001).

Mater. Res. Soc. Symp. Proc. Vol. 1068 © 2008 Materials Research Society 1068-C08-08

InGaN Growth with Indium Content Controlled by GaN Growth Plane

Hisashi Kanie[1], and Kenichi Akashi[2]
[1]Applied Electronics, Tokyo University of Science, Yamazaki 2641, Noda, Japan
[2]Applied Electronics, Graduate School of Tokyo University of Science, Yamazaki 2641, Noda, Japan

ABSTRACT

InGaN crystals yielding blue cathodoluminescence (CL) were grown by the nitridation of the mixture of GaN crystals with indium sulfide powders in an ammonia flow at 1000°C. Grown InGaN crystals have crystal habit with twelve prismatic planes. By a highly spatially resolved CL imagining study on a scanning electron microscope equipped with a monochromator the prismatic planes are grouped into two groups: pentagonal shaped planes yielding blue CL and rectangular planes without luminescence. The pentagonal shaped plane is assigned to the {11-20} a-plane by electron backscattering pattern (EBSP) method. InGaN growth with In content controlled by GaN growth planes is discussed.

INTRODUCTION

High efficient cathodoluminescent phosphors working at a low acceleration voltage are required to develop a low voltage field emission display. Zinc oxide is an efficient green light emitting low voltage phosphor. We displayed indium gallium nitride is a promising material for a low voltage blue light emitting phosphor [1,2]. To obtain luminescent grade InGaN crystals it is necessary to grow crystals at high temperatures. However, dissociation of InGaN starting at the lower temperature as the higher In content resulted in violet emitting InGaN crystals. To prepare high In content InGaN crystals yielding blue CL, we divided growth process into two steps: the high temperature GaN crystal growth step to obtain crystals with high quality and the low temperature InGaN growth step to incorporate indium in InGaN to the extent to yield blue CL. Blue emitting InGaN crystals were grown by the two-step growth method, however, the yield was not high. Some grown InGaN crystals have characteristic crystal habit with a pyramid or a truncated pyramid connected to twelve-faced prism. We observed pentagonal shaped prismatic planes emitting CL and rectangular prismatic planes without CL in highly spatially resolved CL imaging. For InGaN light emitting diode with quantum well structures InGaN layers grown on m- plane or a-plane GaN are more efficient than that grown on c-plane GaN because the a-plane or m-plane is nonpolar and has no piezoelectric internal filed in the well [3,4]. We determined Miller indices of the prismatic planes by electron backscattering pattern (EBSP).

EXPERIMENTAL

Blue light emitting InGaN microcrystals were grown by a two-step growth method. At the first step, GaN crystals were grown by a reaction with ground gallium sulfide placed in a quartz boat in an ammonia gas flow at 1100°C for 5 hours in open tube quartz reactor. At the second step, obtained GaN crystals were mixed with ground indium sulfide and treated in an ammonia gas flow at 1000°C for 5 hours.

SEM imaging and highly spatially resolved CL imaging were performed at 3 to 5 kV on a TOPCON scanning electron microscope (SEM) DS-130 equipped with cathodoluminescent measurement system and normally spatially resolved CL imaging on a Shimazu EPMA 8705. EBSP was performed on a Hitachi SEM SU-70 equipped with Oxford instrument HKL channel 5 EBSP system and analyzed by software called SFC published by Kogure[5].

RESULTS

Miller indices of prismatic planes were assigned by comparing SEM image and EBSP. The exterior angle between the m-plane and the a-plane is 30°. Twelve prismatic planes of InGaN in the SEM image viewed along the c axis are shown in figure 1 (a) and (b). Miller index of the six inclined planes surrounding c plane of truncated pyramid was assigned first. The

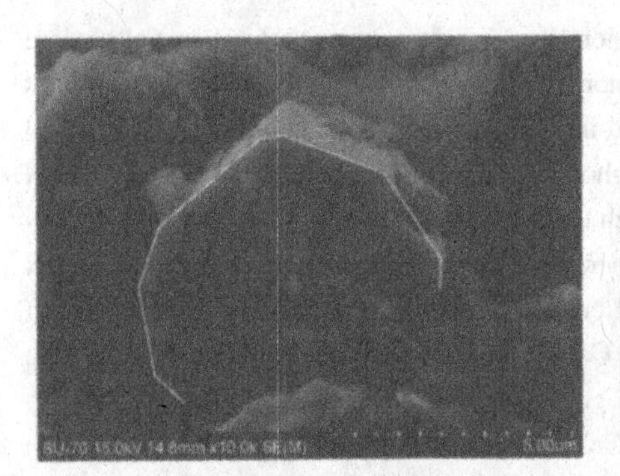

(a) View from flat basal plane

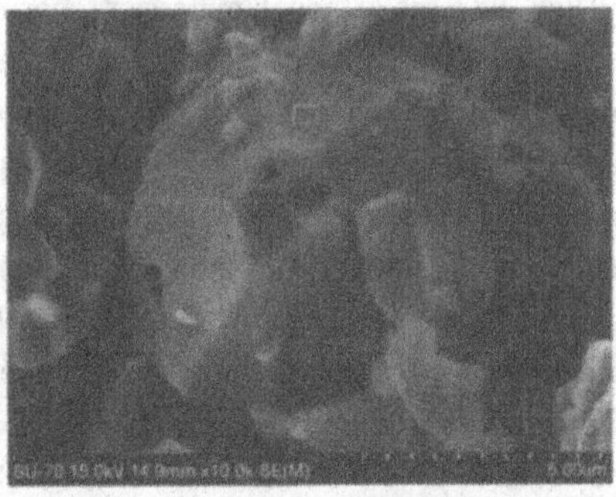

(b) View from the top of truncated pyramid

Figure 1. Prismatic planes of InGaN intersecting at an angle of 30° (a) View from the flat basal plane, (b)View from the truncated pyramid.

c-plane and inclined planes of truncated pyramid belongs to the zone axis <1-100>, as shown in figure 2, so the inclined plane is assigned to {1-10x} plane, where the value x is ambiguous. The inclined planes of a GaN single crystal have the compatible index [6].

As the sample stage for EBSP is tilted at 70° from the horizontal, simultaneously taken SEM images was a slightly tilted elevation view. Kikuchi bands from the prismatic plane at the top in SEM image (figure 3) were projected to the lower center of EBSP image and had characteristics from the <11-20> direction. We indexed the pentagonal prismatic plane to {11-20} (a-plane) and the rectangle plane the {1-100} m-plane.

InGaN crystals with twelve prismatic planes showed luminescence at 420 nm from the

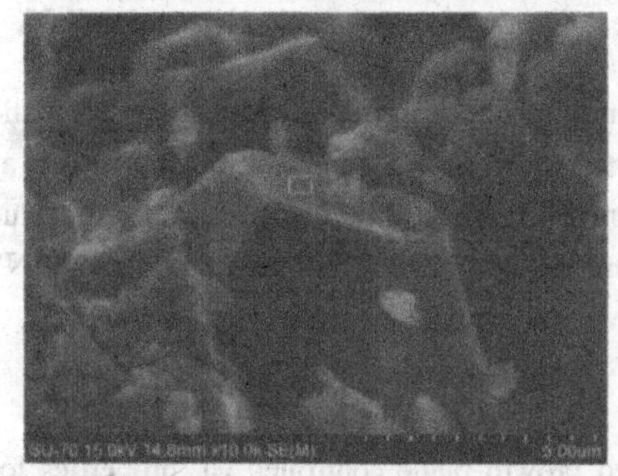

(a) SEM image of truncated pyramidal InGaN. A square indicates the incident beam position.

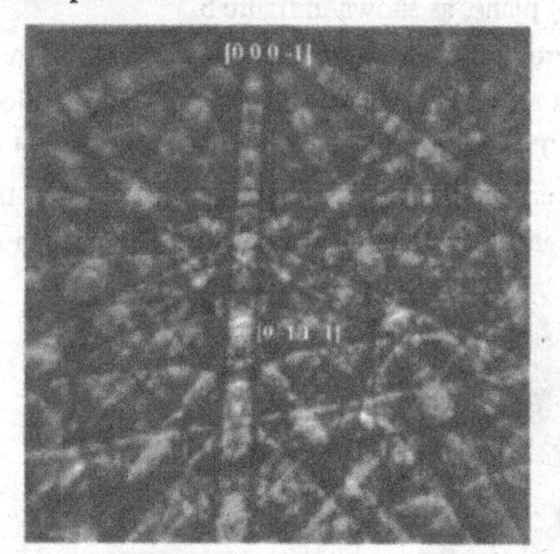

(b) EBSP of {1-10x} plane

Figure 2. SEM image of InGaN. Truncated pyramid crystal and EBSP.

(a) SEM image of twelve prismatic InGaN planes. A square indicates the incident beam position.

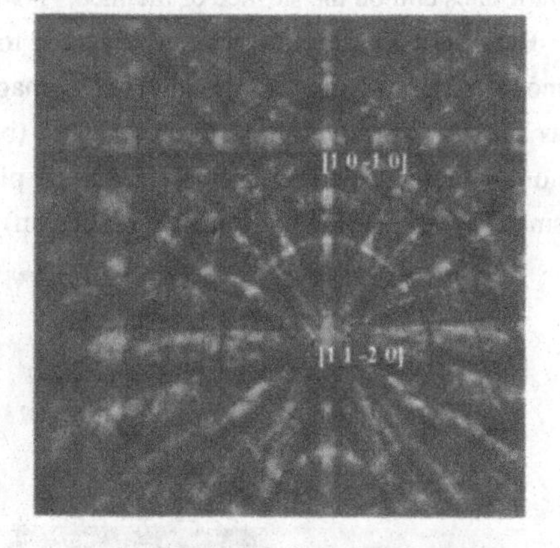

(b) EBSP of prismatic plane

Figure 3. SEM image of InGaN. Prismatic planes and EBSP.

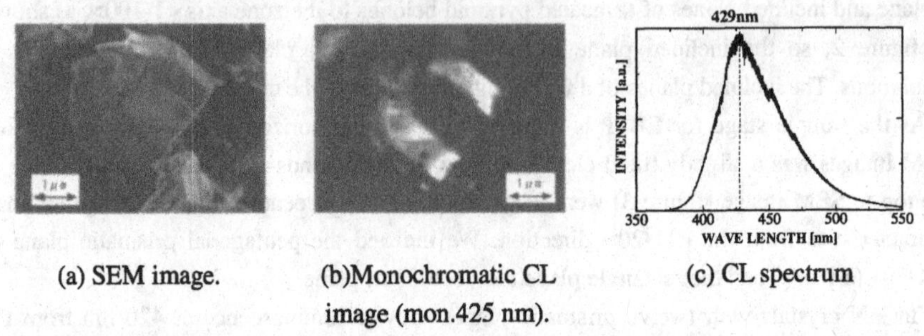

(a) SEM image. (b)Monochromatic CL image (mon.425 nm). (c) CL spectrum

Figure 4. Highly spatially resolved CL images of pentagonal prismatic planes of InGaN.

pentagonal shaped prismatic planes and no luminescence from the rectangular prismatic planes in a highly spatially resolved cathodoluminescent (HRCL) monochromatic image, as shown in figure 4 (a) and (b). The CL spectrum in figure 4 (c) showed characteristic blue emission peak from InGaN and did not show donor-acceptor pair (DAP) emission from GaN.

DISCUSSION AND SUMMARY

Highly spatially resolved CL imaging showed growth plane controlled CL properties for InGaN crystals grown by the two-step growth method. HR CL image of a GaN crystal with 300-500 nm bandpass filter showed dark lines on the basal plane and dark spots where the dark lines end on the surface of inclined {1-10x} plane, as shown in figure 5.

Blue light emitting InGaN was observed to grow also on a GaN basal plane in a different mode comparing monochromatic HRCL images at violet (390nm) and blue (420 nm) region, as shown respectively in figures 6(a) and (b). The black lines in figure 6(a) correspond to nonluminescent defects on the GaN basal plane. The white dots in figure 6(b) showed the small InGaN islands emitting blue (420 nm) light aggregated at the defects. The growth of

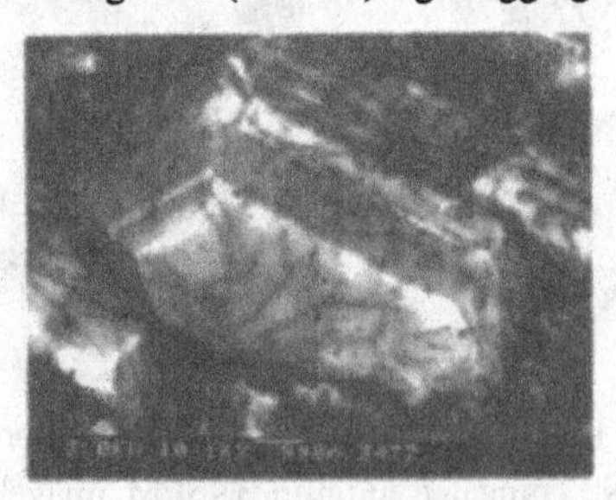

Figure 5. Bandpass filtered CL image of GaN (300-500 nm).

InGaN at the defects is explained by the composition pulling effect [7] by which V-shaped pits formed where the dislocation terminates on the surface of the GaN basal plane pull indium atoms, resulting in preferential growth of InGaN on the walls of the V-shaped pit.

As the two prismatic planes have a nonpolar feature, we stressed the difference of the a-plane and m-plane in the atomic geometry, such as the Ga-N bonds on the surface. For the preferential growth of blue emitting InGaN on the a-planes the existence of two types of the Ga-N bonds in the a-plane plays an important role.

These observations suggest that InGaN grew on the GaN planes with In content controlled to the extent for InGaN emitting blue CL (420 nm). As similar phenomena was observed from MOCVD grown InGaN on GaN [7] and an InN/GaN superlattice structure grown by MBE [8]. The dissociation of GaN and the regrowth of InGaN are the main process to obtain InGaN yielding blue CL in the two-step growth method.

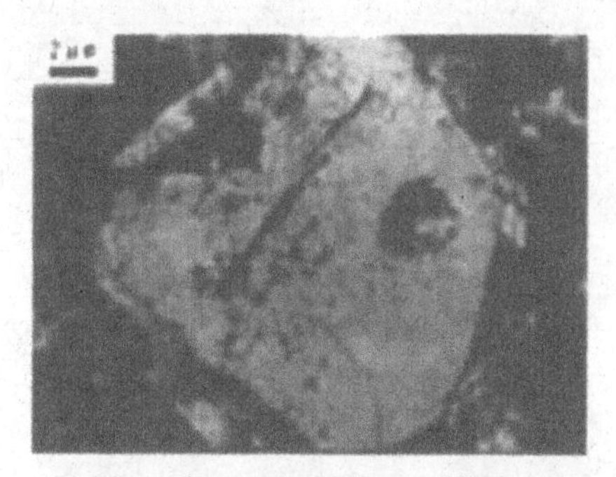

(a) Monochromatic CL image (390 nm)   (b) Monochromatic CL image (420 nm)

Figure 6. HRCL image of InGaN basal plane.

REFERENCES

1. N. Tsukamoto and H. Kanie, Proc. workshop of SiC and related wide band gap semiconductors, Kyoto, Japan 43 (1996) (in Japanese).

2. H. Kanie, T. Kawano, K. Sugimoto, R. Kawai, Proc. MRS spring meeting, Q5.11 (2000).

3. A. Chitnis, C. Chen, V. Adivarahan, M. Shatalov, E. Kuokstis, V. Mandavilli, J. Yang, and M. A. Khan, APL, **84**, 3663 (2004).

4. Y. J. Sun, O. Brandt, S. Cronenberg, S. Dhar, H. T. Grahn, K. H. Ploog, P. Waltereit, and J. S. Speck, Phys. Rev., B **67**, 041306 (2003).

5. T. Kogure, J. Cryst. Soc. Jpn., **45**, 391-395 (2003) (in Japanese)

6.D. Elwell, R. S. Feigelson, M. M. Simkins and W. A. Tiller, J. Crystal Growth, **66,** 45 (1984).

7. H. Hiramatsu, Y. Kawaguchi, M. Shimizu, N. Sawaki, T. Zheleva, Robert F. Davis H. Tsuda, W. Taki, N. Kuwano, K. Oki, MRS Internet J. Nitride Semicond. Res., 2, 6 (1997).

8. A. Yoshikawa, S.B.Che, W. Yamaguchi, H.Saito, X.Q.Wang,Y.Ishitani, and E.S.Hwang, APL, **90**, 073101 (2007).

Mater. Res. Soc. Symp. Proc. Vol. 1068 © 2008 Materials Research Society 1068-C06-09

Electron Traps in n-GaN Grown on Si (111) Substrates by MOVPE

Tsuneo Ito[1,2], Yutaka Terada[1], and Takashi Egawa[1]
[1]Research Center for Nano-Device and System, Nagoya Institute of Technology, Gokiso-cho,Showa-ku, NAGOYA, 466-8555, Japan
[2]DOWA Electronics Materials Co.,Ltd., Sunada 1,Iijima, AKITA, 011-0911, Japan

ABSTRACT

Deep level electron traps in n-GaN grown by metal organic vapor phase epitaxy (MOVPE) on Si (111) substrate were studied by means of deep level transient spectroscopy (DLTS). The growth of n-GaN on different pair number of AlN/GaN superlattice buffer layers (SLS) system and on c-face sapphire substrate are compared. Three deep electron traps labeled E2 (the energy level is not clear), E4 (0.7-0.8 eV), E5 (1.0-1.1 eV), were observed in n-GaN on Si substrate. And the concentrations of these traps observed for n-GaN on Si are very different from that observed for n-GaN on sapphire substrate. E4 is the dominant of these levels for n-GaN on Si substrate, and it is not related to linear array defect like dislocation line, but it behaves like point-defect based on the analysis by electron capture kinetics, in spite of having high dislocation density to the order of 10^{10} cm^{-2}.

INTRODUCTION

GaN and its alloys have found broad application for the visible to ultraviolet optoelectronics, high frequency, high power, and high temperature electron devices. Over a decade, the growth of these nitrides on Si substrate has become a subject of great interest, because of the numerous advantages like significantly lower cost, good thermal conductivity, and availability in large diameter. However, there are some difficulties in the growth of GaN on Si, especially cracking of GaN due to thermal mismatch between them. Recent progress of several kind of buffer layer system enables to grow 2-3 μm thick crack-free GaN layers on Si substrate [1, 2]. And using these techniques, many GaN devices on Si substrate have been demonstrated as reported by us earlier (for example; [3]). Besides, an understanding of the defects in these materials is essential for improving the material quality and device performance. Deep level transient spectroscopy (DLTS) has been used by many device researchers to characterize electronic trap states in GaN grown by MOVPE. However, currently there are no reports available on the study of deep traps in GaN grown on Si substrates. In this paper, we report electron traps in n-GaN grown on Si (111) substrates by MOVPE, and analyzed the crystal quality, electrical properties and investigated the deep traps by means of DLTS measurement.

EXPERIMENT

The growth of Si-doped n-GaN on 4-inch Si (111) substrate and on 2-inch c-sapphire substrate was performed by horizontal MOVPE system. The schematic diagram of these epilayer structures are shown in figure.1. In order to grow 1-μm crack-free n-GaN layers on Si substrate, we have used an i-GaN/i-AlN superlattice buffer layers system. In our experiment, two types of n- GaN on Si substrate samples were prepared. They are the growth of n-GaN on (i) 10 pairs and

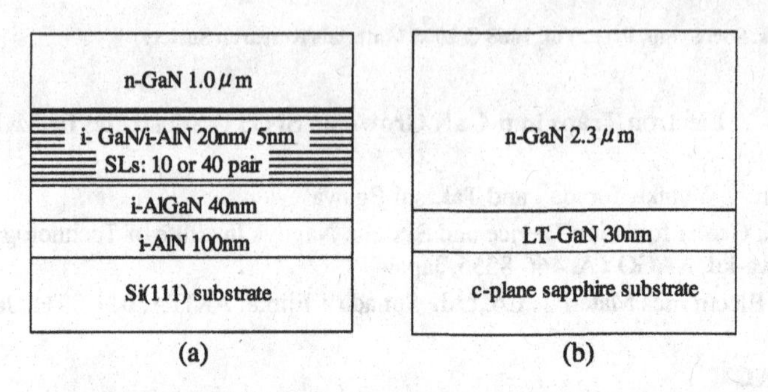

Figure 1. Schematic diagram of n-GaN on (a) Silicon and (b) Sapphire Substrates

(ii) 40 pairs of i-GaN/i-AlN SLS respectively. For comparison, an approximately 2.3-μm-thick n-GaN layer was deposited on c-plane sapphire substrates with a conventional low-temperature 30-nm thick GaN buffer layers(LTBL) grown at 500 ℃. These n-GaN layers on each substrate were grown at 1130 ℃, in atmospheric pressure. Trimethylgallium (TMGa) and trimethylaluminum (TMAl) and ammonia (NH_3) were used as precursors. Hydrogen-diluted silane (SiH_4) was used for n-type dopant, and Si mole flow rates were identical in both the samples. The carrier concentrations in n-GaN are in the range of 6- 9×10^{17} cm^{-3} as calculated from capacitance-voltage(C-V) measurement. For DLTS measurement, Ohmic contacts were formed by evaporating the metals Ti/Al (25 nm/100 nm) using electron beam evaporation technique. The Ohmic metals were annealed at 800 ℃ for 30 sec by rapid thermal annealing in the nitrogen ambient. Schottky contacts of 400 μm in diameter were formed by using the metals Pd/Ti/Au (40 nm/20 nm/60 nm). DLTS measurements were carried out in a Nanometrics DL8000 system in the temperature range of 90-600 K. A quiescent reverse bias voltage and fill biases were -2 V and 0 V, respectively. In this case, the space charge region was less than 100 nm at this reverse bias voltage. The filling pulse times (t_p) from 0.01 to 10 ms was applied to fill the electron traps with an emission time period width (t_w) of 50 ms. And we performed other characterization namely X-ray diffraction (XRD) measurement to study the crystalline nature of n-GaN and Hall effect measurement for electron mobility analysis.

DISCUSSION

Hall, C-V, XRD measurement

The electrical and structural properties of the samples, i.e., Hall mobility, carrier concentration at room temperature (R.T.), C-V, and the full-width at half–maximum (FWHM) of the X-ray rocking curve (XRC) of the n-GaN are summarized in Table I. Dislocation density for each sample was estimated from XRC based on the method described in a previous report [4]. It is generally understood that the growth of n-GaN on multi-pairs of i-GaN/i-AlN SLS system can reduce the dislocation density. It is seen from Table I, when the number of GaN/AlN pairs increase from 10 to 40, the Hall mobility increases, and the value of XRC FWHM decreases respectively. And it shows that the dislocation densities of both the screw type and the edge type decrease. But these dislocation densities of n-GaN on Si are one order of magnitude higher than

Table I. Electrical and structural properties of the studied samples

#	Epi Structure	Hall measurement (R.T.) n (cm^{-3})	Hall measurement (R.T.) Mobility (cm^2/Vs)	C-V (R.T.) Nd-Na (cm^{-3})	XRC FWHM (0004) (arcsec)	XRC FWHM (20$\bar{2}$4) (arcsec)	XRC FWHM (1000) (arcsec)	dislocation density screw ($\times10^8 cm^{-2}$)	dislocation density edge ($\times10^8 cm^{-2}$)
S10	on SLs(10pair)/Si	7.0×10^{17}	186	5.7×10^{17}	865	2134	3067	15	500
S40	on SLs(40pair)/Si	1.1×10^{18}	247	6.3×10^{17}	769	1563	2197	12	256
Sap	on LTBL/Sapphire	8.5×10^{17}	358	9.2×10^{17}	253	606	879	1.3	41

that of the density observed on sapphire.

DLTS characterization

Figure 2 shows the DLTS spectra of n-GaN's for (a) on 10 pairs and on 40 pairs of SLS on Si substrate, and (b) on sapphire substrate. In these spectra, we can observe a dominant peak which has more than two shoulders in n-GaN on Si sample, and two peaks in n-GaN on sapphire sample, respectively. And we labeled three deep levels, E2, E4, and E5 on Si sample, and two deep levels labeled E2, E4 on sapphire sample. Figure 3 shows the Arrhenius plots of DLTS signal of all deep levels in figure 2. Their energy position (E_a), capture cross section (σ), and trap concentration (N_t), are listed in Table II. However, fitting of the transient for E2 did not result in linear Arrhenius plots. Our result corresponds to previous literature reports for deep levels, except for E5 [5-12]. In the DLTS spectra of n-GaN on Si in figure 2(a), the dominant peak is E4 near 400 K. Haase et al. have suggested that E3 in n-GaN grown by MOVPE corresponds to an activation energy of 0.67eV [5], is not clear in the case of n-GaN on Si substrate, and not detected for n-GaN on sapphire substrate. From the result of these DLTS spectra, the concentration of E4 for n-GaN grown on 40 pairs of SLS sample is 60 % that of n-GaN grown on 10 pair SLS, and E4 on sapphire substrate sample is about 14 % that of the 10

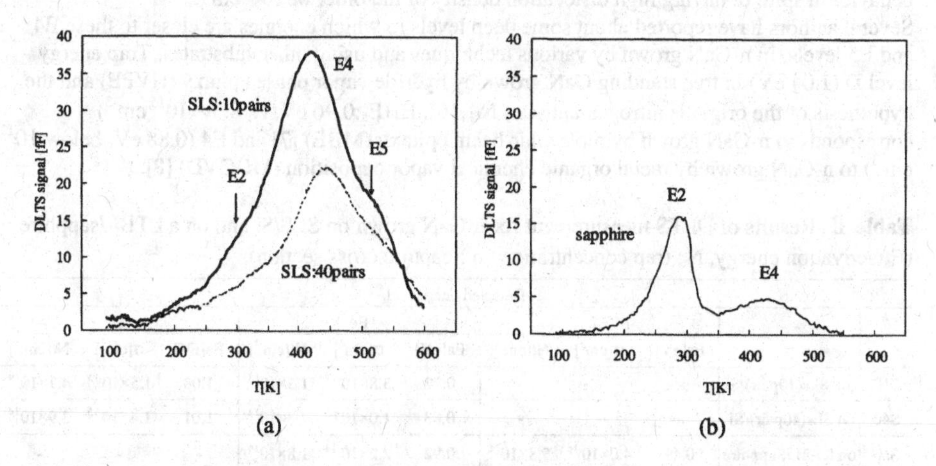

Figure 2. DLTS spectra for n-GaN grown (a) on SLS/Si, and (b) on a LTBL/sapphire

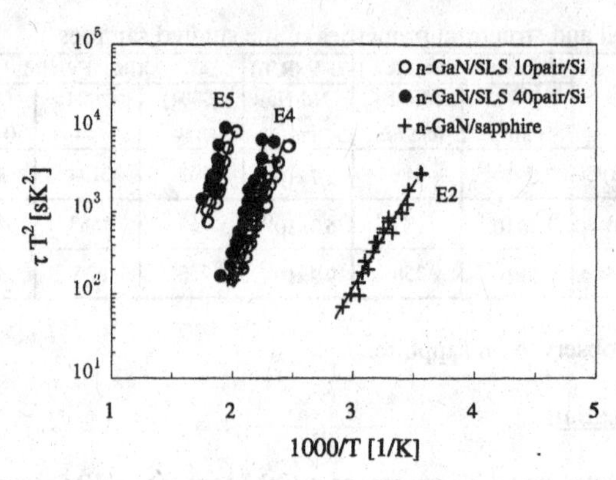

Figure 3. DLTS spectra for n-GaN grown on SLS/Si and on a LTBL/sapphire

pair SLS respectively. The reason for the different concentrations of E4 between on Si samples and on a sapphire sample, is not clear at this moment. In addition to that, it has a possibility that the peak temperature and peak height of E4 are shifted by overlapping several other small peaks. More detailed experiments are in progress to characterize the nature of them.

Electron capture kinetics

Figure 4 shows the dependence of the concentration of electrons trapped at E4 trap states on the logarithm of the filling-pulse duration time (t_p) for n-GaN grown on SLS/Si and on a LTBL/sapphire. It shows a non-linear response but saturated for long filling times for n-GaN on Si substrate, or slight decline for n-GaN on sapphire substrate. This illustrates that E4 of each sample is not related to linear array defect like dislocation line but related to point-defect-like behavior in spite of having high dislocation density of the order of 10^{10} cm^{-2}.
Several authors have reported about some deep levels in which energies are closer to these E4 and E5 levels, in n-GaN grown by various techniques and using other substrates. Trap energy level D (1.04 eV) in free standing GaN grown by hydride vapor phase epitaxy (HVPE) and the hypothesis of the origin is nitrogen antisites N_{Ga} [6]. E_4 (E_a:0.96 eV, N_t:8.3$\times 10^{15}$ cm^{-3}) corresponds to n-GaN grown by molecular-beam epitaxy (MBE) [7] and E4 (0.88 eV, below 10^{13} cm^{-3}) to n-GaN grown by metal organic chemical vapor deposition (MOCVD) [8].

Table II. Results of DLTS measurement for n-GaN grown on SLS/Si and on a LTBL/sapphire. (E_a:activation energy, N_t: trap concentration, σ : capture cross section).

#		Detected deep level								
		E2			E4			E5		
		Ea[eV]	σ[cm²]	Nt[cm⁻³]	Ea[eV]	σ[cm²]	Nt[cm⁻³]	Ea[eV]	σ[cm²]	Nt[cm⁻³]
S10	on SLs(10pair)/Si	-	-	-	0.72	3.8×10⁻¹⁷	1.3×10¹⁵	1.08	1.5×10⁻¹⁵	4.7×10¹⁴
S40	on SLs(40pair)/Si	-	-	-	0.83	1.0×10⁻¹⁵	7.8×10¹⁴	1.01	1.4×10⁻¹⁵	3.9×10¹⁴
Sap	on LTBL/Sapphire	0.49	4.0×10⁻¹⁶	5.8×10¹⁴	0.92	2.2×10⁻¹⁴	1.8×10¹⁴	-	-	-

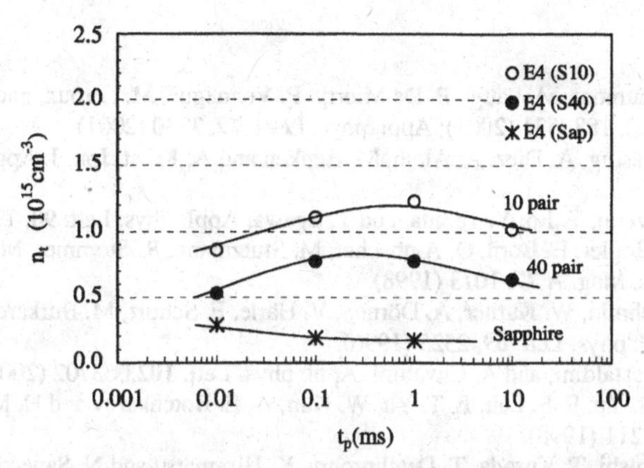

Figure 4. Concentration of electrons trapped at E4 traps vs. filling-pulse duration time for n-GaN grown on SLS/Si and on a LT-BL/sapphire

A1 behaves like a "line defect" in the event of growth by MBE [9], E1 (0.5-0.62 eV) by MOVPE[10]. 0.9 eV trap (0.9 eV, 2.1×10^{13} cm^{-3}) corresponds to a simple point defect in freestanding GaN by HVPE [11], and A1 (1.02 eV, 2.1×10^{14} cm^{-3}), associated with threading dislocations in AlGaN/GaN/SiC schottky barrier diode by MOCVD [12], are shown in previous reports. The energies and concentrations of the deep trap level are very different, and depend on the growth techniques, growth conditions, and substrate materials. Further study on crystal growth condition, thermal history of sample preparation, is needed to understand the reliability of many kinds of optical, electrical devices.

CONCLUSIONS

In this paper, we have grown n-GaN films on SLS buffer layer system on Si (111) substrate by MOVPE. Hall mobility, XRC FWHMs, showed that the crystalline quality was superior on 40 pairs of AlN/GaN superlattice buffer layers rather than the growth on 10 pairs of SLS. And we investigated the electron trap states in n-GaN films on these SLS/Si and on conventional LTBL/c-sapphire, using DLTS measurement. It was confirmed by identifying deep electron traps, such as E4=Ec-(0.72-0.92) eV on Si and on sapphire, E5= Ec-(1.01-1.08) eV on Si, and E2=Ec-0.49 eV on sapphire. The trap level E4 is observed clearly in both n-GaN films grown on Si and on sapphire. E4 of each sample is not related to linear array defect like dislocation line but related to point-defect-like behavior.

ACKNOWLEDGMENTS

We are grateful to Mr. Y. Nomura and Mr. A. Watanabe (Research Center for Nano-Device and System, Nagoya Institute of Technology) for help with epi growth, device fabrication and DLTS measurements.

REFERENCES

1. E. Feltin, B. Beaumont, M. Laügt, P. De Mierry, P. Vennéguè, M. Leroux, and P. Gibart, Phys. Stat. Sol (a), 188, 531 (2001); Appl. phys. Lett. **79**, 3230 (2001)
2. A. Dadgar, J. Bläsing, A. Diez, A. Alam, M. Heuken and A. Krost, Jpn. J. Appl. Phs. **39**, L1183 (2000)
3. S.Lawrence Selvaraj, T. Ito, Y. Terada, and T. Egawa, Appl. Phys. Lett. **90**, 173506 (2007).
4. T. Metzger, R. Hopler, E. Born, O. Ambacher, M. Stutzmann, R. Stommer, M. Schuster and H. Gobel, Philos. Mag. A **77**, 1013 (1998).
5. D. Haase, M. Schmid, W. Kürner, A. Dörnen, V. Härle, F. Scholz, M. Burkard and H. Schweizer, Appl. phys. Lett. **69**, 2525 (1996).
6. L. Polenta, A. Castaldini, and A. Cavallini, Appl. phys. Lett. **102**,063702 (2007).
7. C. D. Wang, L.S. Yu, S. S. Lau, E. T. Yu, W. Kim, A. E. Botchkarev and H. Morkoç, Appl. phys. Lett. **72**, 1211 (1998).
8. P. Hacke, H. Okushi, T. Kuroda, T. Detchprohm, K. Hiramatsu and N. Sawaki, J.Cryst. Growth **189/190**, 541 (1998).
9. Z.-Q. Fang, D. C. Look, P. Visconti, D.-F. Wang, C.-Z. Lu, F. Yun and H. Morkoç, Appl. phys. Lett. **78**,2178 (2001).
10. D. Emiroglu, J. H. Evans-Freeman, M. J. Kappers, C. McAleese and C. J. Humphreys, Physica **B401-402**, 311 (2007).
11. A. Y. Polyakov, N. B. Smirnov, A.V. Govorkov, Z.-Q. Fang, D. C. Look, S. S. Park and J. H. Han, J. Appl. Phs. **92**, 5241 (2002).
12. Z.-Q. Fang, D. C. Look, D. H. Kim, and I. Adesida, Appl. phys. Lett. 87,182115 (2005).

Mater. Res. Soc. Symp. Proc. Vol. 1068 © 2008 Materials Research Society 1068-C03-09

The Influence of Growth Temperature on Oxygen Concentration in GaN Buffer Layer

Ewa Dumiszewska[1,2], Wlodek Strupinski[1], Piotr Caban[1,3], Marek Wesolowski[1], Dariusz Lenkiewicz[1,2], Rafal Jakiela[1,4], Karolina Pagowska[5], Andrzej Turos[1,5], and Krzysztof Zdunek[2]
[1]III-V Epitaxy Department, Institute of Electronic Materials Technology, Wolczynska 133, Warsaw, 01-919, Poland
[2]Faculty of Materials Science, Warsaw University of Technology, Woloska 141, Warsaw, 02-507, Poland
[3]Institute of Microelectronics and Optoelectronics, Warsaw University of Technology, Nowowiejska 15/19, Warsaw, 00-665, Poland
[4]Institute of Physics, Polish Academy of Sciences, Lotnikow 32/46, Warsaw, Poland
[5]Soltan Institute for Nuclear Studies, Swierk/Otwock, 05-400, Poland

ABSTRACT

The influence of growth temperature on oxygen incorporation into GaN epitaxial layers was studied. GaN layers deposited at low temperatures were characterized by much higher oxygen concentration than those deposited at high temperature typically used for epitaxial growth. GaN buffer layers (HT GaN) about 1 μm thick were deposited on GaN nucleation layers (NL) with various thicknesses. The influence of NL thickness on crystalline quality and oxygen concentration of HT GaN layers were studied using RBS and SIMS. With increasing thickness of NL the crystalline quality of GaN buffer layers deteriorates and the oxygen concentration increases. It was observed that oxygen atoms incorporated at low temperature in NL diffuse into GaN buffer layer during high temperature growth as a consequence GaN NL is the source for unintentional oxygen doping.

Keywords: A1. MOVPE; A3 Gallium Nitride.; B1.Oxygen;

INTRODUCTION

Gallium nitride plays a significant role in optoelectronic and high-power, high frequency devices industry. Lack of commercially available free standing GaN substrates makes it necessary to use alternative ones such as sapphire (Al_2O_3), silicon carbide (SiC) or silicon. As the result, GaN layers grown on sapphire by the most commonly used method - metalorganic vapor phase epitaxy (MOVPE) are characterized by high oxygen concentration [1]. Oxygen in GaN layers influences n-type background conductivity and may be responsible for threading dislocation behavior [2,3]. Unfortunately, the origin of oxygen atoms is still not well understood. It was suggested that it was incorporated during the growth [3] or comes from sapphire substrate [4]. There are also numerous works indicating interactions between oxygen atoms and dislocations [3,5]. Liliental-Weber et al. [6] showed the increased nanopipes density at elevated oxygen content. Cherns et al. [7] claimed that oxygen may be responsible for electrical activity of dislocations. Our recent results also indicate that interaction between dislocations and oxygen atoms influences device performance [8]. There is a strong correlation between oxygen concentration and carrier concentration in high temperature (HT) GaN, i.e. the larger oxygen content, the higher carrier concentration.

The aim of this work was the search for the principal source of oxygen in HT GaN. The effects of the growth temperature on oxygen concentration in gallium nitride layers were studied. We have proved that the low growth temperature of the nucleation layer (deposited at 550 °C) is decisive for oxygen content in HT GaN.

EXPERIMENTAL

GaN layers were grown on sapphire substrates in AIX 200/4 RF-S metal organic vapor phase epitaxy low-pressure reactor. The source gases were trimethylgallium (TMGa, 10 ml/min), and ammonia (NH_3, 2000 ml/min). Hydrogen flow of 20 ml/min of through TMGa bubbler and 2000 ml/min of NH_3 was applied. The reactor pressure and temperature were 250 mbar and 1140°C respectively. High purity hydrogen was used as a carrier gas.

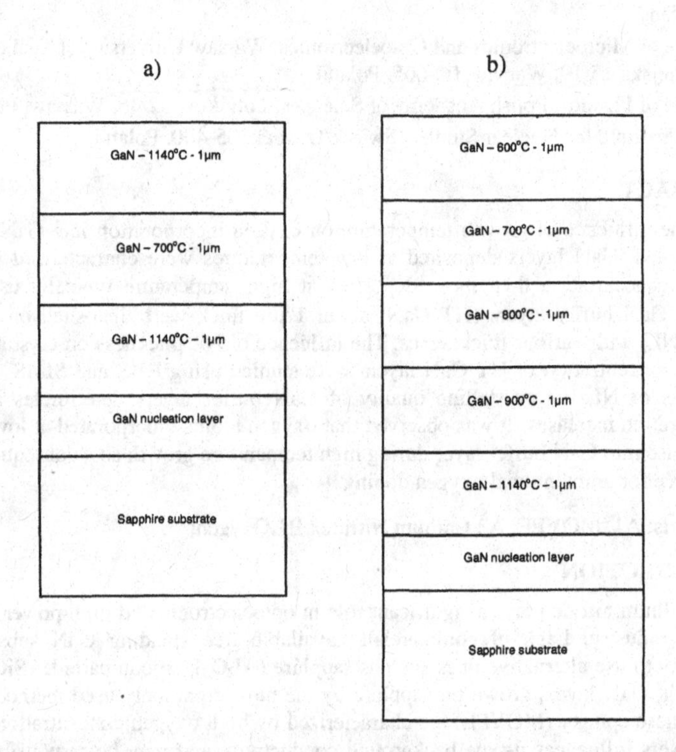

Fig.1. Test structures for oxygen incorporation study. Different layer sequences were grown on sapphire substrate: a) GaN NL, HT GaN layer (1140°C), low temperature GaN layer (700°C), HT GaN layer (1140°C); b) GaN NL, HT GaN layer (1140°C), four GaN layers grown at 900°C-600°C.

Two epitaxial structures showed in Figs. 1a and b were deposited at various temperatures. First one, cf. Fig. 1a, consisted of three layers subsequently deposited at 1140°C at 700°C and finally again at 1140°C. The second structure, cf. Fig. 1b, is composed of GaN HT layer and four GaN layers grown at the temperatures deceasing from 900°C to 600°C with 100°C steps without stopping the growth.

Thickness and crystalline structure of NL and overgrown GaN epilayers was determined by RBS/channeling [9]. Oxygen concentration in these structures was measured by SIMS.

RESULT AND DISCUSSION

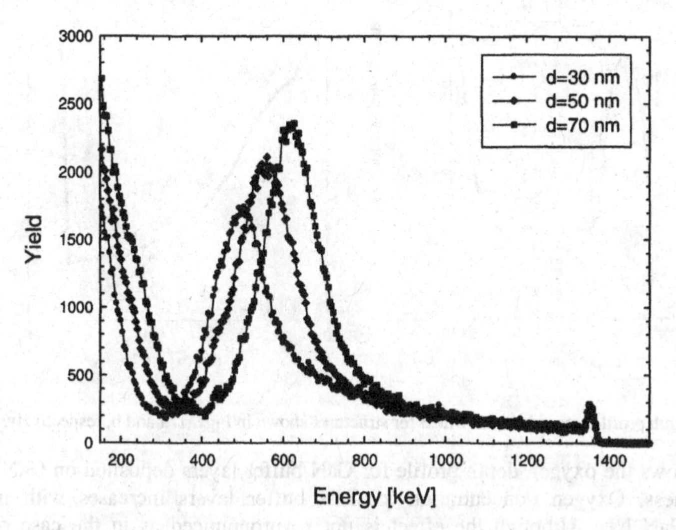

Fig. 2. Aligned RBS/channeling spectra recorded for about 1 μm thick HT GaN layers deposited on NL with various thickness.

Influence of the NL thickness on the grown layers structure was studied by RBS/channeling using 1.7 MeV ^{4}He ions. GaN layers of about 1 μm thickness were grown on NL of varying thickness. As shown in Fig. 2 two regions can be distinguished in the aligned channeling spectra. The first one is delimited by pronounced peaks located in the vicinity of GaN/sapphire interface (around 600 keV of backscattered ion energy) and correspond to the large defect agglomeration, which is composed of defects generated by layer/substrate lattice misfit and remnants of polycrystalline NL. The second region (energies above 900 keV) reflects the top part of layers approximately 600 nm thickness. By applying appropriate evaluation methods layer thickness and their crystalline quality can be determined [10]. Very low scattering yield of aligned spectra in this region amounting to less than 2% of that for random spectra (not shown here) reveal very good quality of layers. Two conclusions can be drown from spectra shown in Fig. 2:

(i) interface defect agglomeration grows with increasing thickness of NL,

(ii) there is no visible influence of this effect on the top layer crystalline quality.

Figs. 3a and b show SIMS profiles recorded for structures shown in Figs. 1. a) and b), respectively. SIMS profile in Fig. 3a clearly demonstrates that there is an enhanced oxygen incorporation into GaN at low temperatures. Oxygen concentration in the middle layer deposited at 700°C contains about two orders of magnitude higher oxygen concentration than the HT GaN layers. SIMS profile showed on fig. 2 b) confirmed this behavior. First layer deposited at 1140°C is characterized by the lowest oxygen concentration. With continuing growth at decreasing temperature oxygen concentration in the subsequent GaN layers increase. Finally, the layer deposited on the top of the structure at 600°C contains the highest oxygen concentration.

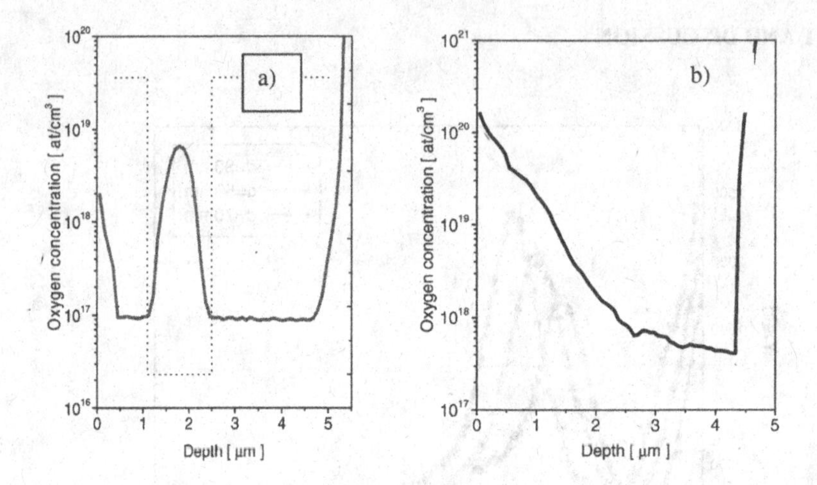

Fig. 3. Oxygen depth profiles recorded with SIMS for structures shown in Figs. 1. a and b, respectively.

Fig. 4 shows the oxygen depth profile for GaN buffer layers deposited on GaN NL with various thickness. Oxygen concentration in GaN buffer layers increases with increasing thickness of GaN NL. Although the effect is not so pronounced as in the case of growth temperature the overall tendency seems to be confirmed.

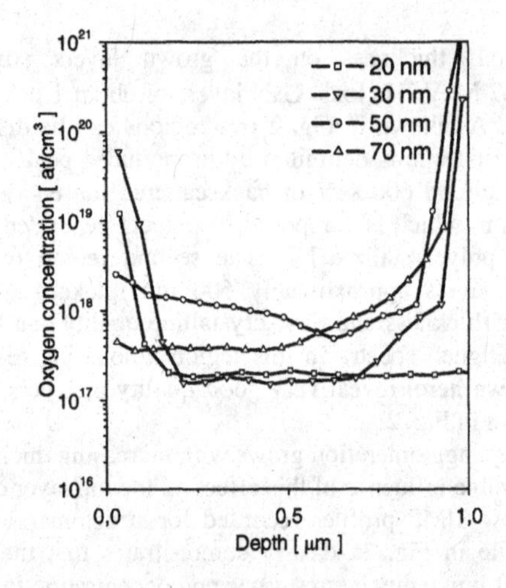

Fig. 4. Oxygen depth profile recorded with SIMS for GaN buffer layers deposited on GaN nucleation layer with various thickness.

Growth temperature has an important influence on properties of GaN epitaxial layers. At low temperatures oxygen incorporation into GaN is facilitated. Most probably the source of oxygen are the trace contaminations of nitrogen precursor: ammonia. Our results confirm observations of Chung and Gershenzon [5] that oxygen can be easily incorporated into GaN at low the temperatures.

Another source of oxygen is the low temperature NL. Our RBS/channeling analysis demonstrated that crystalline quality of the GaN buffer layer/GaN NL interface depends on the thickness of GaN nucleation layer: with increasing thickness of NL the crystalline quality of the interface region strongly deteriorates. This is reflected by the size of defect agglomeration in the vicinity of interface (cf. fig. 2). Also oxygen concentration in HT GaN layer increases with the NL thickness. Since the interface region is the most defective part of a GaN layer it provides a variety of easy paths for oxygen diffusion during the high temperature growth. This is supported by observations of Pearton et al. [11] who proved that the expanded regions around dislocations are easy paths for impurities diffusion. As a consequence, GaN buffer layer deposited on the thickest GaN NL is characterized by the highest oxygen concentration (cf. Fig. 4). One can conclude that nucleation layers deposited at low temperature are effective source of oxygen in GaN buffer layer. At high temperatures large amounts of oxygen incorporated in GaN NL can easily diffuse into GaN buffer layer during its growth.

SUMMARY

In summary, we have studied influence of the growth temperature on oxygen incorporation into GaN layer. Two structures consisted of GaN layers deposited at various temperatures were grown. Layers deposited at lower temperatures were characterized by the higher oxygen concentration. The most probable oxygen source is due to the contamination of precursor gases. Analysis of GaN buffer layers grown on GaN NL with various thickness revealed the presence of large defect agglomerations in the vicinity interface GaN buffer layer/GaN NL that grows with increasing NL thickness. They form easy paths for diffusion of oxygen accumulated in NL. Consequently, oxygen concentration in GaN layers increases with increasing NL thickness.

REFERENCES

[1] C. G. Van de Walle, J. Neugebauer, J. Appl. Phys. **95**, 3851 (2004).

[2] M.E. Hawkridge, D. Cherns, Appl. Phys. Lett. **87**, 221903 (2005).

[3] M. E. Hawkridge, D. Cherns, T. Myers, Appl. Phys. Lett. **89**, 251915 (2006).

[4] G. Popovici, W. Kim, A. Botchkarev et al., Appl. Phys. Lett. **71**, 3385 (1997).

[5] B-C.Chung, M. Gershenzon, J. Appl. Phys. **72**, 651 (1992).

[6] Z. Liliental–Weber, Y. Chen, S. Ruvimov et al., Phys. Rev. Lett. **79**, 2835 (1997).

[7] D. Cherns, C. G. Jiao, H. Mokhtari, J. Cai, and F. A. Ponce, phys. stat. sol. (b) **234**, 924 (2002).

[8] E. Dumiszewska, W. Strupiński, K. Zdunek, J. of Superhard Materials **29**, 174 (2007).

[9] L.C. Feldman, J.W. Mayer: Ion Channeling in Materials Science, Academic Press, New York 1984

[10] L. Nowicki, A. Turos, R. Ratajczak, A. Stonert, and F. Garrido, Nucl.Instr. and Meth. **B240**, 277 (2005).

[11] S. J. Pearton, H. Cho, J. R. LaRoche et al., Appl. Phys. Lett. **75**, 2939 (1999).

Conventional III-V Materials and Devices on Silicon

Mater. Res. Soc. Symp. Proc. Vol. 1068 © 2008 Materials Research Society 1068-C01-01

Formation of III-V Semiconductor Engineered Substrates Using Smart CutTM Layer Transfer Technology

Fabrice Letertre

R&D, Soitec SA, Parc technologique des Fontaines, Bernin, 38190, France

ABSTRACT

Engineered substrates are expected to play a dominant role in the field of modern nano-electronic and optoelectronic technologies. For example, engineered substrates like SOI (Silicon On Insulator) make possible efficient optimization of transistors' current drive while minimizing the leakage and reducing parasitic elements, thus enhancing the overall IC performance in terms of speed or power consumption.

In this paper, we will review wafer bonding and layer transfer technologies with a special emphasis on the Smart Cut™ technology applied to compound semiconductors. Beyond SOI, the innovation provided by substrate engineering will be illustrated by two main examples :

- Silicon and SiC engineered substrate serving as a platform for GaN and related alloys processing,
- Germanium/Si platform for the growth of GaAs/InP materials,

opening the path to monolithic integration of Si CMOS microelectronics with III-V optoelectronics functions.

INTRODUCTION

Integration by bonding of dissimilar materials for the formation of engineered substrates relies essentially on two main steps: wafer adhesion by molecular atomic bonding and thin film formation. Besides these process steps, two other key parameters need to be defined: the donor substrate, which provides the thin film to be transferred and the receiving final substrate on top of which the above thin film is physically integrated. The physical, chemical and electrical properties of both donor and final substrates need to be considered in order to properly design the targeted engineered substrate.

This elegant material integration technology has found interest in the microelectronic community in the beginning of 1980's when Toshiba and IBM research teams pioneered the formation of both bonded thick epitaxial structures and of Silicon On Insulator structures [1,2]. At that time, wafer bonding experiments became possible due to surface quality improvements of the silicon substrates (flatness, surface roughness and cleanliness). Currently, wafer bonding is a generic technological step involving relatively thick pieces of materials with specified surface conditions and bulk thermal expansion properties. Formation of the thin film (or device layer) raises, however, a certain number of technological questions. In the case of SOI formation, critical challenges are mostly film thickness and uniformity, as well as macro and microscopic quality and defectivity. Current SOI substrates require film thickness in the 100 nm range with thickness deviation not more than a few percent across a wafer and surface defectivity similar to silicon bulk substrates. Due to their intrinsic process limitations, initially developed Bonded SOI technologies (BSOI) have rapidly seen their interest being limited to applications requiring thick films (> 5 microns), such as power devices or MEMS. In the case of BSOI technology, the removal of the donor substrate is obtained by a combination of mechanical grinding and polishing, leading to film thickness uniformity challenges [3], and to the complete loss of one substrate, thereby adding cost penalty to this approach. In the case of thin SOI substrates, attempts to solve the thickness and quality issues have led to the development of Bond and Etch technologies (BESOI), in which the final film thickness is determined by an etching step. While SEH company teams explored film thickness etching and control with a dry etching process called PACE (Plasma Assisted Chemical Etch) [4], the use of etch stop layers between the device layer and the substrate has been considered by others. Among the different etch-stop studies, the most popular solutions include Boron [5], Germanium [6], Carbon [7] in the form of continuous films or distributed impurities. Due to selectivity and / or defectivity issues, these solutions were not able to generate satisfying technological solution for the formation of high quality thin SOI structures. In the field of III-V materials, the etch stop benefit, resulting from heteroepitaxial mismatched growth of different compounds, has led to the demonstration of physical transfer of III-V films onto a foreign substrate [8] as well as the formation of GaAs on Silicon [9] and InP on Silicon structures [10] using respectively AlAs and InGaAs as etching or etch stop layers. Again, epitaxial substrate loss is in most cases an economic limitation to the industrial dissemination of these technologies.

The Smart Cut™ technology, introduced in the mid 1990's by M. Bruel [11] is a revolutionnary and powerful thin film technology that solves both thin film formation issues as well as the cost aspect owing to the donor substrate preservation. This is to date the only thin film technological solution that brings industrial maturity to engineered substrates. This technology involves a combination of wafer bonding and layer transfer via the use of ion implantation. It allows multiple high quality transfers of thin layers from a single crystal donor wafer onto other substrates, and it allows integration of dissimilar materials. The Smart Cut™ technology has become today the industrial standard for thin layer transfer [12]. It consists in defining a splitting region within the donor substrate by ion implantation (e.g. H, He) which allows a thin film to transfer to a handle wafer after bonding and splitting. The first industrial implementation of the Smart Cut technology is SOI manufacturing. The simplicity of the concept is schematically illustrated in figure 1.

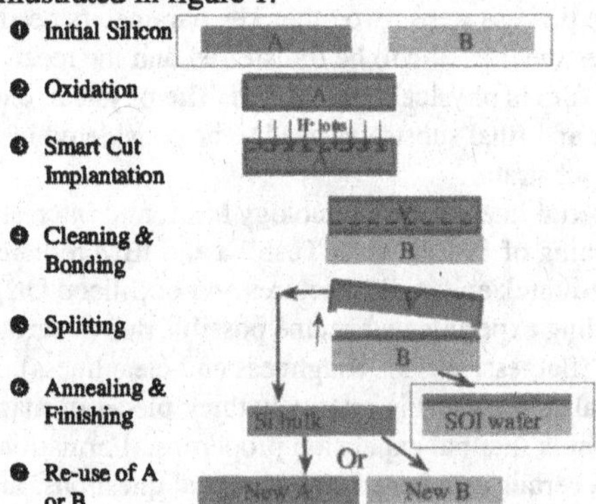

Figure 1 : Schematic Smart Cut process flow.

First, a thermal oxide is formed on the donor wafer, followed by hydrogen implantation with doses typically in the mid $10^{16}cm^{-2}$ range. The implantation energy determines the thickness of the transferred layer, while the oxide thickness fixes the buried oxide (BOX) thickness. After cleaning and surface activation, the donor and handle wafers are bonded together. By splitting at the H implanted region, a thin film is transferred to the handle substrate. Subsequent steps remove the post-splitting surface roughness. Furthermore, the thickness removed from the donor wafer is negligible compared to the total wafer thickness. Consequently, the donor wafer can be reused many times. A range from 10 nm up to a few 10^3 nm in top Si and BOX thickness is easily covered by this technology with standard industrial implanters and furnaces.

The Smart Cut technology is a powerful tool that is applicable to many materials, making it possible to create a wide range of composite substrates and their tailoring to the requirements of the application by properly choosing the active layer, the buried dielectric and the base substrate. As a consequence, it opens a path to the formation of III-V based engineered substrates by integrating, for example, materials like GaAs [13], InP [14], SiC [15], GaN [16], Germanium [17], and Si [18] on a silicon, poly-SiC, sapphire, ceramic, or metal substrates. Compared to previous Bond and Etch technologies, the Smart Cut technology provides a large number of dissimilar integration options, solving at the same time high quality thin film formation and substrate removal issues, while leading to a high volume cost-effective solution.

Beyond SOI technology, the integration of high quality III-V materials on silicon or related substrates is expected to enable a variety of new opportunities in the field of, for example, GaN and related alloys, power, RF and optoelectronic devices that are partly covered by existing epitaxial substrate technologies, as well as in the field of digital CMOS where III-V materials are envisioned as potential candidate for channel material replacement. The following sections of this paper review technical achievements in the area of III-V integration obtained by using the Smart Cut technology to fabricate high quality films of wide band gap materials (SiC, GaN) and Germanium on top of silicon based substrate platforms.

EXPERIMENTS

Si bonding and layer transfer onto polySiC (SopSiC™)

Bulk silicon carbide, sapphire and silicon (111) materials are today used as epitaxial substrates for high quality GaN growth due to their crystalline parameters, physical and chemical properties. In case of large volume applications, bulk SiC wafers suffer from high price, small wafer diameter and lack of industrial availability. In contrast, bulk silicon substrates are perfectly suited for large area applications and low cost solutions.

Considering in particular high power, high frequency microelectronics, one of the main targeted applications here, silicon material is not as good as SiC for thermal dissipation, which is a key figure of merit. Combining these two materials (i.e. silicon and SiC), might be the ideal substrate solution for applications where high device performance and low cost are mandatory. Silicon on insulating, high thermal resistivity poly SiC substrates (SopSiC™) is being investigated to meet such a goal. As shown by thermal simulation results [19], SopSiC substrates demonstrate a strong potential for high power, high frequency applications. The

SopSiC process flow differs slightly from the SOI process flow since the base substrate is a polycrystalline SiC substrate (poly SiC) instead of a monocrystalline silicon substrate. A possible process flow is summarized in figure 2. The starting materials are a high resistivity (111) silicon wafer (HR Si) and a poly SiC substrate. Both are available in different sizes, from 2 inches to 12 inches. Poly-SiC substrates can be made highly resistive (>10^5 Ω.cm), with thermo-mechanical characteristics close to single crystal SiC. Silicon substrate can also be quite highly resistive (above 10^4 Ω.cm

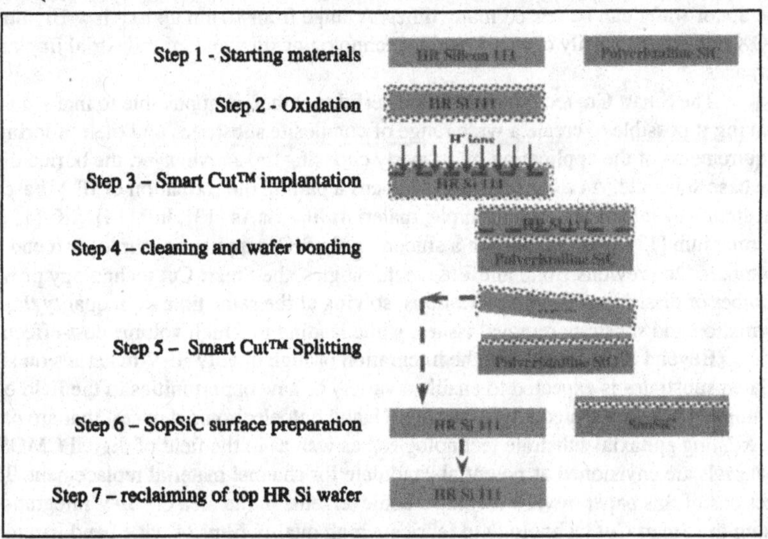

Figure 2 : process flow applied to SopSiC substrate realization

The typical process flow is thermal oxidation and hydrogen implantation of the HR Si substrate, followed by surface preparation and room temperature wafer bonding between HR Si and poly SiC substrates. After wafer bonding, the silicon film is transferred to the poly-SiC base wafer, resulting in a SopSiC substrate. Additional treatments can follow to stabilize the structure and prepare the silicon surface of SopSiC to be epi-ready for the subsequent III-Nitride epitaxy. Typical process parameters are close to standard SOI making possible SopSiC fabrication in an SOI fabrication line. The high resistivity silicon substrate can be reclaimed for subsequent layer transfer.

SiC bonding and layer transfer (SiCOI) :

Silicon carbide bulk material is well suited for the growth of high quality GaN material. However, SiC mechanical hardness and chemical inertia make it difficult to remove these substrates during fabrication of thin film LED structures that are needed for optimum device performance. Bulk SiC substrate removal, although feasible by mechanical and etching means, may greatly increase the cost of the technology. .An alternative is to use a Silicon Carbide On

Insulator (SiCOI) engineered substrate, where a thin single crystal SiC film is bonded onto a removable substrate like silicon and used as a seed layer for further GaN epitaxial regrowth.

<u>SiCOI fabrication</u>

SiCOI engineered substrate was demonstrated by Di Cioccio et al. [19]. Hydrogen implantation conditions were studied extensively to determine splitting process window with different SiC poly-types (6H, 4H, 3C). The effects of main implantation parameters were studied, such as dose, implantation energy and temperature dependency. Dose recommended for SiC splitting is the same range as for silicon, typically between 5e16 and 1e17 H^+/cm^2 depending on the implant energy. Contrary to GaAs or InP, implantation temperature window is quite wide for SiC: it has been reported that H^+ implantation can be performed from 50°C up to 900°C and be efficient to get blistering and fracture [20]. Splitting behavior is similar for different SiC polytypes (6H, 4H and 3C) and doping level (p type, n type and highly resistive). In each case, blistering is observed after implantation and specific annealing treatment depending on the implantation conditions. Defects and micro-cavities created by H^+ in SiC, and their subsequent evolution with annealing, were studied by TEM by Grisolia and al. [21]. It has been found that micro-cavities (also called platelets) nucleation and growth follow an Ostwald-Ripening law before leading to micro-cracks formation and ultimately to crystal splitting. This behavior is similar to what has been observed and recently quantified in the case of silicon [22].

SiC wafer bonding and layer splitting onto a foreign substrate like silicon or poly SiC is demonstrated with no diameter limitation foreseen. As for the case of silicon transfer onto polySiC, the process flow used for making SiCOI is very close to the one for SOI, meaning that SiCOI substrates can be processed using SOI substrate manufacturing lines. As for the SOI case, SiC wafers can be reclaimed many times for subsequent layer transfer, optimizing the use of those materials that are particularly difficult to grow.

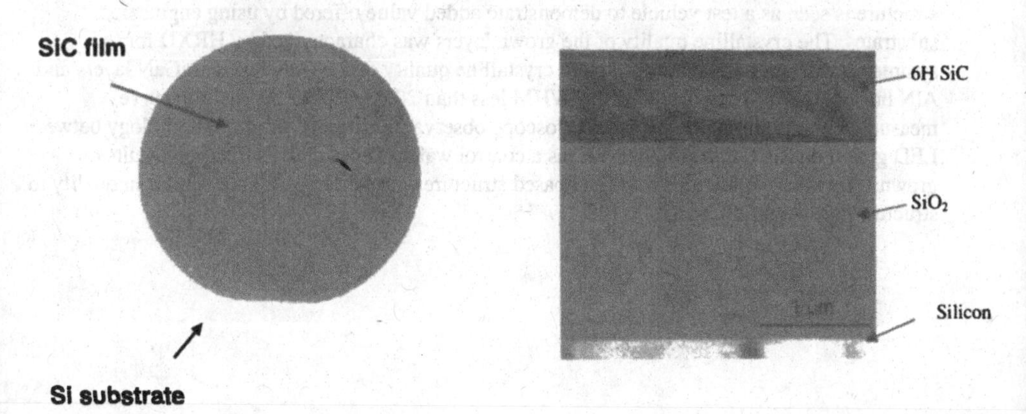

Figure 3 : picture of a 2 inch SiCOI substrate (left) and corresponding cross-section TEM analysis (right).

These substrates are compatible with the formation of high quality GaN layers. First GaN growth on SiCOI was obtained in 2002 [23]. This early Molecular Beam Epitaxy (MBE) result was confirmed by Lahrèche et al. who performed Gas Source MBE growth of an AlGaN / GaN HEMT structure onto SiCOI, with crystalline and electrical results similar to those obtained on bulk SiC [24]. Dikme and al. [25] presented results of GaN epitaxy by Metal Organic CVD (MOCVD) on SiCOI on a silicon substrate. In this case, a high temperature AlN buffer layer was used to start the growth. Some AlGaN interlayers were used to control built-in stress and GaN layers up to 3μm thick were grown despite the CTE (Coefficient of Thermal Expansion) mismatch with silicon substrate used as the base substrate for the SiCOI substrate. PL measurements confirmed good GaN crystalline quality, showing a FWHM of 5meV. Typical FWHM values measured by HXRD in a rocking curve geometry (0002) are as low as 300 arc-sec, confirming the PL measurement result about good structural quality of GaN layers grown on SiCOI. Based on those results, Fieger [26] et al. demonstrated GaN based HEMT structures with good performance. These encouraging results for MBE and MOCVD GaN epitaxy on SiCOI opened the door to GaN based LED and HEMT structure epitaxy and processing.

More recently, Meneghesso et al. and V. Hoel et al. [27, 28] have published very promising characterization results of AlGaN / GaN HEMT epitaxial stacks as well as device results using SiCOI and SopSiC structures. These DC, pulsed, and power RF electrical results indicate a possibility of achieving a low cost, large area substrate solution.

Device demonstration using engineered substrates : GaN based thin film LEDs

The work of Dorsaz et al. [29] who demonstrated the fabrication of InGaN / GaN thin film LEDs bonded onto a silicon substrate, using a SiCOI wafer, illustrates the benefits of using engineered substrates for the sake of improving device performance and facilitating device process integration. Conventional InGaN / GaN LED stack were grown in a MOCVD reactor. The grown stack and related optical characterization are described in figure 4. Although the grown stack was not optimized from device performance point of the view, such device and structure is seen as a test vehicle to demonstrate added value offered by using engineered substrates. The crystalline quality of the grown layers was characterized by HRXD for symmetric diffraction plan (0002). High crystalline quality of the GaN layers, InGaN layers and AlN buffer layer was confirmed with FWHM less than 200arcsec for the rocking curve measurement. Additionally, optical microscope observation showed similar morphology between LED grown on SiCOI and on bulk SiC as a control wafer. Those characterization results on grown LED layers confirmed that GaN based structures grown on SiCOI are similar in quality to structures grown on SiC.

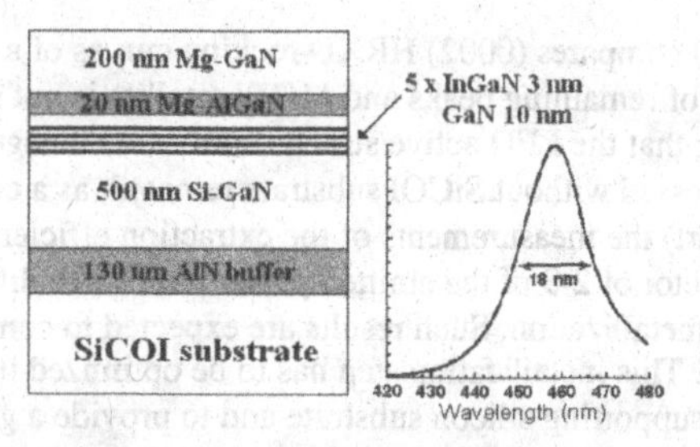

Figure 4 : Cross section schematic of the LED structure (left) and room temperature PL spectrum of the InGaN quantum wells (right)

The LED processing is described in figure 5. Semi-transparent Ni/Au structure (5/5 nm) annealed under O_2 at 600°C for 5 minutes was formed as full surface p type electrical contact. Au layer was deposited on p type contact formed onto GaN LEDs structures and on low resistivity n type silicon substrate. A sealing process at 400 °C under pressure was used to create SiAu alloy as a conductive metallic interface, in order to transfer the full wafer LED stack. The first structure comprises a simple SiAu contact (type 1), while the other two structures comprise an Al mirror directly in contact with p-type contact (type 2) or with W/WN barrier diffusion (type 3). After metallic sealing, SiCOI substrates were easily removed by mechanical grinding and wet etch for Si and SiO2, and by RIE dry etch (standard Cl_2 / Ar chemistry) for SiC and AlN buffer layer. InGaN/GaN LEDs structure transfer at full wafer level was then achieved with good homogeneity and no cracks observed.

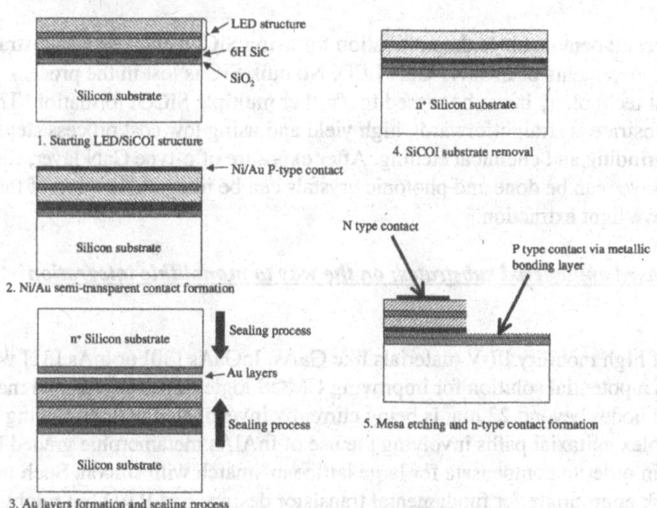

Figure 5 : Schematic of thin film LED processing starting from SiCOI substrate

Figure 6 (left) compares (0002) HRXD rocking curves of a LED layer before and after the transfer. FWHM of remaining peaks and MQW satellite is not affected on transferred structure, confirming that the LED active structure was not damaged by this process. An LED device was also processed without SiCOI substrate removal, as a control wafer. As shown on the same figure (right part) the measurements of the extraction efficiency of the devices show an improvement by a factor of 2-3 of the emitted power with some differences depending on the nature of the buried metallization. Such results are expected to come from an improved extraction efficiency. This metallization step has to be optimized in order to prevent optical absorption from the supporting silicon substrate and to provide a good light reflector.

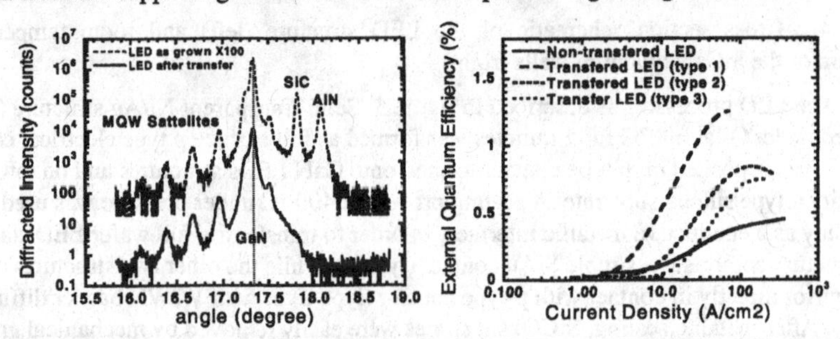

Figure 6 : (left) HXRD rocking curves comparison of the LED, before and after the transfer, (right) External Quantum Efficiency of transferred and non-transferred LEDs

This result demonstrates the motivation for using SiCOI engineered substrate for efficiency improvements of InGaN / GaN LED. No bulk SiC is lost in the process, and thanks to the Smart Cut technology it can be reused for further multiple SiCOI formation. The removal of the silicon substrate is straightforward , high yield and using low cost process steps like mechanical grinding and chemical etching. After exposure of n-type GaN layer, formation of n-type contact layer can be done and photonic crystals can be integrated on top of the surface to further improve light extraction.

SOI on Ge based engineered substrates: on the way to monolithic integration

Integration of high mobility III-V materials like GaAs, InGaAs [30] or InAs [31] with silicon substrates, as a potential solution for improving CMOS logic transistor channel material for technological nodes beyond 22 nm, is being currently investigated with promising device results. Complex epitaxial paths involving the use of InAlAs metamorphic graded buffers are investigated in order to compensate for large lattice mismatch with silicon. Such material platforms look appropriate for fundamental transistor designs and R&D but might not be adequate for monolithic integration and manufacturing as cointegration with silicon transistors and circuitry do not look straightforward in such configuration. One step towards effective cointegration is described in the work of Dorhman et al. [32]. An SOI stack is layer transferred by the Smart Cut technology on top of a Ge-rich SiGe alloy, providing what is described as silicon on lattice-engineered substrate (SOLES). By local etching of the SOI film and localized epitaxial growth, the germanium is expected to serve as a seed for the growth of high quality GaAs film. This work is capitalizing on all the efforts being made in the past years to grow high quality SiGe graded buffers on silicon [33]. As the SOI stack is in the range of 100 nm above the seed and can be easily adjusted, thick SiGe buffer induced topology (couple of microns) is no longer an issue in making surfaces of a silicon layer and of a III-V layer co-planar. This approach might, however, suffer from the underlying quality of the seed on top of the graded buffer. From a technology point of view, the presence of the thick graded buffer and its related wafer bow might be a processing challenge during the wafer bonding step as well as during device processing lithography steps, as the topology of the stack will be governed by the properties of the underlying substrate. One possible alternative to keep the stack as thin as possible, while improving the quality of the Germanium seed film and minimizing technology issues, is to develop a structure with SOI on top of thin Ge-on-silicon wafers with no thick buffer. The elimination of the thick buffer allows to reduce thermo-mechanical stresses in the whole stack during thermal processes as well as the related potential thermal barrier. Such structure uses, for instance, a GeOI (Germanium On Insulator) substrate below the SOI stack. Using such approach, the SOI on GeOI wafer will look very much like a conventional SOI wafer, thus minimizing issues for its processing.

Large area, high quality GeOI substrates have been demonstrated previously [34,35]. Layer transfer of both bulk and epitaxial Ge/ Si donor wafers were demonstrated up to 200 mm, with a possibility of moving to even larger diameters, thus showing a potential path for CMOS logic applications. GeOI have been identified as an enabling substrate platform for Ge PMOSFET development, with co-integration with a III-V material being possible. Processing of GeOI can be made in a state-of-the-art device foundry, avoiding the intrinsic limitations of bulk germanium substrates (price, availability and mechanical fragility). Similar to silicon studies, the formation and evolution, upon thermal treatments, of hydrogen implantation induced extended defects has been studied [36]. GeOI wafers are processed following a conventional SOI process flow with slight adaptations due to the presence of the germanium material. For instance, hydrogen implantations and splitting parameters are adjusted. As described in [34], the nature of the donor wafer (bulk versus on-axis and off-axis epilayers) has no influence on the process conditions. After completing the process, germanium thin films are of high quality and mostly defect free. For instance, threading dislocation densities (TDD) counted using both TEM and defect decoration techniques are below 1E7/cm2, with best results in the 1E4/ cm2 range being obtained with high quality bulk donor wafer. Such high quality is highlighted in figure 7. A generic advantage provided by the Smart Cut technology is that the donor wafer (bulk or epi) can be reclaimed and reused for subsequent layer transfer operation. This unique possibility makes this technology very attractive for future industrial use.

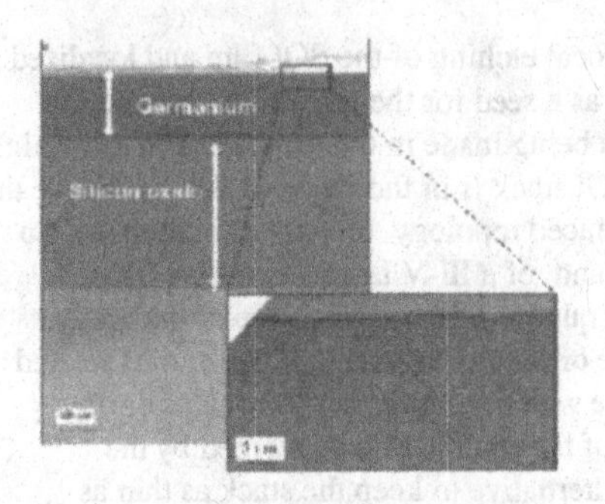

Figure 7 : example of a TEM cross section of GeOI structure. On the left is the Ge / SiO2 / Silicon stack with a HRTEM cross section of the Germanium film and BOX interface. On the right is a photograph of a 200mm GeOI substrate.

Besides digital applications, potential of GeOI wafers has been demonstrated by S.G. Thomas et al. [37] in the field of microwave analog applications. GeOI served as a seed substrate, with on purpose 6° off cut Ge for the growth of high quality InGaP Heterojunction Bipolar Transistor (HBT) structures. Functional transistors on GeOI were found as good as control HBT grown on bulk GaAs and Ge, meaning that epitaxial quality obtained on GeOI was comparable with state-of-the-art MOCVD growth on bulk materials. One interesting feature of GeOI was its lower substrate thermal resistance compared with bulk Ge and GaAs, as indicated by improved electrical characteristics during operation at high power. The decrease of the self heating effect observed in GeOI is attributed to the higher thermal conductivity of silicon compared to Ge and GaAs materials (1.5 W/cm.K versus 0.6 and 0.46 W/ cm.K respectively), with no significant effect of the buried oxide between the Ge thin film and the substrate. This result is the first demonstration of high performance InGaP HBT device grown onto Ge based engineered substrate. More recent direct growth of III-V devices on GeOI is reported by K. Herrick et al [38]

As the quality of the GeOI seed substrate is good enough, SOI formation on top of it has been demonstrated using wafer bonding and layer transfer technique, thus forming the so-called SOLES structure. Due to the presence of buried germanium, the maximum processing temperature is limited around 900°C to avoid germanium melting. The corresponding structures are firstly developped in 100 mm diameter with no other limitation than GeOI and SOI substrates availability to move to larger diameter (200 and 300mm) substrates.s. The SOI on GeOI TEM cross section is shown in figure 8.

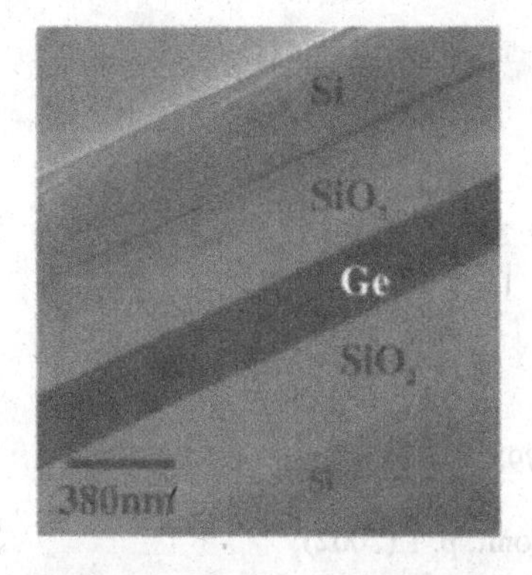

Figure 8: example of a TEM cross section of a SOI on GeOI structure (left : courtesy G. Fitzgerald, MIT). At right is a photograph of a 100 mm diameter SOI on GeOI substrate.

The principle of this multi stack structure can be extended to materials other than germanium (for instance GaAs and InP) to form a high quality seed. Also thermal and electrical properties of each material can be tuned to the specific application.

CONCLUSIONS

The Smart Cut technology opens the path for new structures of engineered substrates for the integration of III-V materials with silicon. Depending on the applications, the integration scheme can be tuned owing to the flexibility of the Smart Cut technology. Besides the generic competitive advantage offered via the multiple use of the donor wafers, new functionalities can be addressed through the unprecedented possibility of material combinations. In the field of high performance LEDs, thin film devices have been demonstrated using SiC based engineered substrates. For power RF GaN applications, silicon on poly SiC (SopSiC) structures show encouraging perspectives of meeting both device performance and cost targets. The extension of the concept of thin film engineering offers a large number of possibilities for the monolithic integration of III-V with the mainstream CMOS logic. SOI on engineered substrates allows potential integration of III-V materials on a silicon substrate, making its further processing affordable in a standard infrastructure, and also enabling the co-integration, on a same platform, of silicon and III-V devices. This engineered material platform might find applications in the field of the System On a Chip.

ACKNOWLEDGMENTS

The work reviewed here is the result of the effort of many teams at different companies, institutes and universities. Special thanks in particular to Bruce Faure, Nicolas Daval, Robert Langer, Ian Cayrefourcq and Carlos Mazuré from SOITEC, Chrystel Deguet and Laurent Clavelier from LETI/CEA, G Fitzgerald from MIT, Nicolas Grandjean from EPFL.

REFERENCES

1. M. Shimbo, et al. JAP, 60, 2987 (1986)
2. JB Lasky, APL, 48,78 (1986)
3. S. Blackstone, Conference proceedings, Vol. 95-7, 1995, p. 56
4. T. Abe et al., Conference Proceedings, Vol. 92-7, ECS 1992, P. 200
5. H Seidel et al. Journal of ECS, 137, 3626 (1990)
6. D. Godbey et al., APL, 56, 373 (1990)
7. V. Lehmann et al., Journal of ECS, 138, L3 (1984)
8. E. Yablonovitch et al., APL, 56, 2419 (1990)
9. JF Klem et al. , JAP, 66, 459 (1989)
10. H. Wada et al, conference proceedings, Vol. 95-7, 1995, p. 579
11. M. Bruel et al., Electron. Lett., vol 31, p. 1201 (1995)
12. A.J. Auberton-Hervé and C. Maleville, IEEE Int. SOI Conf., p. 1 (2002).
13. E. Jalaguier et al., Electron. Lett., 34(4), 408 (1998)
14. E. Jalaguier et al. Proc. llth Intern. Conf. on InP and Related Materials, Davos, Switzerland, (1999)
15. L. Di Cioccio et al., Mat. Sci. and Eng. B Vol. 46, p. 349 (1997)
16. A. Tauzin and al., Semiconductor Wafer Bonding VIII, ECS Proc Vol. 2005-02, pp. 119-127
17. F. Letertre, et al. MRS Symp. Proc., **809**, B4.4 (2004).
18. B. Faure et al., Semiconductor Wafer Bonding VIII, ECS Proc Vol. 2005-02, pp. 106-118
19. L. Di Cioccio et al. , Mat. Sci. and Eng. B Vol. 46, p. 349 (1997)
20. Q.-Y. Tong, U. Gösele, *Advanced Materials,* **11**, pp. 1409–1425 (1999).
21. J. Grisolia, *MRS spring meeting proceedings*, **C3** (1999)
22. S. Personnic et al., JAP 103, 023508 (2008)
23. F. Letertre et al. phys. stat. sol. (c) 0 (7), pp.. 2103 – 2106 (2003)
24. H. Lahrèche et al., Mat. Sci. For., Vols. 457–460 pp.. 1621 – 1624 (2004)
25. Y. Dikme et al., Journal of Crystal Growth, v.272 (1-4), pp. 500-505 (2004)
26 M. Fieger and al., *phys. stat. sol. (c)*, **2 (7)**, pp. 2607 – 2610 (2005)
27. G. Meneghesso et al , IEDM 2007, to be published
28. V. Hoel et al., Electronics Letters, 31st of january 2008, Vol. 44, N°3
29. J. Dorsaz and al., Proceedings, ICNS6 (2005)
30. M.K. Hudait et al., pp. 625-628, IEDM 2007
31. DH Kim et al., pp. 629-632, IEDM 2007
32. C.L. Dohrman et al., Materials Science and Engineering B (135) 235-237
33. E.A. Fitzgerald et al. , APL 59, 93 (1991) 811
34. F. Letertre et al, Mat. Res. Soc. Proc. 809, B4.4, p.153 (2004)
35. C. Deguet et al, ECS Proc., Vol.2005-06 (2005)
36. Akatsu et al., APL 86, 181910 (2005)
37. S.G. Thomas et al., IEEE EDL Vol. 26, July 2005.
38. K.J. Herrick, MRS spring meeting 2008, symposium C, to be published

Mater. Res. Soc. Symp. Proc. Vol. 1068 © 2008 Materials Research Society 1068-C01-06

Transistor Level Integration of Compound Semiconductor Devices and CMOS (CoSMOS)

Kenneth Elliott, Pamela Patterson, James C. Li, Yakov Royter, and Tahir Hussain
HRL Laboratories, LLC, 3011 Malibu Canyon Road, Malibu, CA, 90265-4797

ABSTRACT

In the COSMOS program, HRL Laboratories, LLC. is developing technology for intimate integration of CMOS devices with 400 GHz InP HBTs to form complex integrated circuits. This research is investigating innovative approaches to the transistor-scale integration of compound semiconductor and silicon-based transistors so as to enable revolutionary advances in science, devices, circuits, and systems.

INTRODUCTION

The DARPA COSMOS program is developing new methods to tightly integrate compound semiconductor technologies within state-of-the-art silicon CMOS circuits in order to achieve unprecedented circuit performance levels. In the CoSMOS program, HRL Laboratories, LLC. is developing the process technology to enable intimate integration of CMOS devices and 400 GHz InP HBTs with a goal of demonstrating complex mixed signal integrated circuits with a figure-of-merit up to 100X better than currently obtainable.

Currently, heterogeneous integration of compound semiconductor devices with silicon is achieved through the use of multi-chip modules and similar large-scale assemblies. While adequate for relatively low performance applications (e.g., power amplifiers for cellular telephone handsets), the integration complexity and performance that can be achieved in this manner is extremely limited. The COSMOS program will drive the size-scale at which heterogeneous integration can occur down to that of individual transistors. The net result will be a new integrated circuit infrastructure that can exploit CMOS complexity and III-V performance at a sustainable price.

APPROACH

The HRL approach to the integration of CMOS and InP HBTs is based on aligned die bonding of semiconductor materials and devices onto partially processed Si CMOS obtained from commercial foundries. HRL's approach offers outstanding overlay accuracy, solves thermal expansion and stress issues, and maximizes connectivity between CMOS and InP transistors. The approach will also provide scalability as feature sizes shrink to sub-100 nm and permit rapid technology readiness for emerging technologies into complex integrated circuits. The HRL CoSMOS approach has great potential to solve common issues that could limit transistor level integration. These include:

- Degradation in material quality,
- Limitation to a single material,
- Wafer size incompatibility,
- Poor overlay accuracy (microns vs. nanometers),
- Low yield for complex ICs, and
- Ability to use of Si design cell libraries and design infrastructure.

The HRL approach retains all portions of the CMOS process up to a level that still permits transistor level integration. The constraints are that the bonding surface 1) is reasonably planar over the bonding area and 2) that the wiring can permit a wiring pitch dense enough to connect the InP and CMOS transistors without degradation.

HRL is using its 400 GHz InP HBT fabrication process (1,2) to produce devices for CoSMOS and is currently using a 130 nm CMOS process as a basis for CoSMOS development. In the initial stages of the CoSMOS program, HRL is refining this approach to support a "double flip" process flow that eliminates the need for the complex device material development that would be required to grow layers upside-down in the "single flip" process. HRL had demonstrated essential elements of a single flip material die bonding process using BCB as a bonding material as shown in Figure 2 prior to the start of the CoSMOS program.

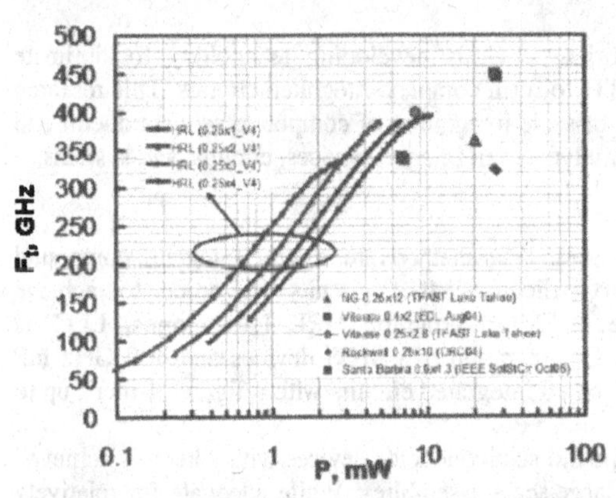

Figure 1 Unity gain cutoff frequency vs. power dissipated per device for HRL 250 nm device family and best published results.

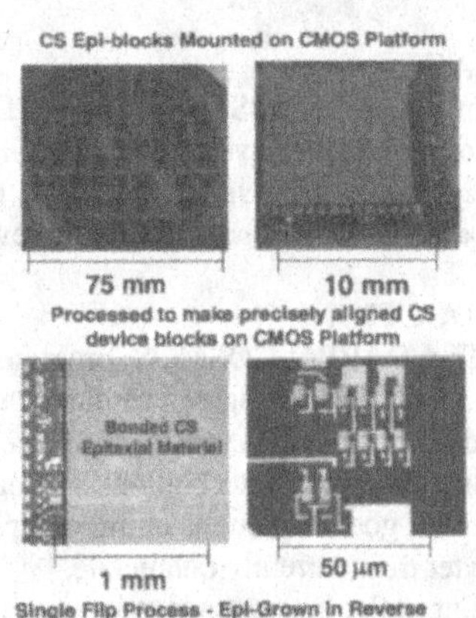

Figure 2 Photos of epi blocks bonded, thinned and patterned with overlay accuracy of 250 nm

In the HRL MatBond "double-flip" process, a very thin epi layer is boned using a temporary BCB adhesive to a silicon handle wafer. The InP substrate is removed leaving active device layers on the order of 1000-2000 nm in thickness as illustrated in Figure 3.

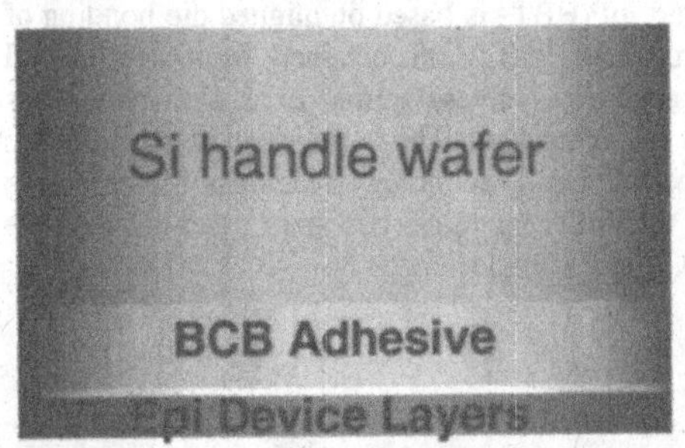

Figure 3 Diagram illustrating "double flip" layer stack

Figure 4 Photo of die -bonded wafer prior to fabrication into large-area HBTs.

The handle wafer is processed into tiles suitable for bonding to a partially fabricated Si wafer. By using a tiled approach, any Si wafer size can be accommodated provided that the InP equipment can handle it. An example is shown in Figure 4. A variation of the process, called DeviceDef, utilizes epilayers processed into transistors prior to bonding. In contrast to the MatBond process, the DeviceDef variation quires a highly aligned bond, but may have yield and other economic advantages. After bonding and BT fabrication, additional process steps provide via and metallization to connect CMOS and InP ansistors. A diagram illustrating a cross section of the final processed structure is shown in Figure 5.

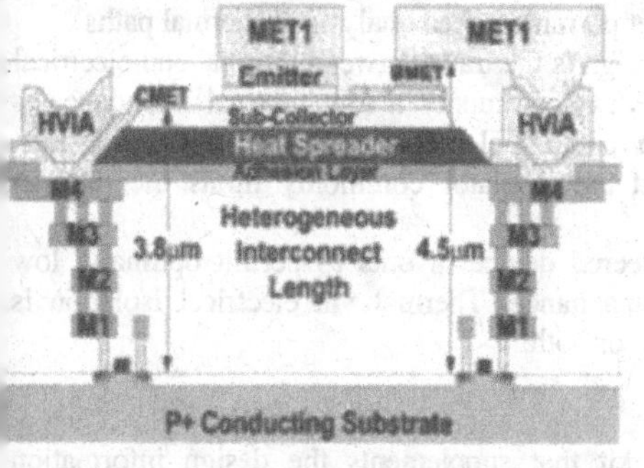

igure 5 Cross sectional diagram illustrating the ıal HRL COSMOS transistor connections

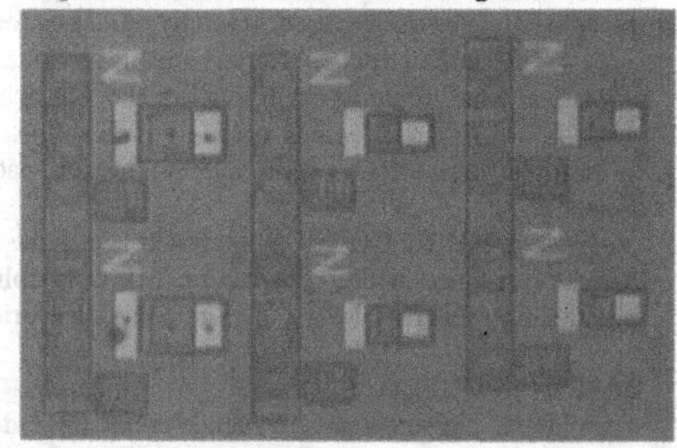

Figure 6 Photos of large area HBTs fabricated on die-bonded material.

BT FABRICATION ON Si

HRL has successfully fabricated large area transistors on Si wafers using the MatBond process. A ıoto of large area transistors is shown in Figure 6 and a forward Gummel plot is shown in Figure 7. The ɔrward Gummel plot exhibits excellent current gain but some minor base-collector leakage. This leakage not expected to have an impact on the designs of interest using CoSMOS technology which typically ɔuld operate at high bias currents.

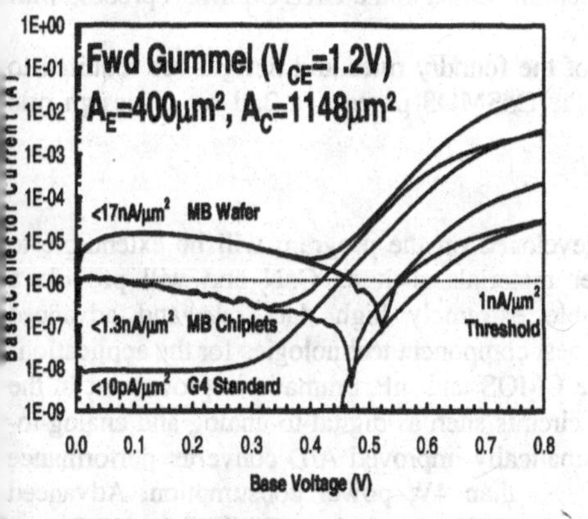

gure 7 Electrical data from large area HBTs bricated on die-bonded material.

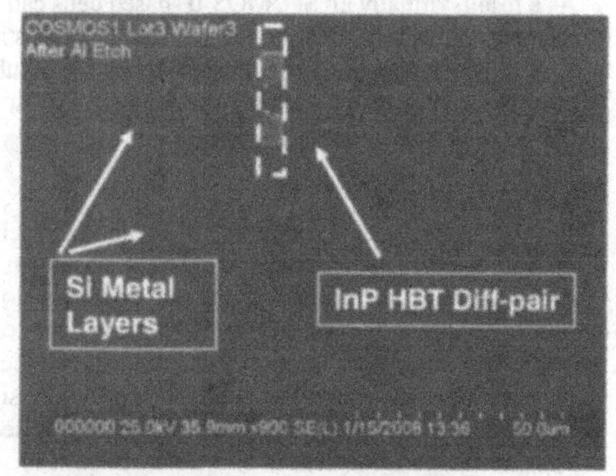

Figure 8 SEM Photo of a pair of InP HBTs bonded to a CMOS wafer.

We are now working to demonstrate small area transistors on CMOS circuit wafers using a atBond "double flip" process. An SEM photo shown in Figure 8 shows a pair of 250 nm emitter HBTs ed in a differential amplifier design that incorporates both InP and CMOS transistors. The CMOS vices are underneath the remaining unetched epi. Etch stops are used strategically to prevent the III-V ocessing from attacking the underlying CMOS circuitry. Additional metal and via layers are used to nnect the Si and InP devices.

THERMAL MANAGEMENT

The materials used for adhesion of very thin epilayers are normally poor thermal conductors. To remedy this limitation efficient thermal management is being obtained through the use of novel heat spreading and thermal designs driven by modeling tools that permit detailed analysis of thermal paths.

A dilemma presented for any method of thermal design is the trade between thermal and electrical properties. This dilemma is driven by the physical fact that commonly a low thermal impedance is obtained by using conducting materials such as metals. To maintain electrical isolation, an insulator must be inserted somewhere within the conduction path and the insulator commonly limits the thermal resistance.

In the HRL COSMOS approach, we have engineered device layouts to permit optimally low thermal resistance while maintaining a low coupling capacitance. Thermal via electrical isolation is provided by a reverse-biased junction, a thin dielectric layer or both.

DESIGN INTERFACE

HRL has created a Cadence-compatible design kit that supplements the design information provided by the CMOS-foundry while maintaining the ability to update the CMOS library with typically no impact on the CoSMOS InP HBT and interface libraries. The designer creates a single design that references the CMOS and CoSMOS libraries, closely mapping common design flows and simulation environments.

A highly desirable if not necessary feature of any CMOS-based integration process with compound semiconductors is to permit an extremely high level of design reuse of CMOS components along with foundry level compatibility. In the current set of designs, 4 thin metal and 2 thick metal Cu-based layers are available in the CMOS design portion with additional metal layers available in the InP HBT portion. As a result, virtually all Si CMOS IP-based cells can be accommodated in the HRL CoSMOS process with minor modifications to the power bus and I/O infrastructure.

The available CMOS design layers are a subset of the foundry rules and designs are required to obey the core foundry design rule checks. To support the CoSMOS process, a 2nd set of design rule checks is also used to verify designs.

APPLICATIONS

The methods for integration technology being developed on the program will be extendible to optical components as well as devices based on other materials such as GaN and will provide a straightforward extension of the Si roadmap to enable extremely high data rate and advanced heterogeneous system-on-a-chip (SoC) using a mix of the best component technologies for the application.

By integrating the two disparate technologies like CMOS and InP, dramatic improvement in the linearity, dynamic range and bandwidth of mixed-signal circuits such as digital-to-analog and analog-to-digital converters can be obtained. We expect to see dramatically improved A/D converter performance with 16-bit performance at 500 MHz bandwidth with less than 4W power consumption. Advanced calibration linearization techniques are expected to improve dynamic range by over 30 dB from combining CMOS with the timing precision of InP in wideband (BW> 500 MHz) converter circuits.

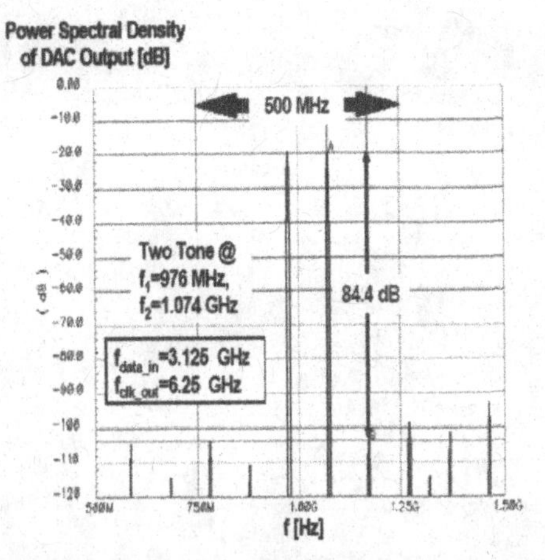

Figure 9 Simulated two-tone spectrum of COSMOS-based DAC using dynamic element matching.

For example, Figure 9 shows a two tone output spectrum of a COSMOS-based DAC. The complexity of the circuit precludes a pure HBT design whereas the performance would not be obtainable using CMOS. COSMOS enables calibration and dynamic element matching to improve linearity. By combining the InP with CMOS on a single chip, both high edge rates can be obtained as well as superior calibration and matching. A pure CMOS implementation would have larger dynamic errors and lower bandwidth. A pure HBT implementation would require 1000 HBTs vs. ~200 for this circuit. An InP HBT configuration would have >30 dB lower SFDR in a similar simulation due to self-heating induced dynamic errors. Our studies also show a significant decrease in power consumption enabled by COSMOS for advanced circuitry that must operate at frequencies > 10 GHz. This reduction largely comes from the ability to eliminate or reduce the need for level shifting circuitry commonly required for bipolar circuits.

CONCLUSIONS

In this paper, we have presented an approach for integration of CMOS with InP HBTs that can enable dramatically increased dynamic range and bandwidth while reducing power consumption for mixed-signal circuits needed for DoD systems.

ACKNOWLEDGEMENTS

This work was supported in part by DARPA through AFRL contract FA8650-07-C-7714 (CoSMOS). The authors would like to the rest of the staff of the HRL Microelectronics group for their assistance in obtaining the results presented in this paper.

REFERENCES

1. D. A. Hitko, T. Hussain, J. F. Jensen, Y. Royter, S. L. Morton, D. S. Matthews, R. D. Rajavel, I. Milosavljevic, C. H. Fields, S. Thomas III, A. Kurdoghlian, Z. Lao, K. Elliott, M. Sokolich, 'A Low Power (45mW/latch) Static 150GHz CML Divider', 2004 Compound Semiconductor IC Symposium.
2. T. Hussain, D. A. Hitko, Y. Royter , R. D. Rajavel, K. Elliott, K. McCalla, M. Madhav, M. Sokolich,' Low Power (51mw per Flip-Flop) CML Static Divider Implemented In Scaled 0.25 μm Emitter-Width InP DHBTs, 2005 InP and Related Materials Conference.

Mater. Res. Soc. Symp. Proc. Vol. 1068 © 2008 Materials Research Society 1068-C02-01

Direct Growth of III-V Devices on Silicon

Katherine Herrick[1], Thomas Kazior[2], Amy Liu[3], Dmitri I. Loubychev[3], Joel M. Fastenau[3], Miguel Urteaga[4], Eugene A. Fitzgerald[5], Mayank T. Bulsara[5], David Clark[6], Berinder Brar[4], Wonill Ha[4], Joshua Bergman[4], Nicolas Daval[7], and Jeffrey LaRoche[2]
[1]Advanced Technology, Raytheon Tewksbury, Tewksbury, MA, 01876
[2]Advanced Technology, Raytheon RF Components, Andover, MA, 01810
[3]IQE Inc., Bethlehem, PA, 08873
[4]Teledyne Scientific Company, Thousand Oaks, CA, 91360
[5]Department of Materials Science and Engineering, Massachusetts Institute of Technology, Cambridge, MA, 02139
[6]Raytheon Systems Limited, Glenrothes, Scotland, United Kingdom
[7]SOITEC, Bernin, France

ABSTRACT

Our direct growth approach of integrating compound semiconductors (CS) and silicon CMOS is based on a unique silicon template wafer with an embedded CS template layer of Germanium (Ge). It enables selective placement of CS devices in arbitrary locations on a Silicon CMOS wafer for simple, high yield, monolithic integration and optimal circuit performance. HBTs demonstrate a peak current gain cutoff frequency ft of 170GHz at a nominal collector current density of 2mA/μm2. To the best of our knowledge this represents the first demonstration of an InP-based HBT fabricated on a silicon wafer.

INTRODUCTION

Combining the best attributes of Compound Semiconductors (CS) and CMOS (complementary metal–oxide–semiconductor) will enable performance superior to that achievable with CS or CMOS alone with CMOS affordability. With an approach that directly integrates the CS into the CMOS wafer, only one wafer is processed to achieve a finished chip. In contrast, device transfer and wafer bonding approaches require fabrication of separate CMOS and CS wafers, followed by assembly.

The starting material, shown in Figure 1, is based on a unique silicon template wafer invented at MIT and fabricated by Soitec [1,2,3]. These silicon-on-lattice-engineered substrates (SOLES) contain an embedded CS template layer of Germanium (Ge). This unique wafer technology enables placement of CS devices in arbitrary locations on the chip, while maintaining co-planarity with the CMOS for simple, high yield, monolithic integration.

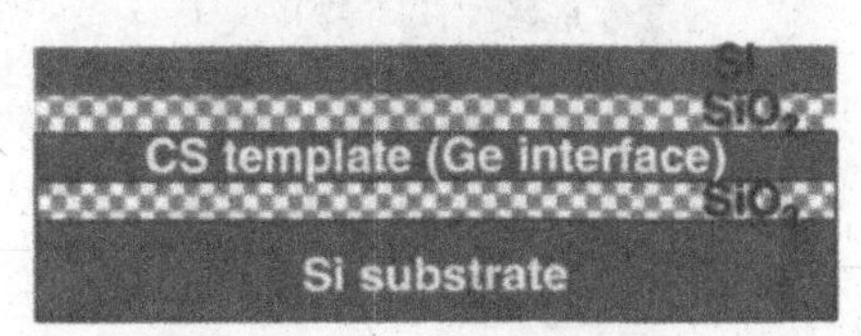

Figure 1. Soitec's Silicon on Lattice Engineered Subtrates (SOLES)

A cross section of CMOS and InP HBT integration on SOLES is shown in Figure 2. The process flow on the SOLES template wafer is as follows 1) front end CMOS device fabrication, 2) selective window opening for the HBT devices, 3) InP HBT epitaxial growth, 4) InP HBT device processing, and 5) interconnects between the InP HBT devices and CMOS.

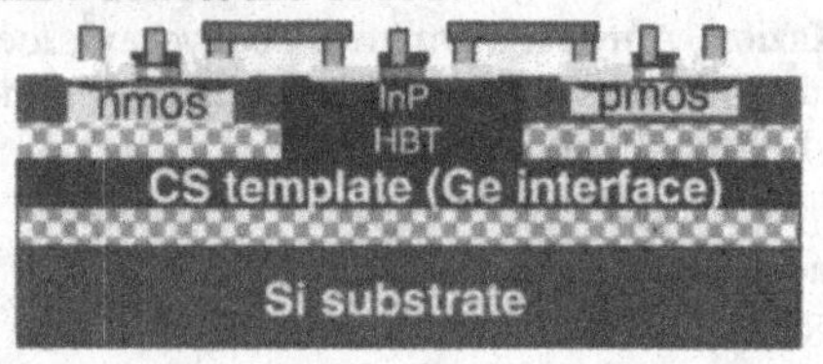

Figure 2. Cross-section of CMOS and InP HBT Integration on SOLES

EPITAXIAL GROWTH ON GEOI SUBSTRATES

The initial process development of InP HBT epitaxy and devices on silicon is with GeOI (Germanium on Insulator) silicon substrates shown in Figure 3. This allows for the development of epitaxial growth and nucleation on Ge without the additional complexity of growing epitaxy within selective windows.

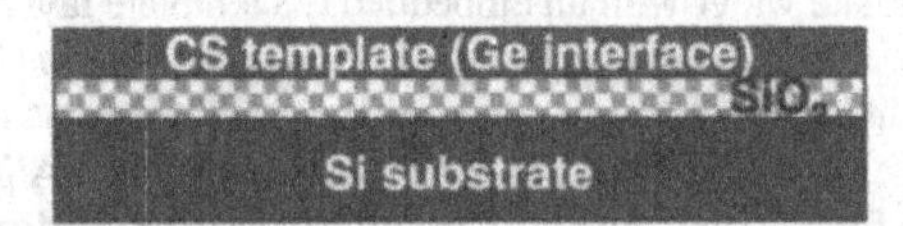

Figure 3. Soitec's Germanium on Insulator (GeOI) Substrate

A schematic diagram of the InP HBT on GeOI structure is shown in Figure 4. The D-HBT structure with type-I band-alignment employs an InP collector and emitter layers and a 400 Å thick InGaAs base layer C-doped at ~4×1019 cm-3. A digital alloy grade was incorporated to suppress current blocking at the base-collector junction. GaAs nucleation conditions on GeOI were optimized to obtain films with mirror-like surfaces. These films exhibit limited formation of anti-phase boundaries (APB), with low dislocation density, and minimize Ge out-diffusion.

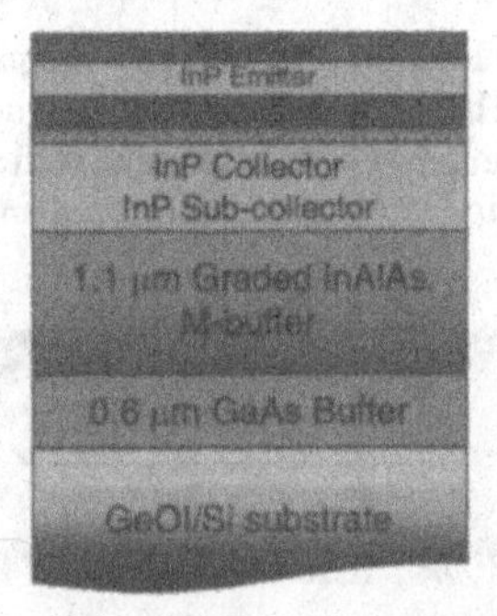

Figure 4: Schematic Diagram of InP HBT on GeOI structure.

Cross-sectional TEM analysis shows minimal APB formation and dislocation density below the TEM detector limit in the top GaAs layer (Figure 5). Well-defined and sharp x-ray diffraction peaks were observed. These baseline parameters suggest that our GaAs-on-GeOI templates are suitable for metamorphic buffer integration.

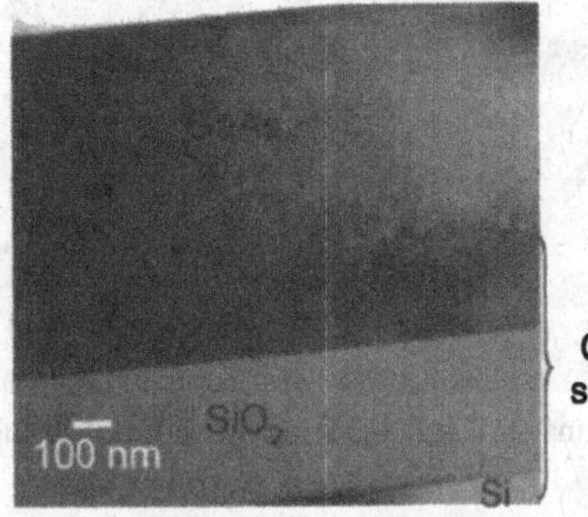

Figure 5. TEM image of GaAs layer grown on GeOI/Si substrate

Further lattice parameter engineering from GaAs to InP was obtained with an InAlAs-based M-buffer. The M-buffer used was a 1.1 μm linearly-graded InAlAs layer including an inverse graded step ending with an alloy composition lattice matched to InP. Since differences in the thermal expansion properties and the surface conditions between the GaAs and GeOI substrates can affect the strain relaxation in the M-buffers, growth conditions for the graded InAlAs were adjusted slightly for each substrate to improve material properties. Typical HRXRD spectrum for InP-HBT structures grown on GeOI/Si substrates are shown in Figure 6.

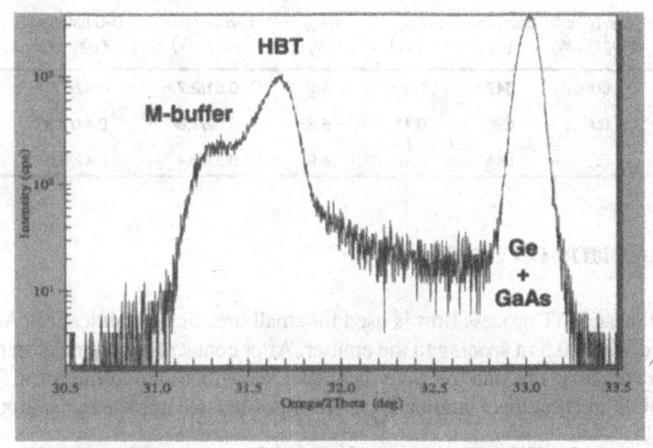

Figure 6. TEM image of InP-HBT grown on GeOI/Si substrate

The TEM image of Figure 7 confirms that density of dislocation networks decreases throughout the InAlAs M-buffer with few dislocations observed at the top, similar to baseline structures grown on GaAs substrates.

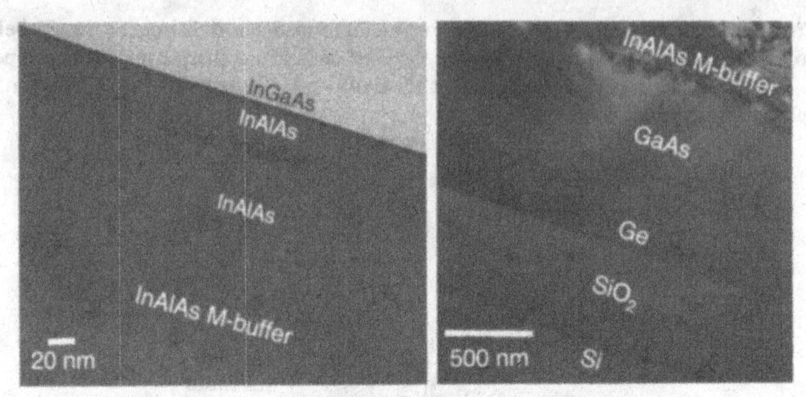

Figure 7. TEM image of InP-HBT grown on GeOI/Si substrate

The characteristics of InP HBT structures compares favorably with baseline structures grown on InP (lattice-matched) and GaAs (metamorphic) substrates as shown in Table 1. To the best of our knowledge, we have demonstrated for the first time the direct epitaxial growth of InP HBT structures on composite InAlAs/GaAs metamorphic buffer (M-buffer) layers on GeOI substrates.

Table 1. Large area device DC characteristics summary of 110x110 μm^2 emitter HBTs.

Substrates	Large-area Device DC Measurements						
	Gain (β) at Ib = 1.2 mA	Base Rsh (Ω)	V_{offset} (V)	BV_{ceo} (V)	B-E junction V_f/V_r (V)	B-C junction V_f/V_r (V)	Ideality factors (n_c/n_b)
InP	60	647	0.10	5.0	0.51/2.7	0.42/7.0	1.08/1.29
Ge	60	695	0.11	5.3	0.52/1.6	0.41/7.3	1.10/1.33
GeOI/Si	55	644	0.12	4.9	0.52/2.4	0.42/6.5	1.10/1.28

INP HBT FABRICATION ON GEOI

A standard mesa-HBT process flow is used for small-area device fabrication. An evaporated base contact is deposited with a 0.5 m spacing to the emitter. After contact formation and semiconductor mesa etches, the devices are passivated and the wafer is planarized with a spin-on-dielectric (BCB). Contact vias are etched in the BCB and first level interconnects are deposited to complete fabrication.

Large area device characteristics are measured for a device with an emitter junction area of 40x40 μm^2. Large area HBTs on the GeOI (Figure 8) and InP (Figure 9) control wafers demonstrate DC current gains β of ~40 and ~50, respectively. The base sheet resistance ρ_{sheet}, extracted from TLM measurements, is 790 Ω/square and 760 Ω/square for the GeOI and InP wafers respectively.

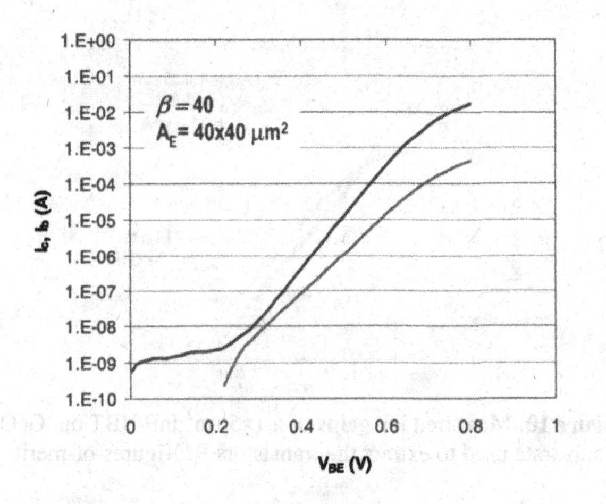

Figure 8. Large area device (LAD) characteristics for InP HBT on GeOI

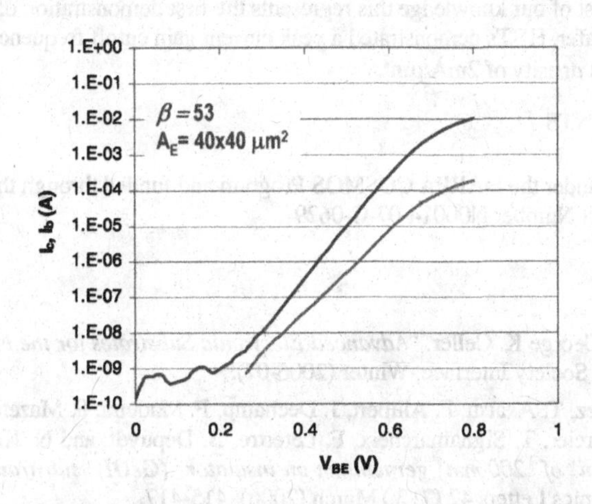

Figure 9. Large area device (LAD) characteristics for InP HBT on InP.

On-wafer s-parameter characterization of the devices is performed from DC-50 GHz. GeOI devices with a 1x5μm² emitter junction area exhibit peak f_t and f_{max} values of 170 GHz and 124 GHz, respectively. Figure 10 shows the measured RF gains of the transistor. The peak f_t of the device indicates good material quality. The f_{max} of the transistor is relatively low due to excessive emitter undercut increasing the base-to-emitter spacing, and hence, the base resistance. The excessive emitter undercut is resolved for future fabrication runs.

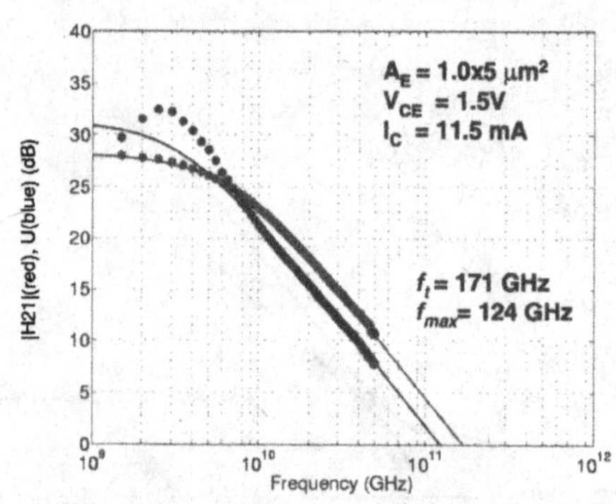

Figure 10: Measured RF gains of a 1x5μm^2 InP HBT on GeOI substrate used to extract the transistors RF figures-of-merit

CONCLUSIONS

The first small-area InP HBTs fabricated on a germanium-on-insulator (GeOI) substrate are demonstrated. To the best of our knowledge this represents the first demonstration of an InP-based HBT fabricated on a silicon wafer. HBTs demonstrated a peak current gain cutoff frequency ft of 170GHz at a nominal collector current density of 2mA/μm^2.

ACKNOWLEDGEMENTS

This work is developed under the DARPA COSMOS Program and funded through the Office of Naval Research (ONR) Contract Number N00014-07-C-0629

REFERENCE

1. Carlos Mazuré and George K. Celler. "*Advanced Electronic Substrates for the Nanotechnology Era.*", The Electrochemical Society Interface, Winter (2006-07).
2. C. Deguet, L. Sanchez, T. Akatsu, F. Allibert, J. Dechamp, F. Madeira, F. Mazen, A. Tauzin, V. Loup, C. Richtarch, D. Mercier, T. Signamarcheix, F. Letertre, B. Depuydt, and N. Kernevez. "*Fabrication and characterisation of 200 mm germanium-on-insulator (GeOI) substrates made from bulk germanium.*" Electronics Letters 42 (7) 30 March (2006): 415-417.
3. C. Maleville and C. Mazure. *"Smart Cut™ technology: From 300 mm ultrathin SOI production to advanced engineered substrates."* Solid-State Electron. 48 (6) une (2004). 1055-1063.
4. *Molecular-beam epitaxy production of large-diameter metamorphic high electron mobility transistor and heterojunction bipolar transistor wafers*, by O. Baklenov, D. Lubyshev, Y. Wu, X.-M. Fang, J.M. Fastenau, L. Leung, F.J. Towner, A.B. Cornfeld, and W.K. Liu, published May/June 2002
5. *Monolithic CMOS-compatible AlGaInP visible LED arrays on Silicon on Lattice-Engineered Substrates (SOLES)*, by Kamesh Chilukuri, Michael J. Mori, Carl L. Dohrmanand, Eugene A. Fitzgerald, published November 2006

Mater. Res. Soc. Symp. Proc. Vol. 1068 © 2008 Materials Research Society 1068-C02-02

Epitaxial and Non-Epitaxial Heterogeneous Integration Technologies at NGST

Augusto Gutierrez-Aitken, Patty Chang-Chien, Bert Oyama, Kelly Tornquist, Khanh Thai, Dennis Scott, Rajinder Sandhu, Joe Zhou, Peter Nam, and Wen Phan
Northrop Grumman Space Technology, Redondo Beach, CA, 90278

Introduction

Present systems requirements are increasingly challenging and complex and this trend is likely to increase rapidly for future systems. This translates into more demanding electronic functions to support these needs and higher performance of the microelectronic technology in which these functions are implemented. There are increased requirements in all aspects of the technology capability such as increased speed, bandwidth, output power, gain, and reduced size, weight and power (SWAP), and also cost and fabrication cycle time. To meet these requirements, it is not enough to use one single semiconductor technology but to combine or integrate several high performance technologies in an efficient and cost effective way. In some cases this involves the integration of semiconductor technologies based on different substrates and in other cases it involves the integration of two types of devices on the same substrate. The combination of two or more dissimilar microelectronic technologies, or heterogeneous integration, leads to a significant higher design flexibility and performance and lower SWAP

There are several approaches to perform this heterogeneous integration. In this paper we describe some of the heterogeneous integration approaches that are being used and developed at Northrop Grumman Space Technology (NGST) that include selective epitaxial growth, metamorphic growth and wafer level packaging (WLP) technology (Figure 1). More recently we are developing a scaled and selective wafer packaging technique to integrate III-V semiconductors with silicon under the Compound Semiconductor Materials on Silicon (COSMOS) Defense Advanced Research Projects Agency (DARPA) program. The integration of compound semiconductors with Silicon adds a new level of complexity due to the significantly different fabrication processes, thermal budgets, wafer sizes and integration scales, but promises an unparalleled capability based on the very high performance of the III-V technology and the large scale integration level of the Si technology.

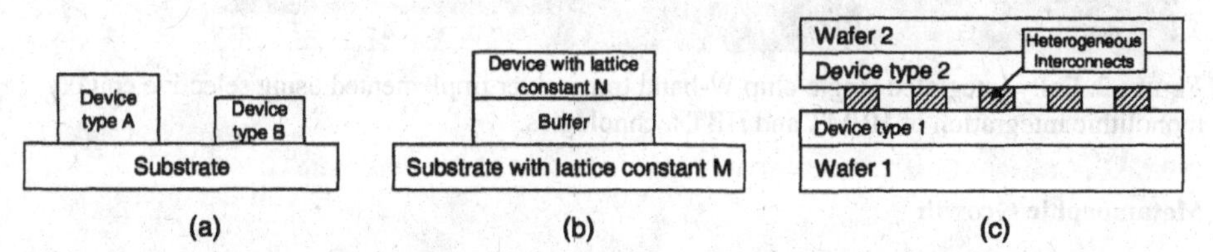

Figure 1. Heterogeneous integration methods used or being developed at NGST. (a) Selective epitaxy, (b) metamorphic growth and (c) wafer level packaging

Selective Epitaxial Heterogeneous Integration

In the selective epitaxy heterogeneous integration technique (Figure 1a), two or more different types of devices are monolithically integrated on one substrate. At NGST, typically these devises are high electron mobility transistors (HEMT's) and heterojunction bipolar transistors (HBT's). HEMT devices provide low noise performance for front-end amplifiers or low noise amplifiers (LNAs) and large RF gain at very high frequencies. On the other hand, HBT devices offer high transconductance, high linearity, high power added efficiency (PAE) in amplifiers and small turn-on threshold variation.

The integration process sequence is as follows. The HBT structure is grown first by molecular beam epitaxy (MBE), then the wafer is patterned with silicon nitride and etched to form HBT islands. Next, the HEMT structure is grown on the patterned HBT material. A high quality epitaxial material is deposited in the areas where the HBT material was removed and the bare substrate is exposed, and polycrystalline material in the areas covered with silicon nitride. Finally the wafer is patterned again to remove this polycrystalline material by wet etching. At this point, the wafer contains some areas with HBT epitaxial layers and other areas with HEMT epitaxial layers and device process can start. The main advantages of this method are capability to achieve transistor scale integration, minimum interconnect parasitics between devices and smaller and simpler packaging since there is only one chip to package instead of two or more. Using this heterogeneous integration technique we demonstrated a fully integrated W-band transceiver (Figure 2), where ten RF functions are integrated closely in a single 6.4 mm x 3.8 mm chip. The size, weight and power, and cost savings are significant compared to a more traditional implementation using multi-chip assemblies.

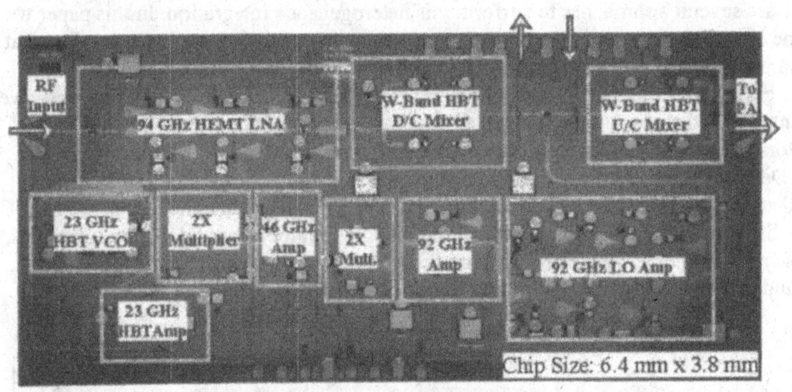

Figure 2. Fully integrated single-chip W-band transceiver implemented using selective epitaxy monolithic integration of HEMT and HBT technologies.

Metamorphic Growth

Metamorphic growth enables the integration of semiconductor devices of one material system on a substrate of a different material system where the device material system lattice constant is significantly different than the substrate lattice constant (Figure 1b). There are several approaches to achieve this that in general use one or more buffer layers between the substrate and the device. The key factor in this approach is to design and implement a buffer layer that provides a high quality "virtual substrate" on which to grow the device. The main advantage of this approach is that it provides large flexibility to the device layer designer by expanding the selection of the material systems and lattice constants available to him or her. The buffer layer should be carefully designed to minimize the increase on thermal resistance due to this added layer. Two examples presented here are Antimonide-based AlSb/InAs HEMTs grown on GaAs substrates and 6.0 Å high Indium content InAlAs HBTs grown on InP substrates, both for ultra-low power applications. These two devices were developed under the Antimonide-Based Compound Semiconductor (ABCS) DARPA program.

The AlSb/InAs structure (Figure 3a) is grown by Molecular Beam Epitaxy on semi-insulating GaAs substrates using an Al(0.7)Ga(0.3)Sb metamorphic buffer. The devices exhibit excellent cut-off frequency (f_T) and maximum oscillation frequency (f_{max}) greater than 220 and 270 GHz, respectively, at drain voltage V_{DS} of 0.3 V. Device fabrication details and additional device performance data can be found elsewhere[1]. Using this technology, two LNAs were demonstrated. One operating at X-band with 18 dB gain and <1.8 dB noise figure (NF) dissipating only 1.34 mW of DC power (Figure 3b)[2] and a W-band LNA with 11 dB gain and 5.4 dB Noise Figure dissipating only 1.5 mW of DC power (Figure 3c)[3].

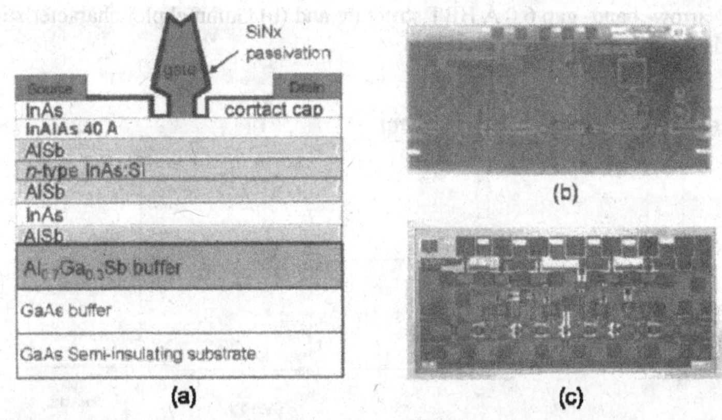

Figure 3. (a) Epilayer structure of an AlSb/InAs HEMT and ultra-low power (a) X-band and (b) W-band LNAs.

In bipolar logic circuits, high indium content in the base layer reduces its energy band gap, which translates into a reduced device turn-on and supply voltage. Our approach to implement this consists of using a metamorphic buffer on semi-insulating InP substrates, with an In(x)Al(1-x)As strain-relieved graded buffer layer (GBL). The HBT structure (Figure 4a) is based on In(0.86)Al(0.14)As/ In(0.86)Ga(0.14)As grown lattice-matched on 6.0 Å. Excellent DC characteristics on small-area devices (1 um emitter width) were observed with high DC gain >40, low leakage at both junctions, and breakdown voltage >3V. A substantial reduction in the turn on voltage was achieved when compared to InP HBT technology (Figure 4b). Peak frequency f_T and f_{max} > 150 GHz were measured without applying any voltage at the base collector junction. As mentioned above, the design and quality of the buffer layer can have a significant impact on the DC and thermal characteristics of the device (Figure 5). Several circuits have been fabricated to evaluate this technology including divide-by-2, divide-by-4, equalizers, delay chains, demultiplexers and a 4-bit flash analog to digital converter (ADC)[4].

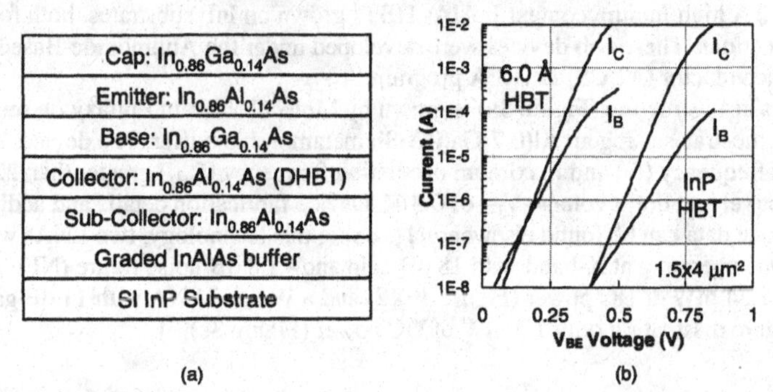

Figure 4. (a) Narrow- band- gap 6.0 Å HBT structure and (b) Gummel plot characteristics of a 6.0 Å HBT and an InP-based HBT.

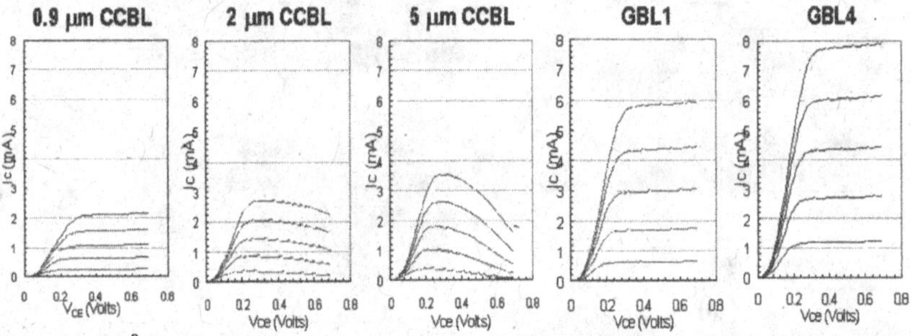

Figure 5. 6.0 Å HBT structure common-emitter DC I-Vs plotted for 30um x 30um emitter area devices fabricated on 0.9, 2, and 5 um thick constant composition buffer layers (CCBL) and two optimized graded buffer layers (GBL) GBL1 and GBL4. IB currents are the same for all five I-V curves (IB = 20, 40, 60, 80 and 100 uA).

Wafer Level Packaging Technology

The wafer level packaging developed at Northrop Grumman Space Technology is a batch fabrication process that starts after the wafers of the individual technologies to be integrated are fully fabricated (Figure 1c)[5]. This technology, also referred as wafer scale assembly, has several advantages that include no device performance degradation, no change in device fabrication process and preservation of existing high-reliability production process. The WLP process combines the desired characteristics of low temperature (<180 °C) solder bonding and thermodynamically stable alloy bonding. It provides the capability of multiple wafer stacks and is compatible with standard assembly processes. In addition, the low bonding temperature results in very low built-in stress during bonding, thereby making it suitable for multi-wafer heterogeneous integration.

The two wafers to be integrated are first processed separately using standard MMIC batch fabrication processes. Next, matching metallic bonding rings are deposited on the two wafers at the end of the front side process. Fabrication of individual wafers is then completed through the backside process before the wafer bonding is performed. The wafer-level bonding process fuses the sealing rings and interconnects of the two wafers together. Cavities are created between the two wafers, encapsulating the front side of the circuits within the cavities at the wafer level (Figure 6). Completed WLP bonded wafers can be diced into individual modules. A variety of circuits and functions were demonstrated with this integration approach. Some examples are vertical WLP interconnects with less than 0.2 dB of insertion loss at W-band and isolation fences that provide 30 dB of isolation[6], a 4-element Q-band linear array[7], LNAs and power amplifiers (PAs)[5]. This technology offers remarkable savings in size and weight that can surpass two or three orders of magnitude compared to traditional implementations

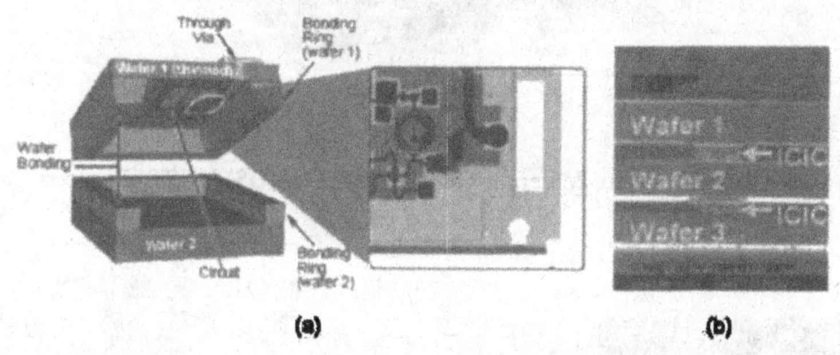

Figure 6. (a) Wafer level packaging integration approach (b) cross section of a WLP stack of three wafer with the intra-cavity interconnects (ICIC) made between the layers during processing.

COSMOS Integration Process

The goal of the COmpound Semiconductor Materials On Silicon (COSMOS) DARPA program is to integrate high performance III-V devices with commercial digital CMOS. This integration will enable significant improvement in ADC dynamic range and bandwidth. Northrop Grumman Space Technology is currently developing an Advanced Heterogeneous Integration process for this program to intimately integrate InP Heterojunction Bipolar Transistors and CMOS. This technology makes possible the optimum partitioning of functional blocks by preferred technology. 0.18 um CMOS will be employed for the digital functions and an advanced 0.25 um planar InP HBT process for the critical high-speed and high-dynamic functions. The NGST AHI approach consists of integrating fully processed Compound Semiconductor "chiplets" on completely processed CMOS wafers using scaled low temperature bond metallic interfaces similar to the interfaces developed at Northrop Grumman Space Technology for wafer level packaging (Figure 7). This integration approach has numerous advantages. The process is largely independent of chiplet technology or chiplet size and complexity, maintains device performance since both technologies are fully processed and the process control monitors can be tested before integration and ensures compatibility with the CMOS and Compound Semiconductor processes.

Figure 8(a) shows an optical photograph of InP HBT chiplets of different sizes integrated onto a 0.18 um CMOS wafer. And Figure 8(b) shows a measurement of a fully functional COSMOS integrated differential amplifier with p-MOS dynamic loads, InP HBT differential amplifier pair and n-MOS current source.

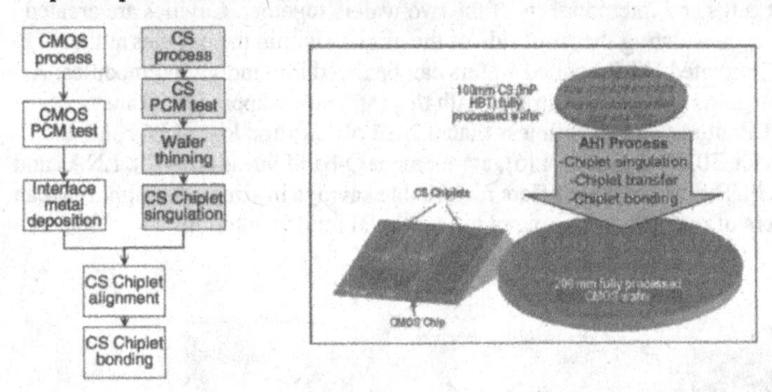

Figure 7. **COmpound Semiconductor Materials On Silicon (COSMOS) NGST advanced heterogeneous integration process.**

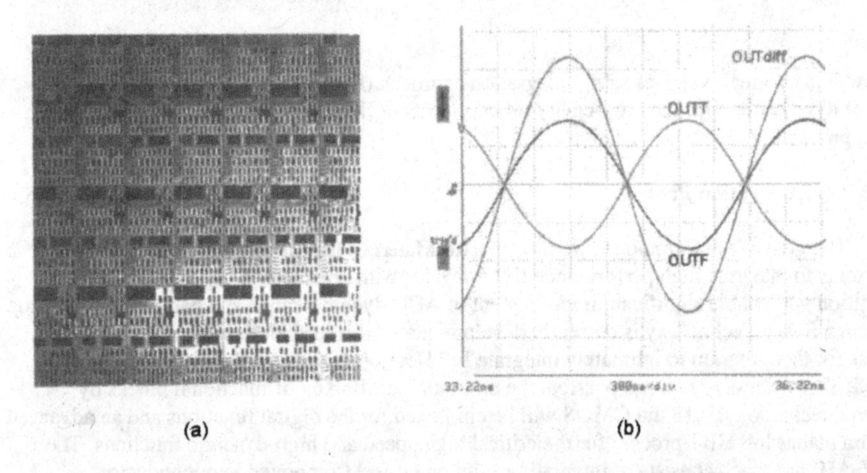

Figure 8. (a) Photograph of a COSMOS wafer with InP HBT chiplets integrated onto a CMOS wafer. (b) Measured COSMOS integrated fully functional differential amplifier.

Summary

Heterogeneous integration offers tremendous reduction and savings in size, weight, power, cost and process cycle time. Furthermore, it provides unparalleled flexibility in circuit and module design for the implementation of increasingly demanding and complex RF, digital and mixed signal functions requiring the integration of several technologies. At Northrop Grumman Space Technology we have demonstrated and are currently developing advanced heterogeneous integration methods to meet these performance challenges.

Acknowledgments

The authors would like to thank DARPA for its support and funding in the programs IRFFE, ABCS, TFAST, ISIS, SMART and COSMOS. The authors also wish to thank the entire engineering, processing and testing team at NGST Microelectronics Product Center.

References

[1] R. Tsai, M. Barsky, J. B. Boos, B. R. Bennett, J. Lee, R. Magno, C. Namba, P. H. Liu, D. Park, R. Grundbacher, and A. Gutierrez "Metamorphic AlSb/InAs HEMT for Low-Power, High-Speed Electronics", 2003 GaAs Integrated-Circuit Symposium, San Diego, CA.
[2] W.R. Deal, R. Tsai, M.D. Lange, J.B. Boos, B.R. Bennett, A. Gutierrez, "A Low Power/Low Noise MMIC Amplifier for Phased-Array Applications using InAs/AlSb HEMT," in *Microwave Symposium Digest, 2006. IEEE MTT-S International*, pp 2051-2054, June 2006
[3] W.R. Deal, R. Tsai, M.D. Lange, J.B. Boos, B.R. Bennett, and A. Gutierrez, "A W-band InAs/AlSb low-noise/low-power amplifier," *IEEE Microwave and Wireless Components Letters*, Volume 15, Issue 4, pp 208 – 210, April 2005.
[4] C. Monier, A. Cavus, R.S. Sandhu, A. Oshiro, D. Li, D. Matheson, B. Chan, and A. Gutierrez-Aitken, "Low Power High-Speed Circuits with InAs-based HBT Technology," Presented at the *2006 Imitational Conference on Compound Semiconductor Manufacturing Technology*, April 24-27, 2006, Vancouver, Canada.
[5] P. Chang-Chien, P. X. Zeng, K. Tornquist, M. Nishimoto, M. Battung, Y. Chung, J. Yang, D. Farkas, M. Yajima, C. Cheung, K. Luo, D. Eaves, J. Lee, J. Uyeda, D. Duan, O. Fordham, T. Chung, R. Sandhu, R. Tsai, "MMIC Compatible Wafer-Level Packaging Technology," in *IEEE 19th International Conference* on *Indium Phosphide & Related Materials*, May 14-18, 2007, Matsue, Japan
[6] D. Farkas, T. Luna, P. Chang-Chien, K. Tornquist, O. Fordham, R. Tsai, "Demonstration of a Low Loss W-band Interconnect and Circuit Isolation Structure for Wafer Scale Assembly," in *IEEE/MTT-S International Microwave Symposium*, June 3-8, 2007.
[7] J. Yang, Y. Chung, M. Nishimoto, M. Battung, A. Long, P. Chang-Chien, K. Tornquist, M. Siddiqui, R. Lai, "Wafer Level Integrated Antenna Front End Module For Low Cost Phased Array Implementation," in *IEEE/MTT-S International Microwave Symposium*, June 3-8, 2007.

Mater. Res. Soc. Symp. Proc. Vol. 1068 © 2008 Materials Research Society 1068-C02-03

An AlGaAs/InGaAs HEMT Grown on Si Substrate with Ge/GexSi1-x Metamorphic Buffer Layers

Edward Yi Chang[1], Yueh-Chin Lin[1], Yu-Lin Hsiao[1], Y.C. Hsieh[1], Chia-Yuan Chang[1], Chien-I Kuo[1], and Guang-Li Luo[2]

[1]Department of Materials Science and Engineering, National Chiao-Tung University, 1001 Ta-Hsueh Rd., Hsin-Chu, 30010, Taiwan

[2]National Nano Device Laboratories, No 26, Prosperity Road 1, Science-based Industrial Park, Hsin-Chu, 30078, Taiwan

ABSTRACT

An AlGaAs/InGaAs HEMT grown on Si substrate with Ge/Ge_xSi_{1-x} buffer is demonstrated. The Ge/Ge_xSi_{1-x} metamorphic buffer layer used in this structure was only 1.0 μm thick. The electron mobility in the $In_{0.18}Ga_{0.82}As$ channel of the HEMT sample was 3,550 cm^2/Vs. After fabrication, the HEMT device demonstrated a saturation current of 150 mA/mm and a maximum transconductance of 155 mS/mm. The well behaved characteristics of the HEMT device on the Si substrate are believed to be due to the very thin buffer layer achieved and the lack of the antiphase boundaries (APBs) formation and Ge diffusion into the GaAs layers.

Index Terms: HEMT, GaAs on Si, SiGe buffer layer.

INTRODUCTION

The heteroepitaxial growth of III-V materials on Si substrate has attracted great attention in recent years due to its potential of integrating Si and III-V based devices on the same platform. The physical gate length of Si transistors used in the current generation node is about 50 nm. The size of the transistor will reach 10 nm in 2011. In order to extend Moore's law well into the next decade, Si-CMOS incorporated with III-V semiconductor compound materials in the device structure is one of the promising solutions for the CMOS technology [1].

However, the two main problems to overcome in growing GaAs on Si by heteroepitaxy are the large lattice mismatch of 4% and the difference in the thermal expansion coefficients (63%) of these two materials. To achieve high quality device structures, it is necessary to reduce the dislocations density in the epitaxial GaAs layer. Si^+ pre-ion-implantation combined with a Ge_xSi_{1-x} metamorphic buffer structure for the growth of Ge layer on Si substrate was proposed [2]. Enhanced strain relaxation of the Ge_xSi_{1-x} metamorphic buffer layer on Si substrate was achieved due to the introduction of the point-defects by heavy dose Si^+ pre-ion-implantation. Because of the strain relaxation enhancement and the interface-blocking of the dislocations in the Ge_xSi_{1-x} metamorphic buffer structure, many threading dislocations was trapped at the heterojunction interface and the total thickness of the Ge/Ge_xSi_{1-x} buffer layers needed for the growth of GaAs on Si was only 1.0 μm. Another problem is the formation of the antiphase boundaries (APBs) at the GaAs/Ge interface, which occur when growing polar GaAs on nonpolar Ge. Several models were reported for the suppression of APBs by using Ge wafer which is 6° off (100) toward <110> direction. This is because as the GaAs epitaxial layer growth proceeds, the initial nuclei grown at the steps coalesce so that APBs-free GaAs can be achieved [3].

In this study, an AlGaAs/InGaAs HEMT grown on Si substrate using the

Ge/Ge_xSi_{1-x} buffer is demonstrated for the first time. The Ge/Ge_xSi_{1-x} layer structure was grown by ultra-high vacuum chemical vapor deposition (UHV/CVD) [4-6] and the AlGaAs/InGaAs HEMT structure was grown by a low pressure metal organic chemical vapor deposition (LP-MOCVD) system [7-8]. The $Al_{0.12}Ga_{0.88}As/In_{.18}Ga_{.82}As$ HEMT structure grown on the Ge/GeSi/Si heterostructure is as shown in Fig 1.

200Å	GaAs	2×10^{18}	n^+
320Å	$Al_{0.12}Ga_{0.88}As$	5.3×10^{17}	n
40Å	$Al_{0.12}Ga_{0.88}As$	undoped	
150Å	$In_{0.18}Ga_{0.82}As$	undoped	
1000Å	GaAs	undoped	
5000Å	$Al_{0.12}Ga_{0.88}As$	undoped	
2μm	GaAs	undoped	
0.55μm	Ge	undoped	
0.45μm	Ge_XSi_{1-X}	undoped	
	Si substrate		

Fig. 1. The layer structure of the InGaAs HEMT structure grown on the Si substrate Ge_xSi_{1-x} metamorphic buffer layers.

EPITAXY AND DEVICE PROCESSING

The growth of the GeSi and Ge buffer layers was carried out using an ultra-high vacuum chemical vapor deposition (UHV/CVD) system with a base pressure of less than 2×10^{-8} torr. First, a 4-inch Si wafer 6° off (100) toward <110> direction was implanted with Si^+ ions with dose of 5×10^{15} cm^{-2} at an acceleration voltage of 50 keV. After implantation, the wafer was cleaned by 10% HF dipping and was baked at 800 °C in the growth chamber for 5 min. Then, a 0.2 μm $Ge_{0.8}Si_{0.2}$, a 0.25 μm $Ge_{0.9}Si_{0.1}$, and a 0.55 μm Ge layer were grown on the substrate at 400 °C in sequence. Between successive layers, growth was interrupted in situ for 15 min annealing at 750 °C. The compositions of the Ge_xSi_{1-x} layers were carefully designed so that the stress at the interfaces can effectively block the dislocations from propagating into the upper layers. After the Ge layer was grown, the dislocation density in the Ge film was measured to be $7.6\times10^6 cm^{-2}$ by X-ray measurement. Then, the sample was switched to a LP-MOCVD system to grow the AlGaAs/InGaAs HEMT structure on the $Ge/Ge_{0.9}Si_{0.1}/Ge_{0.8}Si_{0.2}/Si$ heterostructure. The 150Å undoped $In_{0.18}Ga_{0.82}As$ channel layer was grown at 590 °C, the AlGaAs layer and the GaAs layer were grown at 620 °C and the growth pressure was 40 Torr. Fig. 2 shows the cross-sectional transmission electron microscopy of the HEMT structure grown on the Si substrate. The device structure was composed of 150Å undoped $In_{0.18}Ga_{0.82}As$ channel layer, 40Å undoped $Al_{0.12}Ga_{0.88}As$ spacer layer, 320Å $Al_{0.12}Ga_{0.88}As$ Schottky layer with doping concentration of 5.34×10^{17} cm^{-3} and 200Å GaAs cap layer with doping concentration of 2×10^{18} cm^{-3}. The inserted image is the lattice image of the InGaAs channel of the HEMT grown on the Si substrate.

Fig. 2. The cross section TEM image of the HEMT structure grown on the Si substrate

For the HEMT device fabrication, the mesa isolation was done by wet chemical etching. Ohmic contacts were formed by evaporating Au/Ge/Ni/Au on n^+ GaAs layer and then alloyed at 350°C using RTA (Rapid Thermal Annealing). The T-gate was formed by E-beam lithography with footprint of 0.4 μm. Citrate acid/H_2O/H_2O_2 solution was used for the gate recess process. Ti/Pt/Au was deposited as the Schottky metal for the T-shaped gate. Finally, 100-nm-thick silicon nitride film was deposited by plasma enhanced chemical vapor deposition (PECVD) as the passivation layer.

DEVICE PERFORMANCE

For isolation etch, the HEMT structure was etched 300nm down to the AlGaAs buffer layer. After the isolation etch, the HEMT structure demonstrated a leakage current of 0.016μA/μm when the bias voltage was up to 14.2V as measured by the isolation pattern with a pad spacing of 10μm. If the antiphase boundaries (APBs) formed at the GaAs/Ge interface, Ge atoms could diffuse along the antiphase boundaries into the GaAs layers during the growth process, and act as dopant atoms in the GaAs layers. Both the APBs formation and the Ge diffusion into the GaAs layers will affect the isolation and result in the increased leakage current and influence the pinch off characteristics of the devices [9-10]. The isolation data indicate that the use of 6° off Si substrate had efficiently suppressed the APBs formation and the Ge diffusion into the GaAs layer, and the qualities of the AlGaAs/GaAs buffer layer and the Ge/Ge_xSi_{1-x} buffer layer were good enough for the device application.

The I-V characteristics of the fabricated 0.35 × 100 μm^2 AlGaAs/InGaAs HEMTs on Si with Ge/Ge_xSi_{1-x} metamorphic buffer layers were measured. The drain-to-source saturation current (I_{dss}) was 150 mA/mm at V_{DS} = 1.5V and the pinch off voltage of the device was -1.6V. The device showed very good pinch off characteristics as shown in Fig 3a. The maximum transconductance measured at V_{DS} =1.5V was 155mS/mm as shown in Fig 3b. The drain-to-gate breakdown voltage (V_{BK}) was 3.5V, which was defined at a gate current of 1mA/mm. For the growth of heterostructure materials, the thermal expansion coefficient of the substrate affects the film quality of the heterostructure layers grown when the thermal mismatch between these two materials is large and the epilayer grown is thick [9]. The good performance of the HEMT structure in this study can be attributed to the thin buffer layer achieved between the Si substrate and the HEMT structure by using Si^{+} ion implantation for the Ge_xSi_{1-x} layer strain relaxation and the use of two steps Ge_xSi_{1-x} layer growth to block the dislocation propagation.

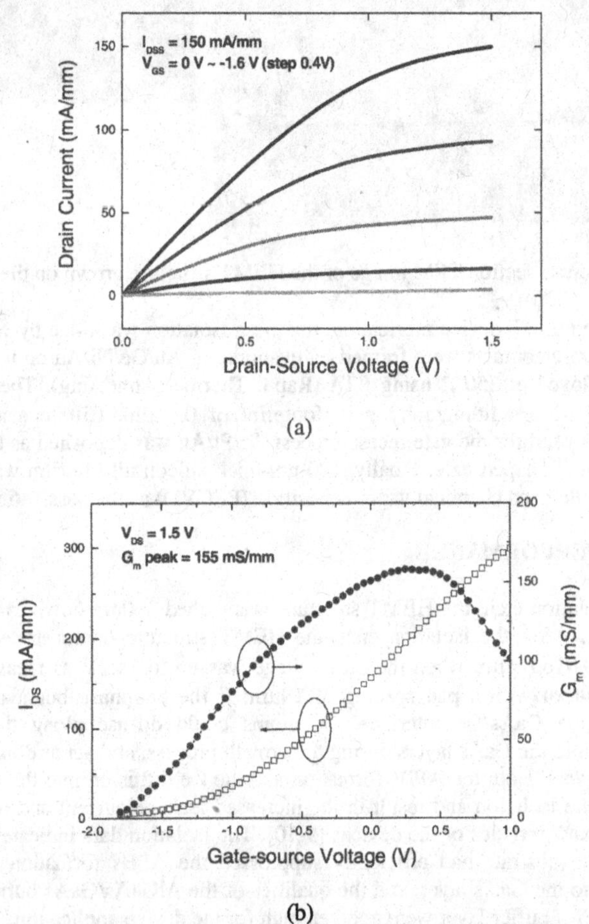

Fig. 3. A 0.35 μm × 100μm AlGaAs/InGaAs HEMT on the Si substrate with Ge/GeSi metamorphic layer: (a) I-V characteristics, (b) G_m and I_{DS} vs. V_{GS}

CONCLUSIONS

In this paper, we have demonstrated a working AlGaAs/InGaAs HEMT device on Si substrate.The 0.4 μm AlGaAs/InGaAs HEMT fabricated on the Si substrate demonstrated a transconductance of 155 mS/mm and a saturation channel current of 150 mA/mm with well behaved pinch off characteristics. The results indicate that the buffer layer quality was good with no Ge diffusion into the GaAs layers and no APBs formation. The good HEMT performance is also attributed to the very thin buffer layer (1μm) achieved by using Si^+ ion implantation and two steps buffer layer growth technique, The thin buffer layer alleviates the stress effect caused by the large thermal expansion coefficient difference between these two materials. Over all, the developed Si_xGe_x metamorphic buffer technology shows great potential for .the integration of III-V and Si devices in the future.

REFERENCE

1. Robert Chau, Suman Datta, Mark Doczy, Brian Doyle, Ben Jin, Jack Kavalieros, Amlan Majumdar, Matthew Metz, and Marko Radosavljevic, IEEE transactions on nanotechnology 4, **153** (2005).
2. Y.C. Hsieh, E.Y. Chang , G.L. Luo , M. H. Pilkuhn, J.Y. Yang, H.W. Chung, and C. Y. Chang, Applied Physics Lett. **90**, 083507 (2007).
3. K. van der Zanden, D Schreurs, Member, IEEE, P. Mijlemans, and G. Borghs, IEEE Electron Device Lett. **21**, 57 (2000).
4. L. Lazzarni, L Nasi, G Salviati, C.Z. Fregonara, Y. Li, L.J. Giling, D.B. Holt, Micron **31**, 217 (2000).
5. Tsung-Hsi Yang, Guangli Luo, Edward Yi Chang, Y. C. Hsien and Chun-Yen Chang, October, J. Vac. Sci.&Techn. B. **22**, L17 (2004).
6. Y. C. Hsieh, E. Y. Chang, G L. Luo, S. H. Chen, Dhrubes Biswas, S. Y. Wang, and C. Y. Chang, J. Appl. Phys. **100**, 064502 (2006).
7. Guangli Luo, Tsung-His Yang, Edward Yi Chang, Chun-Yen Chang and Koung-An Chao, Jpn. J. Appl. Phys. **42**, L517 (2003).
8. Yuan Li, G Salviati, M.M.G Bongers, L. Lazzarini, L. Nasi, L. J. Giling, J. Crystal Growth **163**, 195 (1996).
9. W. Li, S. Laaksonen, J. Haapamaa, M. Pessa, J. Crystal Growth 227, 104 (2001).
10. V. K. Yang, M. Groenert, C. W. Leitz, A. J. Pitera, M. T. Currie, and E. A. Fitzgerald, J. Appl. Phys. **93**, 3859 (2003).

Mater. Res. Soc. Symp. Proc. Vol. 1068 © 2008 Materials Research Society 1068-C02-04

Epitaxial Growth of III-V Nanowires on Group IV Substrates

Erik Bakkers[1], Magnus Borgstrom[1,2], and Marcel Verheijen[1]
[1]Philips Research Labs, Eindhoven, 5656AE, Netherlands
[2]Solid State Physics, Lund University, Lund, Sweden

ABSTRACT

Semiconducting nanowires are emerging as a route to combine heavily mismatched materials. The high level of control on wire dimensions and chemical composition makes them promising materials to be integrated in future silicon technologies as well as to be the active element in optoelectronic devices.
In this article, we review the recent progress in epitaxial growth of nanowires on non-corresponding substrates. We highlight the advantage of using small dimensions to facilitate accommodation of the lattice strain at the surface of the structures. More specifically, we will focus on the growth of III–V nanowires on group IV substrates. This approach enables the integration of high-performance III–V semiconductors monolithically into mature silicon technology, since fundamental issues of III-V integration on Si such as lattice and thermal expansion mismatch can be overcome. Moreover, as there will only be one nucleation site per crystallite, the system will not suffer from antiphase boundaries.
Issues that affect the electronic properties of the heterojunction, such as the crystallographic quality and diffusion of elements across the heterointerface will be discussed. Finally, we address potential applications of vertical III–V nanowires grown on silicon.

INTRODUCTION

"Epitaxy" is derived from the Greek word meaning "ordered upon" and is the crystalline deposition of material on a substrate with identical lattice structure and orientation. For heteroepitaxial growth, materials with different lattice parameters are combined. If the lattice mismatch between the deposited film and the substrate is large, typically a few percent, misfit dislocations can be incorporated near the interface. Alternatively, adlayers may form three-dimensional nuclei to release the strain in order to minimize the energy in the system, at the cost of creating more surface.
With the development of the vapor–liquid–solid (VLS) wire growth method in 1964,[1] it has become possible to vary the chemical composition within a nanowire. With this method thin semiconducting nanowires are grown from a catalytic metal particle. The small diameter (typically some tens of nm) results in effective strain release circumventing the usual requirement of lattice matching. Several groups have recently shown that such heteroepitaxial junctions can be defined in nanowires in the longitudinal (axial)[2–6] direction as well as in the radial[6–11] direction for different III–V compound semiconductors and for the Si/Ge system. These different materials can have a substantial lattice mismatch and the crystal lattices will be elastically deformed near the heterointerface. Such crystal deformations can be relieved at the surface for small enough wire diameters. If the lattice strain is totally accommodated by elastic deformations, the introduction of misfit dislocations at the heterointerface will be avoided. Such strain relief in, for instance, a heterostructured InAs/InP (~4% mismatch) nanowire has been

supported by a detailed analysis of the electrical characteristics.[12,13] Importantly, this method allows us to combine heavily mismatched semiconductors, maintaining perfect crystallinity. This concept of strain engineering can be extended to grow nanowires on a substrate with different lattice parameters, or in other words, to have the heterointerface at the nanowire/substrate junction. Epitaxy of VLS grown (sub)micrometer-sized whiskers on corresponding bulk substrates was pioneered in the 1970s by Wagner[14] and Givargizov[15–17] for the Si/Ge system, and Hiruma[18,19] investigated VLS growth of compound semiconductor nanowhiskers in the early 1990s. In recent years, Ge nanowires have been heteroepitaxially grown on silicon,[20,21] GaN and ZnO nanowires on oxidic substrates,[22–24] and ZnSe on GaP.[25] More recently, growth of III–V nanowires on silicon[26–28] and germanium[29] substrates has been presented.

In this context the nanowire-epitaxy approach might be interesting for the monolithic integration of III–V semiconductors into silicon technology. This combines the best parts of two worlds: the superior electronic properties of the III–V semiconductors with the highly advanced and relatively inexpensive silicon process technology. Most of the III–V materials intrinsically have higher electron mobilities and a direct bandgap, whereas silicon has a higher thermal conductivity and favorable mechanical properties as compared to the III–V substrates.

This article highlights recent developments that have been made with regard to the structural and chemical characterization of epitaxial III–V nanowires on silicon or related substrates.

NANOWIRE EPITAXY

Nanowires grown by the VLS mechanism are nucleated from a nanometer sized metal seed or catalyst particle that collects the precursor material from the vapor phase, establishing local supersaturation. There are a few examples showing that the metal particle acts as a catalyst for the decomposition of the precursor molecules.[6,30] A vapor pressure of the precursor components can be created by means of molecular beam epitaxy (MBE),[31–34] metal organic vapor phase epitaxy (MOVPE)[14–19] and related techniques,[35] pulsed laser ablation[36–39] or by simple evaporation of the precursor material.

When supersaturation has been reached, crystalline material precipitates underneath the metal particle. In order to enable epitaxial growth of a nanowire it is of crucial importance to have a clean and crystalline substrate surface. Methods to clean Si substrates are well established and an etching step with hydrofluoric acid is almost always included. The catalytic metal particles are deposited directly after substrate cleaning. So far, Au has mainly been used as the active particle, but alternative metals,[14,16,30,38] oxides[40] and silicides,[41] being compatible with standard silicon processing, have been used to some extent. Importantly, any exposure to air affects the epitaxy, since Au catalyzes the oxidation of Si; an SiO_2 layer tens of nm thick forms on top of the Au particle within days at room temperature.[42] Alternatively, Si samples provided with a thin Ge layer or bulk Ge samples have been used,[29] since any germanium oxide is readily desorbed by a mild *in situ* thermal treatment.

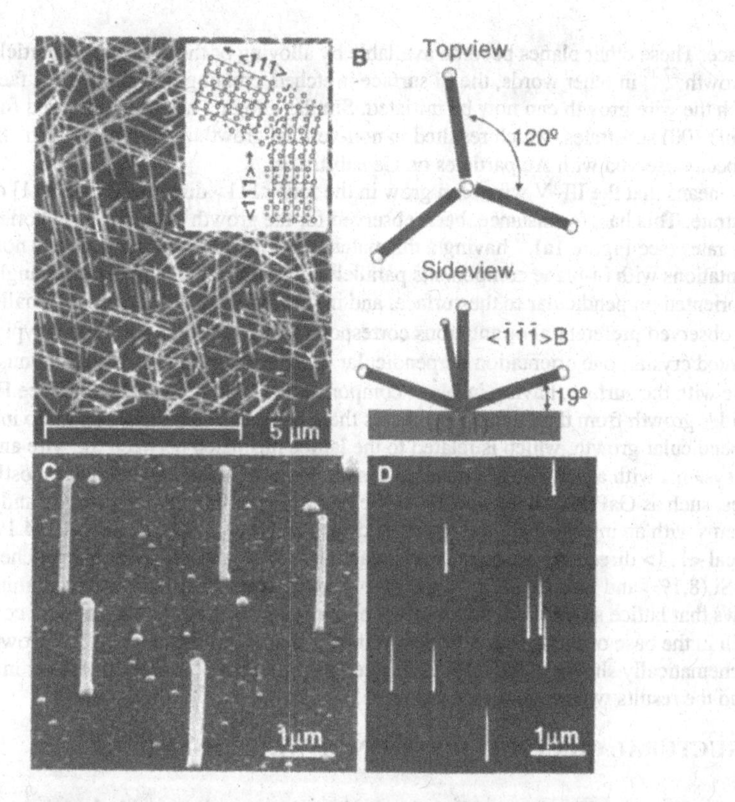

Figure 1. Scanning electron microscopy images of (a) InP wires grown by MOVPE on a Si(111) substrate (top view). Inset in (a) shows a schematic of the mechanism to relieve compressive strain in the nanowire base by a change into another <111> growth direction without a rotational twin dislocation. (b) Overview of expected wire orientations on a (111)-oriented silicon or germanium substrate. (c) SEM image of GaP wires grown by laser ablation on Si(111) (side view, 30º). (d) GaAs wires grown by MOVPE on Ge(111) (side view, 30º).

Scanning electron microscopy (SEM) can give a first indication of nanowire epitaxy. Epitaxial growth results in alignment of the nanowires in specific directions determined by the crystal symmetry of the substrate. However, orientation of one-dimensional structures can also be obtained during growth by other mechanisms, such as interaction with external fields[43] or with a gas flow.[44,45] In general, the III–V wires tend to grow in the [111]B direction[16,18], which means that the wire crystal is terminated at the wire/catalyst interface by a group V plane. Therefore, vertical growth can be induced on [111]B-oriented III–V substrates. We should note however that the elemental semiconductors, such as silicon, are not polar, and all {111} facets are chemically equivalent. When growing III–V wires on a silicon (111) substrate, non-vertical wire growth is expected if the growth is nucleated from {111} facets other than those at the substrate surface. These other planes become available by alloying of the Au catalyst particle with Si prior to growth;[27,28] in other words, the Si surface is etched, creating the other {111} facets from which the wire growth can now be initiated. Similar alloying has been observed for Au particles on InP(100) substrates, which resulted in non-vertical growth directions.[46] Such alloy etching has not been observed with Au particles on Ge substrates.

This means that the III–V wires can grow in the four <111> directions on a [111] oriented Si substrate. This has, for instance, been observed for the growth of InP on (111) oriented silicon substrates (see Figure 1a),[27] having a mismatch of 8.1% There are three clearly noticeable orientations with in-plane components parallel to the sides of an equilateral triangle. Some wires are oriented perpendicular to the surface, and in this top view they appear as small bright spots. The observed preferential orientations correspond to the four ⟨111⟩ directions typical for a (111) oriented crystal; one orientation perpendicular to the surface and three orientations forming a 19° angle with the surface, having in-plane components at 120° from each other (see Figure 1b). Besides growth from the other {111} facets there is an additional mechanism to induce non-perpendicular growth, which is related to the lattice mismatch between the wire and substrate. For systems with a small lattice mismatch, samples have been prepared with mostly vertical wires, such as GaP/Si (0.4%), and GaAs/Ge (0.1%), as is shown in Figures 1c and 1d. For systems with an intermediate mismatch, such as InP/Ge (3.7 %) and GaAs/Si (4.1%) the four typical <111> directions are equally abundant and for the most heavily mismatched systems, InP/Si (8.1%) and InAs/Si (11.6%) the non-perpendicular growth directions dominate. This shows that lattice strain has a major effect on the wire orientation. The residual compressive strain in the base of the wire can be relieved by a change into another <111> growth direction, as is schematically shown in the inset of Figure 1a. We studied these strain effects in more detail[27-29] and the results will be discussed below.

STRUCTURAL CHARACTERISATION OF WIRE ENSEMBLES

The fact that nanowires have well-defined orientations consistent with the substrate's crystal symmetry is a clear indication of epitaxial growth. X-ray diffraction (XRD) and cross sectional TEM are the established techniques to substantiate this. We studied the crystallographic relation between the substrate and a large number of nanowires by XRD pole figure measurements.[27,29] Pole figures were measured for the (111) reflections of the substrate and the wires. In the reference pattern of the silicon substrate, plotted in Figure 2b, (111)-spacings were found at four orientations typical for a (111)-oriented single crystal. One set of reflections is in the center of the pole figure, corresponding to the substrate normal; the other three sets have ψ=19° and an in-plane angle of 120° with respect to each other (these orientations are reflected by the wire orientations as observed with SEM).

The pole figure for InP wires grown by MOVPE on Si(111) is shown in Figure 2c. The signals associated with the majority of the InP wires, labeled A (i.e., about two-thirds of the total signal), matches the pole pattern of the substrate, providing an unambiguous signature of heteroepitaxial growth. With such pole figures we confirmed the epitaxial relation between a range of III–V nanowires, such as GaAs, InP, and InAs, with the Si(111) substrate.[27] The difference in lattice spacings for the GaP/Si system was too small to be resolved in the pole figures.

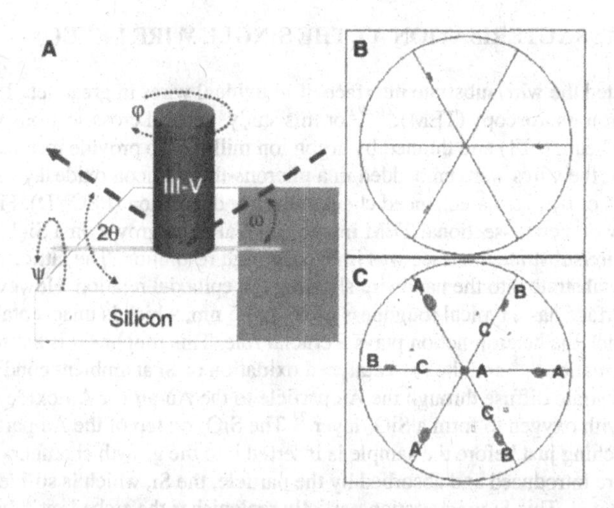

Figure 2. X-ray diffraction pole measurements on InP wires grown on Si(111). (a) schematic of the experimental geometry. To record the pole figure, the detector was set at a 2θ angle corresponding to one of these lattice spacings and the substrate was rotated continuously around φ, and stepped around ψ, (b) pole figure of the Si(111) reflections of the Si(111) substrate, and (c) InP(111) reflections originating from the wires.

The most commonly found crystal defects in all types of nanowires are rotational twin dislocations, which occur specifically around a [111] axis.[14] In terms of electron-wave function propagation, a twin behaves as a junction of two essentially different materials and electrons can scatter at these interfaces.[47] The observed twin density in nanowires seems to be a function of growth temperature and wire diameter and strongly depends on the material system. Typically, the twin density observed is much lower in wires grown epitaxially on germanium and silicon than in wires grown on silicon oxide.

With x-ray pole measurements it is possible to investigate the twin density for a large ensemble of wires. For example, the three spots in the pole figure (in Figure 2c) that have a 180° in-plane rotation with respect to the peaks from the epitaxial wires, labeled with a B, arise from wires that have a rotational twin dislocation around the substrate surface normal vector. The twin boundaries can be in the wire or at the wire/substrate interface. An example of such a twin boundary at the heterointerface is indicated in the TEM cross section (Figure 3b) with a white dotted line. The small signals appearing closer to the center of the pole figure (labeled C in Figure 2c) originate from wires that have grown in one of the non-perpendicular <111> directions, having an angle of 19° to the surface, and that have a twin dislocation orthogonal to their longitudinal axis. The fact that the mirrored orientations give a much smaller signal than the orientation identical to the substrate reveals that the density of twinning defects in the wires is very low (one or two per wire).

INTERFACE CHARACTERISATION AT THE SINGLE WIRE LEVEL

We have investigated the wire/substrate interface of individual wires in great detail by transmission electron microscopy (TEM).[27-29] For this study, vertical cross sections were sliced with a focused ion beam (FIB) and thinned by argon ion milling. To provide mechanical support during this process, the wires were embedded in a microns-thick silicon oxide layer, deposited by the spin-on process or by plasma-enhanced chemical vapor deposition (PECVD). Figure 3a shows an overview of a cross-sectional TEM image of a GaP wire grown on a Si(111) substrate. In Figure 3b the wire/substrate interface was imaged at high resolution. The lattice planes continue from the substrate into the nanowire showing the epitaxial relation. However, the wire/substrate interface has a typical roughness of about ~5 nm, which is unacceptable for applications in which the heterojunction plays a crucial role. This roughness is a direct result of the Au/Si alloy formation[27,28] and the Au-catalyzed oxidation of Si at ambient conditions. Si atoms from the substrate diffuse through the Au particle to the Au-air (or Au-oxide) interface, where they react with oxygen to form a SiO_2 layer.[27] The SiO_2 on top of the Au particle is removed by HF-etching just before the sample is inserted into the growth chamber. When the III–V precursors are introduced and absorbed by the particle, the Si, which is still left in the particle, is precipitated. This Si precipitation partially replenishes the etched pit, (the other part has been oxidized and removed), and the rest is filled up with III–V material resulting in a rough interface.

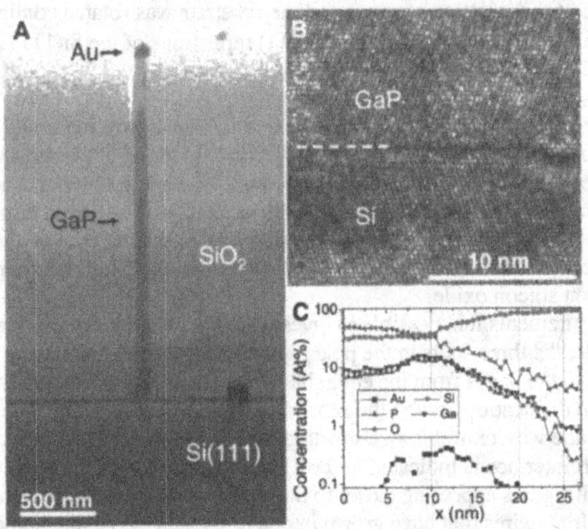

Figure 3. Cross-sectional transmission electron microscopy (TEM) study of (a) GaP wire on Si(111). The red vertical line indicates the position of the EDX line scan(b) High-resolution image of the GaP/Si interface. The dotted line indicates the position of the heterointerface and a twin boundary. (c) Energy-dispersive x-ray analysis line scan across the GaP/Si junction. The blue area indicates the position of the heterointerface.

Such a rough interface has never been observed for InP wires grown on a Ge(111) surface (3.7 % mismatch),[29] although the Ge concentration (28%) in the eutectic Au-Ge mixture is even higher than the Si concentration (18.6%) in the Au-Si system.[48] A typical TEM image of an InP wire grown on Ge(111) is shown in Figure 4a. The flatness of the InP/Ge interface shows that uptake of substrate material to form the eutectic during sample heating and the subsequent precipitation is not the main origin for the surface roughening. This confirms that the rough wire/Si interface is mainly induced by the Au-catalyzed oxidation of Si at ambient.[28,42]
The InP/Ge epitaxy was investigated in more detail by taking the fast Fourier transformation (FFT) of the TEM image of the InP/Ge interface (see Figure 4b).[29] Double spots were observed corresponding to the InP and Ge reciprocal lattices along the [011] zone axis. The lattice spacings corresponding to the InP wire, as extracted from the FFT pattern, are 0.5% lower than the literature values for InP.[49] This shows that for this wire the InP lattice is almost but not fully relaxed within a few nanometers from the Ge substrate. High-resolution XRD measurements revealed that the crystal lattice parameters of the bulk of the wires correspond to the literature values. In addition, diffraction measurements have shown that the crystal lattice orientation of these wires is identical (± 0.09°) to that of the substrate.

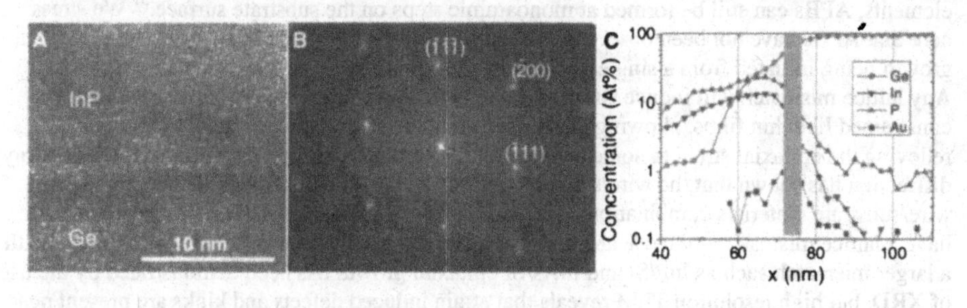

Figure 4. Cross-sectional transmission electron microscopy (TEM) study of (a) an InP nanowire/Ge substrate interface, (b) Fast Fourier transformation of the image shown in (a), and (c) EDX line scan across the InP/Ge junction. The blue area indicates the position of the heterointerface.

In addition to crystal defects, diffusion of atoms across the interface might affect the electronic properties of the wires or of the active areas defined underneath the wires. Group III elements (In, Ga) are *p*-type dopants for Si and Ge, whereas the group V elements (P, As) induce *n*-type doping. For certain applications, a low resistive substrate/wire contact is essential and the interdiffusion could be employed in order to obtain highly doped regions close to the interface. On the other hand, the diffusion of species across the interface might be suppressed by a blocking layer to avoid degradation of highly advanced electronic structures. The interdiffusion has been studied in a TEM with energy dispersive x-ray analysis (EDX).[28] EDX line scans were taken across the interfaces as indicated by the red line in Figure 3a. The heterointerface is in the range of x=10–15 nm for the GaP/Si (blue area in Figure 3c) and around x = 70 nm for InP/Ge (blue area in Figure 4c). The line scans for GaP/Si and InP/Ge show that both group III and V elements diffuse into the substrate to at least some tens of nm below the substrate surfaces. For the GaP/Si system we consistently observed[28] a slightly higher Ga than P concentration in the Si, whereas for InP/Ge the P concentration in the Ge substrate was significantly higher than that of In. These trends are consistent with the solid solubility's of these elements in Si and Ge.[50] Within the detection limit of EDX (0.2%) we could not detect any Au contaminant from the catalyst particle neither in the Si and Ge substrates nor in the nanowires. A small amount of Au was detected close to the interface. This signal originates from particles that have not initiated wire growth after breaking up the thin Au film.[28]

ISSUES AT THE HETERO INTERFACE

The crystallographic quality of the heterointerface might be of crucial importance for certain applications. Crystal imperfections at the interface can be induced by the polar/non-polar nature of the materials and by the lattice mismatch.

Regarding growth of polar III–V semiconductors on a non-polar elemental semiconductor, such as Si or Ge, nucleation can occur with either the group III or the group V element, resulting in opposite polarity. When two nuclei having opposite polarities merge, an antiphase boundary (APB) is formed. Alternatively, when the surface is uniformly covered with one of the precursor elements, APBs can still be formed at monoatomic steps on the substrate surface.[51] We stress here that APBs have not been observed in any of the nanowires. This could be explained by the growth being initiated from a single nucleation site per wire.

Any lattice mismatch will induce strain near the heterointerface. Obviously, wires are not constrained like thin films, allowing elastic deformation of the lattice parallel to the interface, relieving the epitaxial strain to some extent. Detailed analysis of high resolution TEM and x-ray diffraction has shown that the wire/substrate interface is practically free of defects for wire/substrate materials combinations, such as GaP/Si, GaAs/Si, GaAs/Ge and InP/Ge, which have a lattice mismatch up to 4% and a wire diameter typically around 20 nm. For systems with a larger mismatch such as InP/Si and InAs/Si epitaxial growth has been demonstrated by the use of XRD, but high resolution TEM reveals that strain induced defects and kinks are present near the interface. For instance, InAs wires grown on Si(111) contain a complex pattern of twin boundaries at their bottom part, which are oriented parallel to various {111} planes, resulting in a multiple twinned lump of material from which a single wire emerged, growing along its <111> axis. Due to the many twins separating the wire from the substrate, the long wire axis is not always exactly parallel to one of the substrate <111> vectors. For a number of InP wires grown on Si(111), we observed a change into another <111> growth direction as a mechanism to relieve the lattice strain, as indicated in the inset of Figure 1a. Often, such a change was not induced by a twin defect, implying that the growth is continued into a <111> direction with opposite polarity to that of the vertical part.

It seems that a maximum mismatch of about 4 % is allowed for the typical wire dimensions without suffering from defects and kinking. An obvious way to integrate heavily mismatched materials into silicon is to engineer the lattice constant within the wire. A gradual increase of the lattice mismatch with respect to silicon can, for instance, be realised by starting with a GaP (0.4%) segment at the base of the wire, followed by sections of GaAs (4.1%) , InGaAs and InAs (11.6%).

APPLICATIONS & CHALLENGES

Monolithic, or single-crystalline, integration of optically active and high-mobility semiconductor nanowires into the mature silicon technology could boost the performance of existing devices, but could moreover introduce new functionalities. Epitaxial nanowires on (relatively cheap) silicon substrates open up promising applications in the fields of electronics, optoelectronics (light emission[5,19] and detection[52]), sensors (gas and bio sensors[53]), energy scavenging (thermoelectrics[54] and solar cells[55]), and field emission.[17] For instance, nanowire-based *p–n* diodes electronically integrated in the silicon circuitry could enable fast optical inter-chip communication or could distribute and sense the chip's clock frequency optically. Optical signals do not suffer from resistive delays as in electronic circuits and higher frequencies might become available. So far, there are just a few reports showing deliberate *n*- and *p*-type doping of III–V nanowires[5,19,56,57] and an effort should be focused toward obtaining quantitative control of doping levels in the wires.

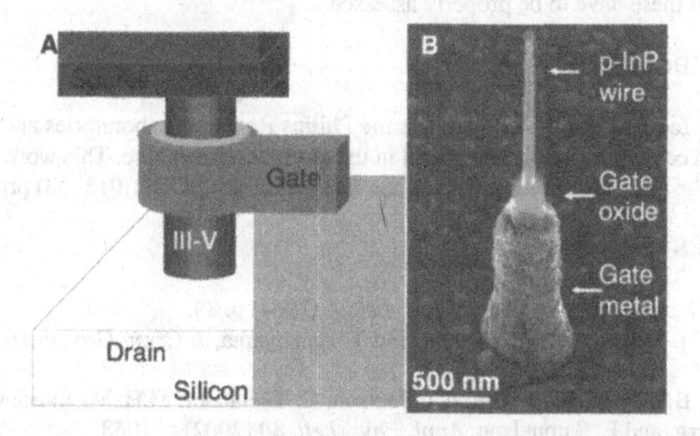

Figure 5. (a) Schematic layout of a vertical nanowire transistor. The active channel is a III–V nanowire epitaxially connected to the silicon substrate, and the gate is wrapped around the channel. (b) SEM image (side view, 30°) of a nearly finished vertical device, consisting of a* p-*type InP wire covered with gate oxide and gate metal.

The possibility to engineer the electronic structure within a nanowire and at the nanowire/substrate junction in combination with a vertical geometry allows optimization of already existing devices, such as a field effect transistor.[58] For instance, InAs is a very interesting material since it has a high electron mobility and it tends to form 'ideal' ohmic contacts[59,60] to many metals. The highest reported values for the electron mobility in InAs wires (up to 10,000 cm^2/Vs)[61] are still below that of the bulk material (33,000 cm^2/Vs). The carrier mobility in wires is most probably limited by surface scattering, which could be suppressed by radial overgrowth of the core with a wide-bandgap semiconductor or by an organic capping.[62] In addition to the freedom of choice of material, a vertical geometry facilitates a wrap-around gate configuration, increasing the electrostatic coupling of the gate to the channel. A vertical FET device with a high mobility channel and a gate-around geometry, as shown in Figure 5, is expected to show an enhanced transconductance.[58] Several groups have recently shown vertical FET devices with

promising characteristics.[61,63–65] However, device performance needs further improvement in order to be able to contribute to current-day nanoscale CMOS devices.

Great progress has been made over the last few years in understanding and controlling nanowire growth and device fabrication. Nanowire heteroepitaxy represents an important step toward the final goal of bringing new materials and novel device layouts into silicon-based integrated circuits.

From a technological point of view an important issue that still remains to be addressed is the vertical growth of III–V wires on currently used Si(100) substrates. Therefore, the nanowires must grow in the [100] direction. Considering the III–V materials, InAs[59,66] and InP[46] wires have been grown in this direction by using homoepitaxy.

Finally, epitaxial growth of nanowires on silicon enables electrical characterization of the junction between heavily mismatched semiconductors. So far, only a few papers have reported preliminary data on the electronic properties of the III–V/Si wire/substrate junction.[27,29] The effects of polar/non-polar interfaces and strain on the electronic properties are important for any application and these have to be properly assessed.

ACKNOWLEDGMENTS

We would like to thank our collaborators at the Philips Research Laboratories and the Delft University of Technology who participated in the work described here. This work was partially supported by the Marie Curie program and the European FP6 NODE (015783) project.

REFERENCES

1. R.S. Wagner and W.C. Ellis, *Appl. Phys. Lett.* **4** (1964) p. 89.
2. K. Hiruma, H. Murakoshi, M. Yazawa, and T. Katsuyama, *J. Cryst. Growth* **163** (1996) p. 226.
3. M.T. Björk, B.J. Ohlsson, T. Sass, A.I. Persson, C. Thelander, M.H. Magnusson, K. Deppert, L.R. Wallenberg, and L. Samuelson, *Appl. Phys. Lett.* **80** (2002) p. 1058.
4. Y. Wu, R. Fan, and P. Yang, *Nano Lett.* **2** (2002) p. 83.
5. M.S. Gudiksen, L.J. Lauhon, J. Wang, D.C. Smith, and C.M. Lieber, *Nature* **415** (2002) p. 617.
6. M.A. Verheijen, G. Immink, T. deSmet, M.T. Borgström, E.P.A.M. Bakkers, *J. Am. Chem. Soc.* **128** (2006) p. 1353.
7. L.J. Lauhon, M.S. Gudiksen, D. Wang, and C.M. Lieber, *Nature* **420** (2002) p. 57.
8. J. Goldberger, R. He, Y. Zhang, S. Lee, H. Yan, H.-J. Choi, and P. Yang, *Nature* **422** (2003) p. 599.
9. J. Hu, Y. Bando, Z. Liu, T. Sekiguchi, D. Goldberg, and J. Zhan, *J. Am. Chem. Soc.* **125** (2003) p. 11306.
10. O. Hayden, A.B. Greytak, and D.C. Bell, *Adv. Mater.* **17** (2005) p. 701.
11. F. Qian, S. Gradecak, Y. Li, C.-Y. Wen, and C.M. Lieber, *Nano Lett.* **5** (2005) p. 2287.
12. E. Ertekin, P.A. Greaney, T.D. Sands, and D.C. Chrzan, in *Mater. Res. Soc. Symp. Proc.* **737** (2003) F10.4.1–6.
13. M. Zervos and L.F. Feiner, *J. Appl. Phys.* **95** (2004) p. 281.
14. R.S. Wagner, in *Whisker Technology*, edited by A.B. Levitt (Wiley Interscience, New York, 1970) p. 47 .

15. E.I Givargizov *J. Cryst. Growth* **20** (1973) p. 217.
16. E.I. Givargizov, *J. Cryst. Growth* **31** (1975) p. 20.
17. E.I. Givargizov, *J. Vac. Sci. Technol. B* **11** (1993) p. 449.
18. M. Yazawa, M. Koguchi, and K. Hiruma, *Appl. Phys. Lett.* **58** (1991) p. 1080.
19. K. Hiruma, M. Yazawa, T. Katsuyama, K. Ogawa, K. Haraguchi, M. Koguchi, and H. Kakibayashi, *J. Appl. Phys.* **77** (1995) p. 447.
20. T.I. Kamins, X. Li, R.S. Williams, and X. Liu, *Nano Lett.* **4** (2004) p. 503.
21. J.W. Dailey, J. Taraci, T. Clement, D.J. Smith, J. Drucker, and S.T. Picraux, *J. Appl. Phys.* **96** (2004) p. 7556.
22. M.H. Huang, S. Mao, H. Feick, H. Yan, Y. Wu, H. Kind, E. Weber, R. Russo, and P. Yang, *Science* **292** (2001) p. 1897.
23. Z. Zhong, F. Qian, D. Wang, and C.M. Lieber, *Nano Lett.* **3** (2003) p. 343.
24. T. Kuykendall, P.J. Pauzauski, Y. Zhang, J. Goldberger, D. Sirbuly, J. Denlinger and P. Yang, *Nature Mater.* **3** (2004) p. 524.
25. Y.F. Chan, X.F. Duan, S.K. Chan, I.K. Sou, X.X. Zhang, and N. Wang, *Appl. Phys. Lett.* **83** (2003) p. 2665.
26. T. Mårtensson, C.P.T. Svensson, B.A. Wacaser, M.W. Larsson, W. Seifert, K. Deppert, A. Gustafsson, L.R. Wallenberg, and L. Samuelson, *Nano Lett.* **4** (2004) p. 1987.
27. A.L. Roest, M.A. Verheijen, O. Wunnicke, S. Serafin, H. Wondergem, and E. P.A.M. Bakkers, *Nanotechnology* **17** (2006) p. S271.
28. M.A. Verheijen, E.P.A.M. Bakkers, A.R. Balkenende, A.L. Roest, M.M.H. Wagemans, M. Kaiser, H.J. Wondergem, and P.C.J. Graat, in *Proc. MSM XIV: Microscopy of Semiconducting Materials, Springer Proc. Physics,* Vol. 107, edited by A.G. Cullis and J.L. Hutchison (2005) p. 295.
29. E.P.A.M. Bakkers, J.A. Van Dam, S. De Franceschi, L.P. Kouwenhoven, M. Kaiser, M. Verheijen, H. Wondergem, and P. Van der Sluis, *Nature Mater.* **3** (2004) p. 769.
30. G.A. Bootsma and H. Gassen, *J. Cryst. Growth* **10** (1971) p. 223.
31. L. Schubert, P. Werner, N.D. Zakharov, G. Gerth, F.M. Kolb, L. Long, U. Gösele, and T.Y. Tan, *Appl. Phys Lett.* **84** (2004) p. 4968.
32. R. Calarco, M. Marso, T. Richter, A.I. Aykanat, R. Meijers, A.v.d. Hart, T. Stoica, and H. Lüth, *Nano Lett.* **5** (2005) p. 981.
33. Y. Ohno, T. Shirahama, S. Takeda, A. Ishizumi, and Y. Kanemitsu, *Appl. Phys. Lett.* **87** (2005) p. 43105.
34. M.C. Plante and R.R. LaPierre, *J. Cryst. Growth.* **286** (2006) p. 394.
35. M.T. Bjork, B.J. Ohlsson, T. Sass, A.I. Persson, C. Thelander, M.H. Magnusson, K. Deppert, L.R. Wallenberg, and L. Samuelson, *Appl. Phys. Lett.* **80** (2002) p. 1058.
36. A.M. Morales and C.M. Lieber, *Science* **279** (1998) p. 208.
37. Y.F. Zhang, Y.H. Tang, N. Wang, D.P. Yu, C.S. Lee, I. Bello, and S.T. Lee, *Appl. Phys. Lett.* **72** (1998) p. 1835.
38. X. Duan and C.M. Lieber, *Adv. Mater.* **12** (2000) p. 298.
39. E.P.A.M. Bakkers and M.A. Verheijen, *J. Am. Chem. Soc.* **125** (2003) p. 3440.
40. D.D.D. Ma, C.S. Lee, F.C.K. Au, S.Y. Tong, and S.T. Lee, *Science* **299** (2003) p. 1874.
41. T.I. Kamins, R.S. Williams, D.P. Basile, T. Hesjedal, and J.S. Harris, *J. Appl. Phys.* **89** (2001) p. 1008.
42. A. Hiraki, E. Lugujjo, and J.W. Mayer, *J. Appl. Phys.* **43** (1972) p. 3643.

43. A.M. Cassell, N.R. Franklin, T.W. Tombler, E.M. Chan, J. Han, and H. Dai, *J. Am. Chem. Soc.* **121** (1999) p. 7975.
44. S. Huang, X. Cai, and J. Liu, *J. Am. Chem. Soc.* **125** (2003) p. 5636.
45. L. Gangloff, E. Minoux, K.B.K. Teo, P. Vincent, V. Semet, V.T. Binh, M.H. Yang, I.Y.Y. Bu, R.G. Lacerda, G. Pirio, J.P. Schnell, D. Pribat, D.G. Hasko, G.A.J. Amaratunga, W.I. Milne, and P. Legagneux, *Nano Lett.* **4** (2004) p. 1575.
46. U. Krishnamachari, M.T. Borgström, B.J. Ohlsson, N. Panev, L. Samuelson, W. Seifert, M.W. Larsson, and L.R. Wallenberg, *Appl. Phys. Lett.* **85** (2004) p. 2077.
47. Z. Ikonic, G.P. Srivastava, and J.C. Inkson, *Phys. Rev. B.* **52** (1995) p. 14078.
48. J.H. Westbrook, Ed., *Moffatt's Handbook of Binary Phase Diagrams* (Genium Group, New York, 2004).
49. D.R. Lide, Ed., *Handbook of Chemistry and Physics* (CRC Press, Boca Raton, Fla., 1995).
50. W. Scott and R.J. Hager, *J. Electron. Mater.* **8** (1979) p. 581.
51. S.F. Fang, K. Adomi, S. Iyer, H. Morkoc, H. Zabel, C. Choi, and N. Otsuka, *J. Appl. Phys.* **68** (1990) p. R31.
52. O. Hayden, R. Agarwal, and C.M. Lieber, *Nature Mater.* **5** (2006) p. 352.
53. Y. Cui, Q. Wei, H. Park, and C.M. Lieber, *Science* **293** (2001) p. 1289.
54. R.F. Service, *Science* **306** (2004) p. 806.
55. M. Law, L.E. Greene, J.C. Johnson, R. Saykally, and P. Yang, *Nature Mater.* **4** (2005) p. 455.
56. X. Duan, Y. Huang, Y. Cui, J. Wang, and C.M. Lieber, *Nature* **409** (2001) p. 66.
57. M.H.M. van Weert, O. Wunnicke, A.L. Roest, T.J. Eijkemans, A. Yu Silov, J.E.M. Haverkort, G.W.'t Hooft, and E.P.A.M. Bakkers, *Appl. Phys. Lett.* **88** 043109 (2006).
58. H.S.P. Wong, *IBM J. Res. Dev.* **46** (2002) p. 133.
59. Y.-J. Doh, J.A. van Dam, A.L. Roest, E.P.A.M. Bakkers, L.P. Kouwenhoven, and S. De Franceschi, *Science* **309** (2005) p. 272.
60. J.A. van Dam, Y.V. Nazarov, E.P.A.M. Bakkers, S. De Franceschi, and L.P. Kouwenhoven, *Nature* **442** (2006) p. 667.
61. T. Bryllert, L.-E. Wernersson, L.E. Fröberg, and L. Samuelson, *IEEE Electron Dev. Lett.* **27** (2006) p. 323.
62. L.K. van Vugt, S.J. Veen, E.P.A.M. Bakkers, A.L. Roest, and D. Vanmaekelbergh, *J. Am. Chem. Soc.* **127** (2005) p. 12357.
63. H.T. Ng, J. Han, T. Yamada, P. Nguyen, Y.P. Chen, and M. Meyyappan, *Nano Lett.* **4** (2004) p. 1247.
64. J. Goldberger, A.I. Hochbaum, R. Fan, and P. Yang, *Nano Lett.* **5** (2006) p. 973.
65. V. Schmidt, H. Riel, S. Senz, S. Karg, W. Riess, and U. Gösele, *Small* **2** (2006) p. 85.
66. W. Seifert, M. Borgström, K. Deppert, K.A. Dick, J. Johansson, M.W. Larsson, T. Mårtensson, N. Sköld, C.P.T. Svensson, B.A. Wacaser, L.R. Wallenberg, L. Samuelson, *J. Cryst. Growth.* **272** (2004) p. 211.

Mater. Res. Soc. Symp. Proc. Vol. 1068 © 2008 Materials Research Society 1068-C02-05

Growth and Characterization of InSb Films on Si (001)

Lien Tran, Julia Dobbert, Fariba Hatami, and W. Ted Masselink
Department of Physics, Humboldt-Universität zu Berlin, Newtonstr. 15, Berlin, 12489, Germany

ABSTRACT

In this paper we describe the growth of InSb on Si (001) using molecular-beam epitaxy and discuss the structural and electrical properties of the resulting InSb films. The samples were characterized in terms of background electron concentration, mobility, deep level traps, Hall sensitivity, and x-ray rocking curve. We have investigated samples grown at temperatures between 300°C and 420°C. To prevent the formation of the defects we introduced in some samples GaSb/AlSb superlattice buffer layer. The best structural quality has been achieved at a growth temperature of 420°C using GaSb/AlSb superlattice buffer layer, resulting in our best electron mobility of 26500 cm^2/Vs for a 2μm film at room temperature. This sample has the narrowest x-ray rocking curve width. Deep level noise spectra indicate the existence of the deep levels in all samples in a temperature range of 80 to 300 K. The sample with the best crystal quality and highest mobility, however, has the lowest trap density. The deep levels have a temperature dependent behavior.

INTRODUCTION

The replacement of native oxides with deposited oxides in CMOS technology opens the door to replacing the Si with semiconductors without high-quality native oxides. For example, the use of InSb in logic applications could allow much lower operating voltages and power dissipation due to the InSb channels reaching saturation at significantly lower electric fields. Epitaxy of InSb onto Si could be done directly or using an intermediate layer such as GaP, GaAs, or InP. Furthermore, InSb, the narrowest bandgap III-V material, has potential applications in high speed devices, infrared (IR) detectors and magnetic sensors. Especially for IR application, InSb thin films should be grown on semi-insulating IR transparent substrates [1].
Although, there is a large lattice mismatch between GaAs and InSb (14.6%), GaAs is still an attractive substrate material due to its chemical stability and high resistivity. The epitaxial growth of InSb on GaAs substrates have been demonstrated by several groups [2-6]. Compared to the growth on GaAs, less has been reported on the heteroepitaxy of InSb on Si [7-8]. The difficulties involved in growth of high quality InSb films on Si include the large lattice mismatch (19%), the different thermal expansion coefficients ($\alpha_{InSb} \sim 2\alpha_{Si}$ at room temperature) and antiphase domain formation due to the growth of polar semiconductor on nonpolar semiconductor. Some problems have been eliminated using tilted substrates [9], indium pre-deposition and the insertion of various buffer layers [10–12].
In the current paper, we describe the detailed growth of InSb on Si (001) using molecular-beam epitaxy (MBE) and discuss the structural and electrical properties of the resulting InSb films. Our results are based on X-Ray diffraction, Hall-effect measurement and the deep level noise spectroscopy.

EXPERIMENT

Samples were grown using molecular-beam epitaxy in a Riber-Compact 21T system. Antimony was supplied with a Veeco valved cracker cell and was cracked at temperature between 800- 900°C. The indium beam was provided by conventional effusion cell. The structures were grown on p-type Si substrates with 4° off-cut towards (110) and resistivity of about 50-80 Ωcm. Substrates were prepared by repeated oxidation and oxide-removal and then loaded into the MBE system. After the substrate temperature had been increased to about 820°C, the surface shows a clear 2x4 reconstruction and appears to be free of oxide. This reconstruction remains until the substrate temperature reaches 1015°C and a 2x1 appears, indicating a dominance of double-height steps. After cooling the substrate to the intended growth temperature for InSb, the substrate exposed to Sb, resulting in a change in surface reconstruction from 2x1 to 1x1. The 1x1 reconstruction remains throughout the subsequent InSb deposition. InSb was deposited with an Sb/In flux ratio of about 5 and a growth rate of 0.2 nm/s. We have investigated growth temperatures between 300 and 420°C. To prevent the formation of the defects we introduced in some samples GaSb/AlSb superlattice. In those samples, after cooling the substrate temperature to the InSb growth temperature, a 30 period strained layer superlattice consisting of 1nm GaSb/1nm AlSb periods was incorporated between substrate and 2μm InSb layer. After the growth the InSb films the temperature was decreased to about 200°C under the antimony flux. The quality of the films has been examined using in-situ RHEED, X-ray diffraction (XRD), and atomic-force microscopy. XRD measurements have been carried out in a Bede scientific QC1a diffractometer. Atomic-force micrographs were taken using a Veeco CP-R ThermoMicroscopes in contact mode. Hall mobility and carrier concentration have been studied using the van der Pauw method on 4x4 mm^2 samples. Deep level noise spectroscopy has been carried out using Greek cross geometry between 80 and 300 K.

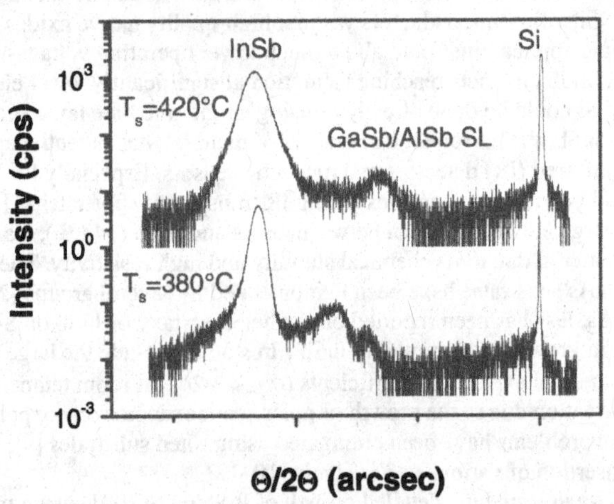

Fig. 1. Double-crystal x-ray rocking curves of InSb films grown on GaSb/AlSb SL on Si Substrate.

DISCUSSION

The crystal quality of the InSb films has been investigated using the X-Ray diffraction measurements. To improve the crystal quality, we applied several techniques. One of the techniques was two-step temperature growth; first growing of an InSb layer at 350°C and then increasing the substrate temperature and growth the second layer at this temperature. We also used the insertion of various buffer layers between substrate and active InSb layer such as AlSb, GaAs and GaSb/AlSb superlattice to suppress the formation of defects due to the lattice mismatch. Structures grown on GaSb/AlSb superlattice show the best crystal quality. Figure 1 shows the (004) double-crystal X-ray (DCXR) patterns of two samples grown at 380°C and 420°C with GaSb/AlSb superlattice. The right shoulder of the InSb peak is due to the strain between the layers. The InSb peaks related with (004) plane appeared with quite high intensity. For the sample grown at 420 °C, the FWHM is found about 951 arcsecs, which is the narrowest bandwidth among the InSb/Si samples grown at different conditions. The lattice constant of the InSb film determined from the XRD peak is 0.648 nm. This value matches closely the lattice constant of the InSb bulk. According to the data obtained from the atomic-force microscopy the root-mean-squared roughness (RMS) for areas around 2x2 μm^2 is lower then 16 nm. Our results show that good quality InSb epilayers can be grown on Si substrates using the GaSb/AlSb superlattice layer.

In order to assess the electrical properties, Hall measurements with Van der Pauw geometry were performed on samples. These measurements provide apparent carrier mobility and concentration. Figure 2 shows the Hall mobility as a function of temperature. The temperature dependencies of the electron mobility in all samples are almost identical. The highest Hall mobility of 26500 cm^2/Vs (corresponding to $n = 2.5e16\ cm^{-3}$), can be observed in the sample grown at 420°C on GaSb/AlSb superlattice.

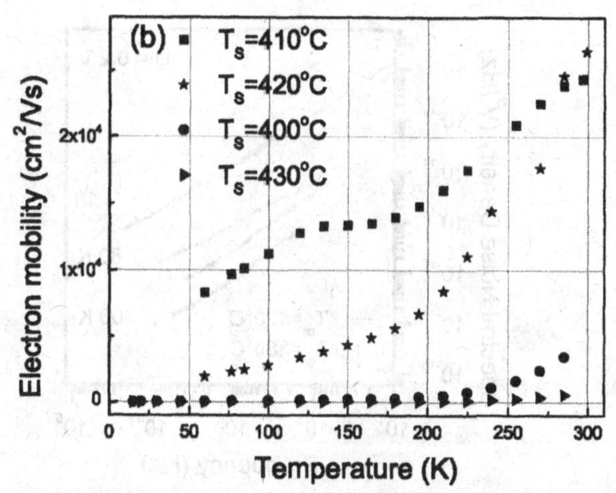

Fig. 2. The temperature dependence of the electron mobility for InSb films grown on Si substrate. The thickness of InSb films is 2 μm. The sample grown at 420°C has GaSb/AlSb SL buffer layer.

The electron mobility increases with increasing temperature. According to Matthiessen's rule, the electron mobility of InSb layer, μ, can be related to μ_b and μ_d as follows:
$1/\mu = 1/\mu_b + 1/\mu_d$
where μ_b is the intrinsic electron mobility of the bulk material and μ_d is the mobility limited by dislocation scattering. Electrons can be scattered by the depletion potential around the dislocations and the lattice dilatation associated with the dislocations. Using the Pödor and Dexter-Seitz models [13-14], Weng and Rudawski [2] found that the dominant factor limiting the electron mobility was the lattice dilatation scattering, which is apparently consistent with the predictions of the Dexter-Seitz model. According to this model, the mobility limited by deformation potential scattering μ_d is inversely proportional to the edge dislocation density D. These results, therefore, show a quantitative correlation between increasing dislocation density and decreasing electron mobility. The dislocation scattering increases as the temperature decreases since the screen effect of free carriers decreases. Therefore, the degradation of electron mobility is observed at low temperature.
To further investigation of dislocation in the samples, noise spectroscopy was carried out on the samples with GaSb/AlSb superlattice layer on Si. Figure 3 shows noise spectra for both samples using Greek cross geometry of 17 µm square size. The measurements were accomplished at 300 and 80 K and a bias voltage of 200 mV. At 300 K both samples exhibit similar flicker noise ($\propto 1/f$) features. The Hooge factor α_H is defined by $S_V=\alpha_H V^2/Nf$, where S_V is the spectral noise density, V the bias voltage, N the number of carriers and f the frequency. For both samples the Hooge factor is about 5e-5 at room temperature, which is quite usual for III-V materials. The flicker noise increases strongly with decreasing temperature for the sample grown at 380°C, which can be ascribed to the reduction of free carriers, since the Hooge factor is nearly constant with 7e-5 at 80 K. The carrier density of sample grown at 420°C does not change with decreasing temperature.

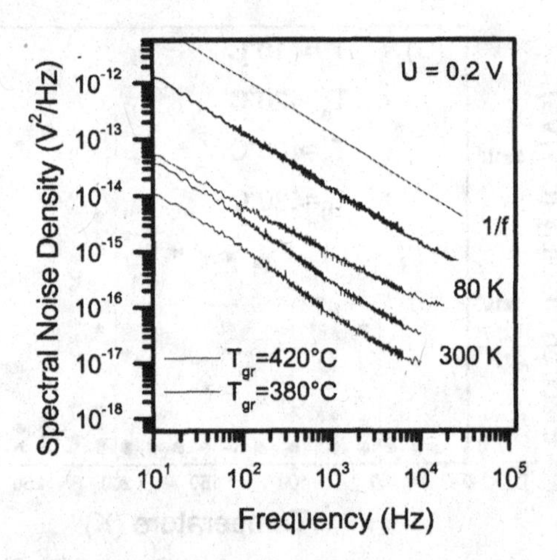

Fig 3. Noise spectra of InSb samples grown on GaSb/AlSb superlattice layer on Si substrate at two different substrate temperatures. Spectra were measured at 80 and 300 K for Greek cross geometry with a bias voltage of 200 mV.

The rising flicker noise must here be ascribed to the rise of the Hooge factor. At 80 K α_H is 3e-4. It shows together with the low mobility of the electrons that the transport may happen in an environment rich of defects. Both samples exhibit several deep levels, which could be examined by the deep level noise spectroscopy. Pronounced Generation-Recombination noise has been detected in sample grown at 380°C between 120 K and 140 K and between 170 K and 230 K, and in sample grown at 420°C between 160 K and 230 K.

CONCLUSIONS

To conclude, we have demonstrated the growth of high quality InSb films on (001) Si substrate using MBE in spite of very large lattice mismatch between InSb and the substrate. The highest room temperature electron mobility is 26500 cm^2/Vs for 2μm-thick InSb film grown on GaSb/AlSb superlattice buffer layer at 420 °C. Our results indicate a highly correlation between the crystal quality of InSb films and the growth condition such as substrate temperature, growth rate, Sb/In flux ratio, and the substrate tilting. Deep level noise spectra indicate the existence of the deep levels in all samples. The sample with the best crystal quality and highest mobility has the lowest trap density.

ACKNOWLEDGMENTS

The authors would like to thank O. Bierwagen and M. P. Semtsiv at Humboldt-Universität zu Berlin, and Was. P. Kunets at University of Arkansas (USA) for helpful discussions and assistance.

REFERENCES

1. A. Rogalski, R. Ciupa, and W. Larkowski, Solid-State Electronics 39(11), 1593 (1996).
2. X. Weng, N. G. Rudawski, P. T. Wang, and R. S. Goldman, Journal of Applied Physics. 97, 043713 (2005).
3. E. Michel, G. Singh, S. Slivken, P. Bove, I. Ferguson, and M. Razeghi, Applied Physics Letters. 65, 26 (1994).
4. V. K. Dixit, Bhavtosh Bansal, V. Venkataraman, H. L. Bhat, G. N. Subbanna, K. S. Chandrasekharan, and B. M. Arora Applied physics letters. 80, 12 (2002).
5. D. L. Partin, L. Green and J. Heremans, Journal of Electronic Materials. 23, 2 (1994).
6. T. L. Tran, F. Hatami, and W.T. Masselink, to be published in Journal of Electronic Materials (2008).
7. J.-I. Chyi, D. Biswas, S. V. Iyer, N. S. Kumar, H. Morko, R. Bean, K. Zanio, H.-Y. Lee, and Haydn Chen, Appl. Phys. Lett. 54, 11 (1989).
8. S. V. Ivanov, A. A. Boudza, R. N. Kutt, N. N. Ledentsov, B. Ya. Meltser, S. S. Ruvimov, S. V. Shaposhnikov, P. S. Kop'ev, Journal of Crystal Growth. 156, 191-205 (1995).
9. H. C. Lu, H. R. Fetterman, C. J. Chen. C. Hsu, and T. M. Chen, Solid-State Electron. 36, 533 (1993).
10. D. M. Li, M. Yamazaki, T. Okamoto, T. Tambo, C. Tatsuyama, Applied Surface Science. 130-132, 101-106 (1998).
11. W. K. Liu, J.Winesett,Weiluan Ma, Xuemei Zhang, M. B. Santos, X. M. Fang and P. J. McCann, J. Appl. Phys. 84, 4 (1997).

12. M. Mori, N. Akae, K. Uotani, N. Fujimoto, T. Tambo, C. Tatsuyama, Applied Surface Science. 216, 569-574 (2003).
13. B. Pödör, Phys. Status Solidi 16, K167 (1966).
14. D. L. Dexter and F. Seitz, Phys. Rev. 86, 964 (1952).

Mater. Res. Soc. Symp. Proc. Vol. 1068 © 2008 Materials Research Society 1068-C02-06

A Study of Conformal GaAs on Si Layers by Micro-Raman and Spectral Imaging Cathodoluminescence

Oscar Martínez[1], Luis Felipe Sanz[1], Juan Jiménez[1], Bruno Gérard[2], and Evelyn Gil-Lafon[3]
[1]Física de la Materia Condensada, Universidad de Valladolid, Paseo de Belen 1, Edificio de I+D, Valladolid, 47011, Spain
[2]Corporate Research Laboratory, THALES, Orsay Cedex, 91404, France
[3]LASMEA UMR CNRS 6602, Université Blaise Pascal, Les Cézeaux, Aubiére Cedes, 63177, France

ABSTRACT

Spectral imaging cathodoluminescence and micro-Raman spectroscopy studies of GaAs layers grown on Si substrates by the conformal method allow to reveal a great variety of physical features of the layers, such as the complete stress distribution, self-doping effects, or the incorporation of dopants. We present herein the characterization of GaAs conformal layers grown by hydride vapor phase epitaxy, where the main issues concerning the distribution of defects and stresses are revealed. Also, intentionally doped layers were analyzed, revealing the main aspects of the incorporation of dopant impurities during growth.

INTRODUCTION

GaAs is a material of choice for optoelectronic applications due to its specific properties, with some important advantages over silicon, like the fact that it posses a direct band gap of 1.42 eV at room temperature [1]. In particular, this makes it the ideal semiconductor for the production of solar cells in terms of conversion efficiency [2]. It is also used for the fabrication of light emitting diodes and laser diodes [3]. However, it is rather expensive compared to Silicon, which prevents its introduction in some niches of volume production. GaAs based optoelectronic devices present a very low tolerance to the presence of dislocations. Epitaxial GaAs on Si substrates, that could benefit from drastic cost reduction, suffers form the large lattice mismatch (4%) and the large difference of thermal expansion coefficients (56%) between Si and GaAs [1], which results in dislocation densities, $>10^6$ cm^{-2}, unacceptable for optoelectronic devices; e.g. the luminescence emission is dramatically quenched [4]. Therefore, one needs to explore alternative ways to prepare high quality GaAs epitaxial layers on Si. Among the alternatives to the direct GaAs/Si growth, the conformal growth, in which the GaAs layers grow laterally from GaAs seeds previously deposited on a Si substrate, have shown to produce a material with optical quality very improved, because the dislocation density is drastically reduced. A complete description of the growth method can be found elsewhere [5].

In this paper we investigated by means of spectral imaging (SI) cathodoluminescence (CL) in the scanning electron microscope (SEM) and micro-Raman spectroscopy (μ-R) the optical properties of GaAs on Si layers grown by the conformal method. Both characterization techniques provide for a fast and non destructive investigation of the material in terms of the crystal quality and the identification of defects.

EXPERIMENT

Nominally undoped conformal samples were prepared on 2° misoriented (001) silicon substrates on which 1.5 μm-thick antiphase domain-free GaAs layers were grown by metalorganic vapor-phase epitaxy. The GaAs layer is covered with a poly-Si dielectric cap layer in which [110] oriented stripes are periodically opened (typically 10 μm wide every 200 μm). The GaAs layer is then selectively underetched using an $H_2SO_4/H_2O_2/H_2O$ etchant solution, so as to realize 125 μm-wide oriented <110> GaAs seed stripes. The conformal growth step is then carried out by hydride vapor-phase epitaxy (HVPE) at 730°C, using gaseous GaCl molecules and arsine (AsH_3) precursors. The lateral growth rate was about 8 μm/h, with a total width of the conformal layer reaching more than 40 μm [5].

Spectrally resolved CL measurements were carried out with a XiCLOne system from Gatan attached to a JEOL 820 SEM. The luminescence detection was done with a Peltier cooled charge coupled device (CCD) camera allowing the acquisition of the full spectrum at each scanned pixel. Further data treatment allows the spectral imaging (SI), giving the distribution of the different spectral parameters [intensity, peak wavelength, and full width at half maximum (FWHM)] of the different luminescence band emissions. The measurements were carried out at liquid nitrogen temperature (~80 K). μ-R spectra were recorded with a HR LabRam spectrometer (Jobin-Yvon) attached to a metallographic microscope. The excitation was done with an Ar^+ laser through the microscope objective, which also collected the scattered light, thus conforming a nearly backscattering geometry. The nominal spatial resolution, corresponding to the beam diameter at the focal plane, was ~0.7 μm in our usual experimental conditions (λ=514 nm, 100X objective, numerical aperture=0.95). The detection was done with a liquid nitrogen cooled CCD camera.

DISCUSSION

Panchromatic CL images give a first overview of the luminescence distribution of the conformal layers, the contrast is mostly governed by the radiative/non radiative recombination ratio, figure 1a. The luminescence intensity is almost killed in the GaAs seed, due to the high density of dislocations. On the other hand, the intensity is highly increased in the conformal layer, thus accounting for a drastic reduction of the non radiative recombination centers (NRRCs), among other the dislocations. However, the emission intensity is not uniform over the layer, but presents fluctuations both parallel and perpendicular to the growth direction. The spectrum of each individual pixel (of a matrix of pixels selected at convenience) is obtained in our CL equipment. Spectra at points (1) to (3) (as indicated in figure1a) have been represented (inset of the figure), showing important changes in amplitude, peak position and FWHM of the main band around 875 nm, related to band to band transitions [1]. Accompanying the increase in intensity from point (1) to (2) (respectively dark and bright areas in the image), the peak position shifts to the blue, while the FWHM decreases. Point 3, near to the seed, shows a quite different behavior, with a decrease of intensity respect to point (2), a high blue shift and an increase of the FWHM. The SI capability of the CL system allows mapping not only the whole luminescence but the different spectral features, such as the amplitude of the different bands, their peak wavelengths and their FWHM, which allows visualizing the local inhomogeneities and to correlate them with the structure of the samples.

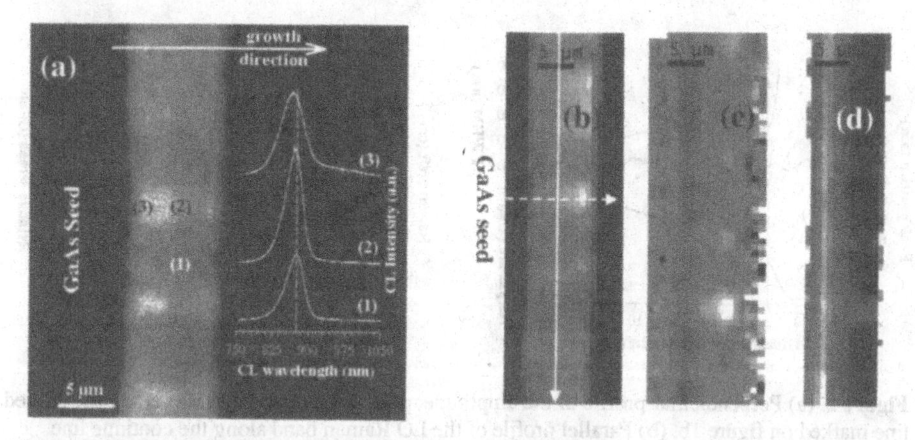

Figure 1. (a) PanCL image of a conformal layer. The inset shows the CL spectra taken at points (1) to (3) as indicated. Spectral imaging maps of (b) the amplitude, (c) peak position and (d) FWHM of the band to band emission, in the studied area.

Stress distribution

Figures 1b-d shows the maps of the amplitude, peak position and FWHM of the 875 nm band in the studied area. The amplitude shows again the quasi-periodically distributed fluctuation of the luminescence intensity perpendicular to the growth front. The peak position and FWHM also reproduce such fluctuations. A dark fringe parallel to the growth front is also observed in the amplitude map, and much better observed in the peak position and FWHM images.

Regarding the fluctuations perpendicular to the growth front, there is a quite good correlation in the fluctuations of the amplitude, peak wavelength and FWHM, the peak shifts to the blue in the areas of higher amplitude, with a drop of the FWHM. Both the peak wavelength and FWHM can be correlated to the stress level. The valence band is degenerated for stress free GaAs crystals, while this degeneracy is lifted in the presence of stress, which results in two band to band transitions corresponding to conduction band to both heavy hole (e-hh) and light hole (e-lh) valence bands [6]. These two bands are not resolved at 80 K in our experimental setup, which means that the residual stresses are below the threshold for a complete splitting out of the two bands. However, if the stress level increases, the peak position of the convoluted band is expected to change, shifting to the red in the case of increasing tensile stress, with an increase in the FWHM. We observed that the dark areas are under higher tensile stress levels respect to the bright areas, as observed from the red shift of the peak position and the increase of the FWHM,

figure1b-d. The fluctuations are well observed for the whole conformal layer. However, a continuous increase of the amplitude and peak wavelength values, and a continue drop of the FWHM is observed as the growth proceed, see figure 2a (apart from the abrupt change at the position corresponding to the fringe parallel to the seed, discussed later on). This indicates a continue decrease of the tensile stress level as the growth proceed.

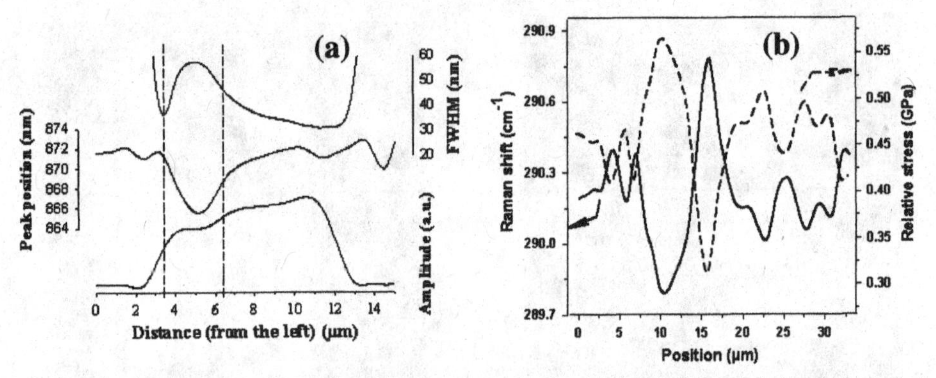

Figure 2. (a) Perpendicular profile of the amplitude, peak position and FWHM along the dashed line marked on figure 1b. (b) Parallel profile of the LO Raman band along the continue line marked on figure 1b. The relative stress distribution (right labels) has been calculated from it (see text).

The observed features might be attributed to the specific characteristics of the conformal growth. The GaAs conformal layer grows from GaAs seeds that are not fully released. The GaAs conformal layer is matched to the seed, close to its lattice parameter; going far in the growth direction the influence of the substrate becomes more important and the lattice expands. This expansion is not uniform, but it finds energetically favorable a quasiperiodic distribution, as observed. The fact that the luminescence intensity decreases in the areas under higher tensile stress levels should be due to the accumulation in these areas of nonradiative recombination centers, probably As_{Ga} related defects, or Oxygen impurities.

Raman spectroscopy was also used for characterizing the conformal layers. The longitudinal optic (LO) Raman band shifts obtained along a line parallel to the conformal layer has been plotted in figure 2b. The LO band is shifted to high wavenumbers for the bright areas respect to the dark areas, which also evidences the stress relaxation in the bright regions of the conformal layers. According to the observed shift [7], the stress levels fluctuations between the stressed and relaxed regions are of the order of 0.3 MPa, with a total stress level of 0.6 MPa, which are lower than the yield strength in GaAs [8], which accounts for the very low dislocation density observed in these layers.

Self-doping

The fringe parallel to the seed observed in the images of figures 1b-d has a behavior different from the quasi-periodic modulation of the intensity discussed before. In fact, the intensity is partially reduced, with an important blue shift of the peak wavelength and an important increase of the FWHM (see figure 2a and the spectrum at point (3), figure 1a). Such a behavior is not ascribed to a variation of the stress level. In this case, the spectral changes correlate with the incorporation of dopants in this region, probably Si coming from the substrate. The hypothesis here is that Si diffuses from the substrate due to the presence of dislocations in this part of the conformal layer. It could be observed that superimposed to the bright fringe the quasi-periodic fluctuation discussed before is also observed. This modulation is not well defined for the amplitude and FWHM, while it is well observed for the peak wavelength. The mean value of the peak wavelength is lowered respect to the other areas of the conformal layer (see figure 2a), corresponding to the blue shift due to the incorporation of Si dopants. In this case, the luminescence band is contributed by stress and dopant incorporation. As observed, the Si incorporation follows again the stress distribution, showing that stress distribution plays a relevant role in the distribution of defects and impurities in these layers.

Intentional doping

Intentionally doped samples, which are of interest for the realization of junctions, were also prepared and studied by CL spectral imaging and Raman spectroscopy. Figure 3a shows the PanCL image of a conformal layer with three fringes, two non intentionally doped (corresponding to the bright fringes) and one intentionally doped with Si in between the two previous ones (corresponding to the dark fringe). The CL spectra are different for the different regions of the layer, see the inset on figure 3a. In the non doped areas, it mainly consists of the band to band transition at 875 nm. On the other hand, the intentionally doped fringes present a strong drop of the band to band emission, while a band centered at ~930 nm, related to Si complexes [9], peaks up. Figure 3b shows the profiles parallel to the fringes of the amplitude of the 875 nm band for the non doped fringes and the amplitude of the 930 nm band for the doped one. The quasi-periodic modulation due to stress is well observed in the profile of the amplitude of the band to band emission. Moreover, the amplitude of the Si-complex related band follows exactly these fluctuations, indicating that the incorporation of Si complexes is also mediated by the stress distribution.

The Raman spectrum obtained in the intentionally doped fringes (lower inset on figure 3a) presents LO phonon-plasmon coupled (LOPC) modes due to n-doping. The carrier concentration was estimated using a hydrodynamic approach [10], giving values of 4.10^{18} cm^{-3}. The Raman data show abrupt interfaces between doped and undoped stripes.

CONCLUSIONS

GaAs conformal layers grown on Si have been studied by spectral imaging cathodoluminescence and Raman spectroscopy. The samples present an intense luminescence, related to the high crystal quality and the low dislocation density of the conformal layers. A quasiperiodic fluctuation of stress is revealed from the CL spectrum images. Raman spectroscopy shows the same fluctuation in the peak position of the LO band, indicating stress level fluctuations of 0.3 MPa, with a total stress level of 0.6 MPa. A self-doping effect, related to the incorporation of Si atoms coming from the substrate, was observed. Intentionally doped samples (Si doped) showed abrupt interfaces between doped and undoped stripes.

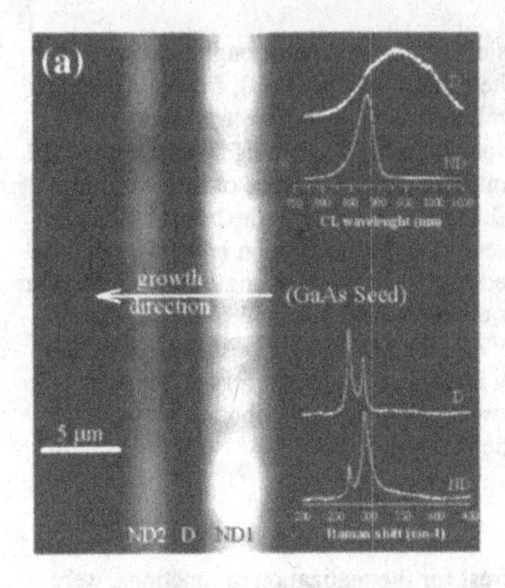

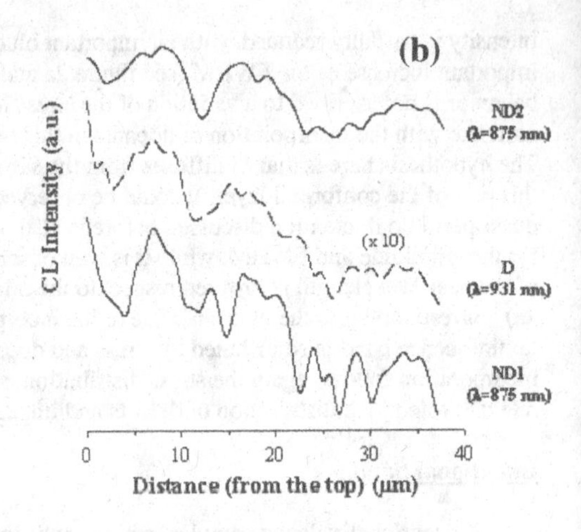

Figure 3. (a) PanCL image of a conformal layer with three fringes, two non intentionally doped (ND) and one intentionally doped (D). The upper inset shows the CL spectra of both ND and D areas. The lower inset shows the Raman spectra of both ND and D areas. (b) Parallel profile of the amplitude of the band at 875 nm (non doped fringes) and of the band at 930 nm (doped fringe).

ACKNOWLEDGMENTS

This work was done within the framework of an EU contract (Contract No. BRPR CT97-0512, CONFORM). The Spanish group of authors was funded by the Consejería de Educación de la Junta de Castilla y León under research Project No. VA103A06.

REFERENCES

1. S.F. Fang, K. Adomi, S. Iyer, H. Morkoc, H. Zabel, C. Choi, and N. Otsuka, J. Appl. Phys. **68**, R31 (1990)
2. M. Tsugami and K. Mitsui, Optoelectronics Devices and Technologies **4**, 59 (1989)
3. T. Egawa, T. Jimbo, and M. Umeno, Jpn. J. Appl. Phys., Part 1 **32**, 650 (1993).
4. T. Yodo and M. Tamura, Jpn. J. Appl. Phys., Part 1 **34**, 3457 (1995).
5. E. Gil-Lafon, J. Napierala, D. Castelluci, A. Pimpinelli, R. Cardonet, and B. Gerard, J. Crys. Growth **222**, 482 (2001)
6. C.M.N. Mateo, A.T. Garcia, F.R.M. Ramos, K.I. Manibog, and A.A.Salvador, J. Appl. Phys. **101** (2007) 073519
7. M. Holtz, M. Seon, O. Brafman, R. Manon, and D. Fekete, Phys. Rev. B **54**, 8714 (1996)
8. P. Wickboldt, E. Anastassakis, R. Sauer, and M. Cardona, Phys. Rev. B **35**, 1362 (1987)
9. K.N. Hong, L. Pavesi, D. Araújo, J.D. Ganière, and F.K. Reinhart, J. Appl. Phys. **70**, 3887 (1991)
10. J. Ibáñez, R. Cuscó, and L. Artús, Phys. Status Solidi (b) **223**, 715 (2001).

Mater. Res. Soc. Symp. Proc. Vol. 1068 © 2008 Materials Research Society 1068-C07-02

Epitaxial Growth of Single Crystalline Ge Films on GaAs Substrates for CMOS Device Integration

Hock-Chun Chin, Ming Zhu, Ganesh Samudra, and Yee-Chia Yeo
Department of Electrical and Computer Engineering, National University of Singapore, Silicon Nano Device Laboratory, Singapore, 119260, Singapore

ABSTRACT

We report a novel chemical vapor deposition (CVD) process for epitaxial growth of Ge film on GaAs substrate. The resultant layer exhibits device level quality, as shown by high-resolution transmission electron microscopy (HRTEM), Raman spectroscopy, high-resolution X-ray diffraction (HRXRD). In addition, atomic force microscopy (AFM) scanning indicates low RMS surface roughness of 5 Å. Secondary ion mass spectrometry (SIMS) reveals negligible out-diffusion of Ga and As into the Ge epilayer. By employing silane passivation, Ge p-MOSFET with TaN/HfO_2 gate stack was fabricated on Ge/GaAs heterostructure for the first time, showing excellent output and pinch-off characteristics. A GaAs channel n-MOSFET was also fabricated, using similar SiH_4 treatment during gate stack formation. These results reveal a potential solution to integrate Ge p-channel and GaAs n-channel MOSFET for advanced CMOS applications.

INTRODUCTION

High mobility materials have recently been actively explored to replace conventional Si or strained Si channels for advanced CMOS technology beyond the 22 nm technology node. III-V compound semiconductors and Ge are attractive candidates for n-MOSFETs and p-MOSFETs due to their high electron and hole mobility, respectively. GaAs and Ge MOSFETs have been demonstrated recently [1-3]. In addition, progress has been reported on improving the quality of metal gate/hafnium-based high-κ dielectric stacks on GaAs and Ge by using surface passivation method such as surface nitridation [4-6] and silane passivation [7-8]. However, the integration of III-V n-MOSFET with Ge p-MOSFET remains as a major challenge for CMOS application. One easy way to integrate GaAs and Ge MOSFETs on a single platform is to hetero-epitaxial growth of a thin Ge layer on selected regions of the GaAs substrate. The Ge/GaAs heterostructure has been investigated for several applications, such as solar cells, hetero-bipolar transistors, and waveguides for electro-optic integrated circuits. However, the integration of MOS devices on this Ge/GaAs heterostructure has not been reported due to the stringent requirements on the Ge epi-layer. For MOS applications, this Ge epilayer should meet the following criteria, including (1) excellent single crystalline quality and free of defects and dislocations; (2) minimal surface roughness to reduce mobility degradation by surface roughness scattering; (3) relatively thin such that the step height between Ge layer and GaAs is within the depths of focus of lithography tools. Recently, we have demonstrated effective passivation of GaAs surface by employing silane (SiH_4) treatment in a multiple chamber metal-organic chemical vapor deposition (MOCVD) cluster system [8]. This SiH_4 passivation is also useful for Ge [7]. In this paper, we report an effective method for hetero-epitaxial growth of a high-quality single crystalline Ge layer on GaAs substrate using a multiple chamber CVD system. High mobility Ge p-MOSFET fabricated on this Ge/GaAs heterostructure was demonstrated for the first time. In addition, we also report a GaAs n-MOSFET on the underlying GaAs substrate.

EXPERIMENTAL DETAILS

Fabrication of Ge/GaAs substrates

Starting semi-insulating GaAs (100) wafers were first degreased in acetone and isopropanol before native oxide and elemental As removal using HCl and NH_4OH. The wafers were then quickly loaded into a high vacuum multiple chamber MOCVD cluster tool. The native oxide of GaAs is very unstable at moderate temperature (≥ 580 °C) [9]. In the cluster tool, a 600°C vacuum anneal (pressure 1×10^{-7} Torr) was carried out in the first process chamber for thermal desorption of native oxide. The As evaporation rate is very low at this moderate temperature [10]. Nevertheless, the duration of vacuum annealing was limited to only 60 s in the experiment in order to reduce excessive As evaporation from the GaAs surface. After vacuum baking, the wafers were transferred to a second chamber for CVD deposition of Ge, without breaking vacuum using our high vacuum transfer module. The CVD process employed GeH_4 (10% GeH_4 + 90% Ar) and N_2 gases at temperature of 400°C and process pressure of 500 mTorr. For comparison, Ge films were also deposited on another GaAs wafers by using magnetron sputtering process. In the PVD process, the GaAs wafers received Ar sputtering for surface cleaning before Ge deposition at temperature of 350 °C. After deposition of Ge films on GaAs, post-deposition anneal (PDA) was performed in a rapid thermal processor at 600 °C for 30 s in N_2 ambient to improve the quality of the Ge film. Extensive material analysis was performed to investigate the Ge/GaAs heterostructure. High-resolution transmission electron microscopy (HRTEM), Raman spectroscopy and high-resolution X-ray diffraction (HRXRD) were employed to examine the crystalline quality of the Ge layers. The surface morphology of the samples was also evaluated by atomic force microscopy (AFM). In addition, the distribution of Ge, Ga, and As in a Ge/GaAs heterostructure was investigated by time-of-flight secondary ion mass spectroscopy (TOF-SIMS).

Fabrication of Ge p-MOSFET

For Ge p-MOSFET fabrication, Ge/GaAs substrates with ~100 nm Ge film were used. After n-well formation by arsenic implantation and activation, the wafers were loaded into the MOCVD system. In the gate cluster system, *in situ* SiH_4 treatment at 400 °C for 60 s was conducted for Ge surface passivation before HfO_2 deposition. After PDA at 500 °C for 60 s, TaN metal was deposited by reactive sputtering and defined by optical lithography and plasma etching. Self-aligned p^+ source/drain (S/D) regions were formed by boron implantation and dopant activation. Finally, Al contact metallization and forming gas annealing were implemented to complete the fabrication.

Fabrication of GaAs n-MOSFET

The fabrication of GaAs n-MOSFET involved undoped semi-insulating GaAs substrates. Wet cleaning process, which was also used in the fabrication of Ge/GaAs substrates, was conducted. The GaAs wafers were then quickly loaded into the MOCVD cluster, where 600 °C vacuum anneal, SiH_4 treatment and HfO_2 high-κ dielectric deposition were performed without breaking vacuum during wafer transfer in three different chambers. PDA at 500 °C for 60 s was conducted before reactive sputter deposition of TaN metal gate. After gate patterning, n^+ S/D of the transistors were formed by silicon and phosphorus co-implantation and dopant activation. Low resistance PdGe contacts were formed by lift-off process to complete the device fabrication.

RESULTS AND DISCUSSION

The crystalline quality of the Ge films was first investigated by HRTEM. Figure 1 (a) reveals that the as-deposited Ge layer by CVD method is crystalline. This is likely attributed to the high growth temperature of 400 °C during the CVD process, providing sufficient energy for epitaxial growth of Ge on the lattice-matched GaAs substrate. However, dislocation loops were observed in this Ge layer, as shown in Figure 1 (b). After annealing at 600 °C for 30 s, the entire Ge layer exhibits single crystalline quality (Figure 1 (c)). No structural defects such as dislocations or stacking faults can be observed in the Ge epilayer. For comparison, the as-deposited Ge film by PVD method was found to be amorphous (Figure 1 (d)). HRTEM image shown in Figure 1 (e) reveals the polycrystalline nature of the Ge film after annealing at 600 °C for 30 s.

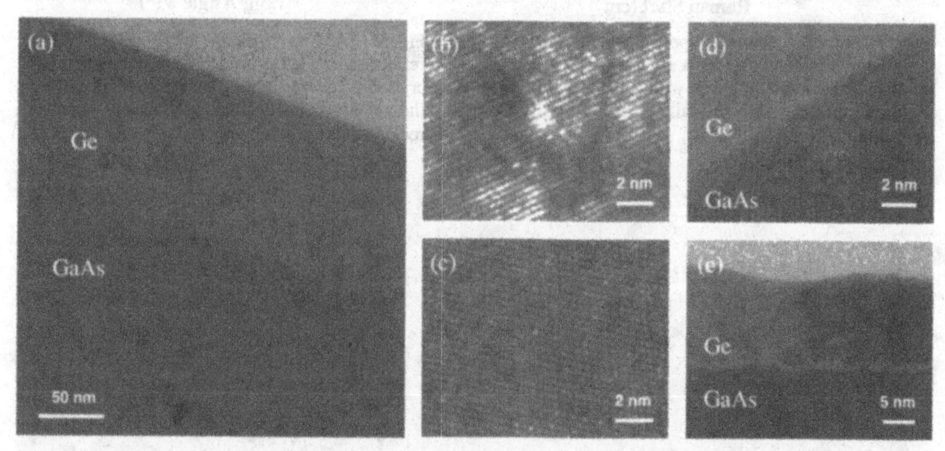

Figure 1. High resolution TEM micrographs of Ge layer grown on GaAs substrate by CVD deposition (a, b, c) and sputter deposition (d, e). (a) As-grown Ge/GaAs heterostructure by CVD deposition is crystalline. (b) Dislocation loops were observed in the as-grown Ge layer. (c) Structural defects were removed after annealing at 600 °C for 30 s. (d) As-grown Ge layer by PVD method is amorphous. (e) Polycrystalline Ge film was observed after.

Figure 2 shows the Raman spectrum of Ge/GaAs heterostructure after anealing at 600 °C for 30 s. The thickness of the Ge layer is ~ 120 nm. The Raman spectrum exhibits sharp peak at 300.2 cm^{-1} with FWHM of 4.3 cm^{-1}, indicating the crystallization of Ge in the film. Two small peaks at 269 cm^{-1} and 292 cm^{-1} are due their respective TO phonon and LO phonon in the underlying GaAs. The high crystalline quality of CVD Ge layer was also confirmed by the X-ray rocking curve of the (004) reflection of Ge/GaAs heterostructure, as shown in Figure 3. An interference pattern (Pendellösung oscillations) is clearly observed in the rocking curve. This interference originates from the beating of two X-ray wave fields inside the crystal. These two wave fields are generated at the interface between GaAs and Ge as well as from the surface of Ge layer. As a result, interference can only be observed in crystals which have almost perfectly parallel boundaries.

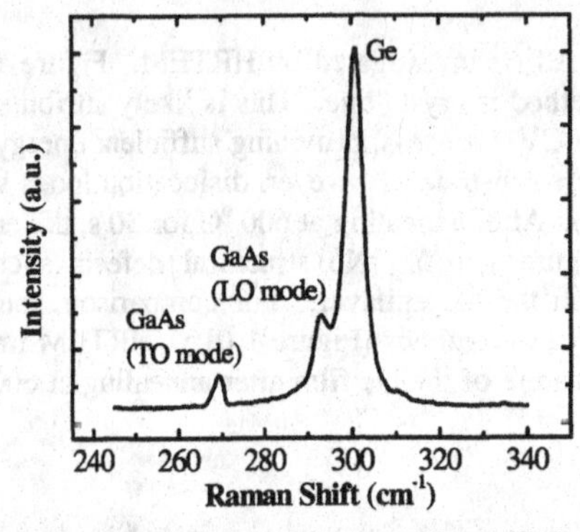

Figure 2. Raman spectrum of Ge/GaAs heterostructure after PDA at 600 °C for 30 s. The sharp peak at 300.2 cm^{-1} with FWHM of 4.3 cm^{-1} indicates the high crystalline quality of the Ge film on GaAs.

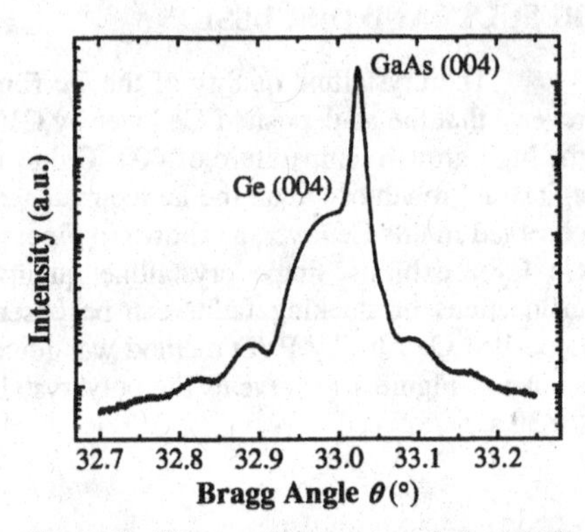

Figure 3. X-ray rocking curve of the (004) reflection of Ge/GaAs heterostructure. The high crystalline quality of the Ge layer is confirmed by the clear Pendellösung oscillations in the rocking curve.

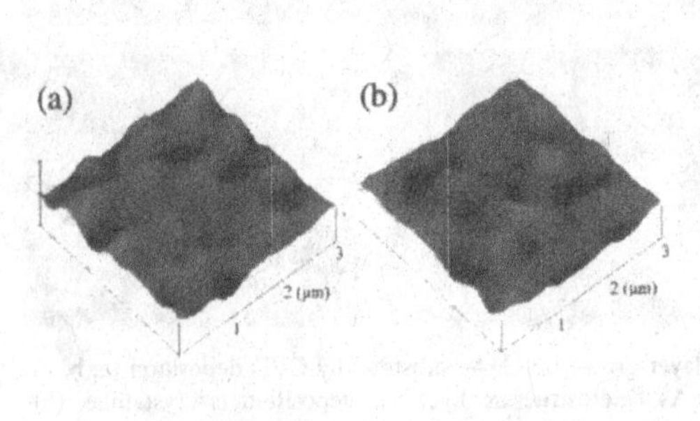

Figure 4. AFM images of Ge/GaAs heterostructures which are (a) as-grown (RMS = 4.9 Å), and (b) after annealing at 600 °C for 30 s (RMS = 5.0 Å). The surface of the Ge epitaxial layers is reasonably smooth and uniform for device fabrication.

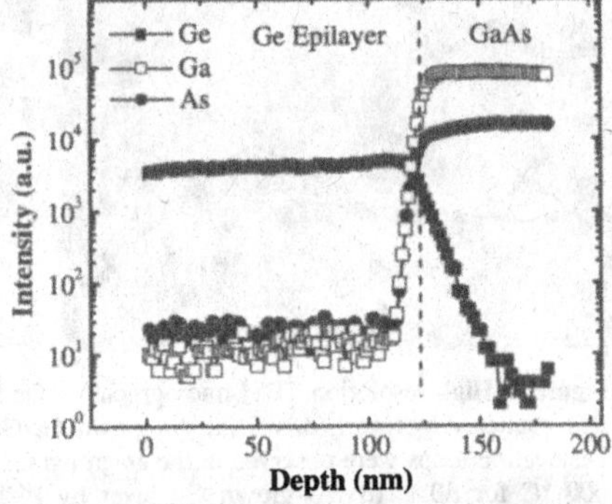

Figure 5. SIMS depth profile analysis showing the distribution of Ge, Ga, and As in a Ge/GaAs heterostructure. The concentration of Ga and As at the surface is below the detection limit of SIMS.

The surface roughness of Ge epilayer is very critical in high performance MOSFET applications as rough surface can degrade the high-field mobility of a transistor. AFM was employed to examine the surface morphology of the CVD Ge layers (Figure 4). AFM analysis reveals RMS roughness of 4.9 Å and 5.0 Å in the as-deposited and as-annealed CVD Ge films. To our knowledge, RMS roughness of 5 Å in this work is one of the smoothest Ge surface ever reported for Ge/GaAs heterostructures. The distribution of Ge, Ga, and As in the Ge/GaAs heterostructure was further investigated by SIMS analysis, as shown in Figure 5. Out-diffusion of Ga and As into the Ge epilayer is negligible as the concentration of Ga and As is below the detection limit of SIMS.

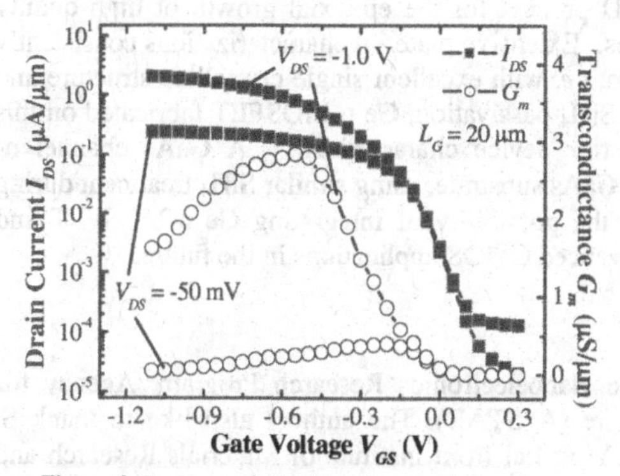

Figure 6. I_{DS}-V_{GS} and G_m- V_{GS} characteristics of a Ge p-MOSFET formed on the SiH_4 passivated Ge/GaAs heterostructure.

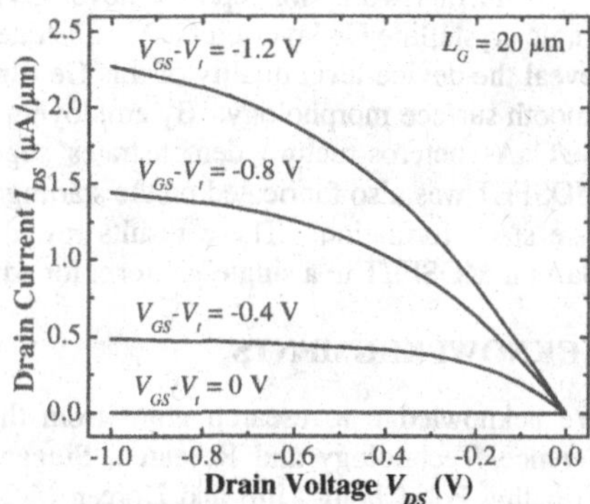

Figure 7. I_{DS}-V_{DS} curves of a Ge channel p-MOSFET at various gate overdrives.

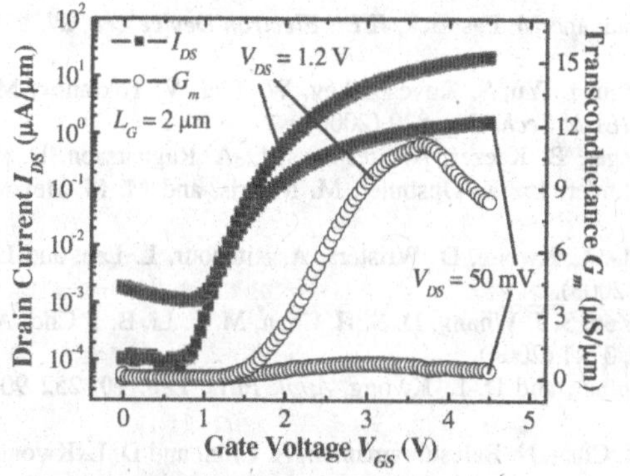

Figure 8. I_{DS}-V_{GS} and G_m- V_{GS} plots of a GaAs n-MOSFET fabricated on undoped semi-insulating GaAs substrate. The SiH_4 passivation was introduced during the gate stack formation.

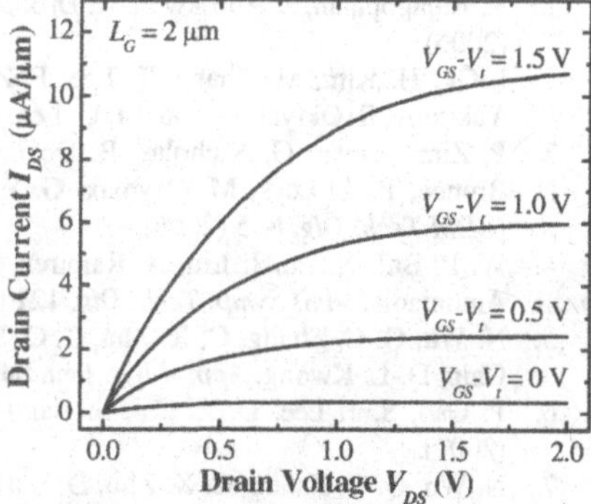

Figure 9. Output characteristics of a GaAs n-channel transistor at various gate overdrives.

Figure 6 demonstrates the I_{DS}-V_{GS} and G_m-V_{GS} characteristics of a Ge p-MOSFET that was fabricated on this Ge/GaAs heterostructure. This SiH_4 passivated Ge device exhibits excellent transfer behaviors with high I_{on}/I_{off} ratio and subthreshold swing, *SS*, of 88 mV/decade. I_{DS}-V_{DS} curves of the Ge transistor at various gate overdrives are plotted in Figure 7. The I_{DS}-V_{GS} and G_m-V_{GS} characteristics of a GaAs n-MOSFET with a gate length of 2 μm are shown in Figure 8. Using the similar SiH_4 passivation during gate stack formation, this GaAs channel device exhibits good transistor characteristics with high I_{on}/I_{off} ratio. I_{DS}-V_{DS} characterictics of the GaAs transistor are also plotted in Figure 9.

CONCLUSION

In this work, we report a novel CVD process for the epitaxial growth of high quality single crystalline Ge layer on GaAs substrates. Extensive material characterizations consistently reveal the device level quality of this Ge film, *i.e.* with excellent single crystalline structure and smooth surface morphology. By employing SiH_4 passivation, Ge p-MOSFET fabricated on this Ge/GaAs heterostructure demonstrates superior device characteristics. A GaAs channel n-MOSFET was also fabricated on the starting GaAs substrate, using similar SiH_4 treatment during gate stack formation. These results reveal the possibility of integrating Ge p-MOSFET and GaAs n-MOSFET in a single platform for advanced CMOS applications in the future.

ACKNOWLEDGMENTS

We acknowledge a research grant from the Nanoelectronics Research Program, Agency for Science, Technology and Research, Singapore (A*STAR). The authors also like to thank S. Tripathy, Poh-Chong Lim, and Doreen Mei-Ying Lai from Institute of Materials Research and Engineering (IMRE), Singapore and Chih-Hang Tung from Institute of Microelectronics (IME), Singapore for material characterization.

REFERENCES

1. K. Rajagopalan, J. Abrokwah, R. Droopad, and M. Passlack, *IEEE Electron Device Lett.* **27**, 959 (2006).
2. I. Ok, H. Kim, M. Zhang, T. Lee, F. Zhu, L. Yu, S. Koveshnikov, W. Tsai, V. Tokranov, M. Yakimov, S. Oktyabrsky, and J. C. Lee, *IEDM Tech. Dig.* 829 (2006).
3. P. Zimmerman, G. Nicholas, B. De Jaeger, B. Kaczer, A. Stesmans, L.-A. Ragnarsson, D. P. Brunco, F. E. Leys, M. Caymax, G. Winderickx, K. Opsomer, M. Meuris, and M. M. Heyns, *IEDM Tech. Dig.* 655 (2006).
4. W. P. Bai, N. Lu, J. Liu, A. Ramirez, D. L. Kwong, D. Wristers, A. Ritenour, L. Lee, and D. Antoniadis, *VLSI Symp. Tech. Dig.* 121 (2003).
5. N. Wu, Q. C. Zhang, C. X. Zhu, C. C. Yeo, S. J. Whang, D. S. H. Chan, M. F. Li, B. J. Cho, A. Chin, D.-L. Kwong, *Appl. Phys. Lett.* **84**, 3741 (2004).
6. F. Gao, S. J. Lee, D. Z. Chi, S. Balakumar, and D.-L. Kwong, *Appl. Phys. Lett.* **90**, 252 904 (2007).
7. N. Wu, Q. C. Zhang, C. X. Zhu, D. S. H. Chan, N. Balasubramanian, A. Chin, and D. L. Kwong, *Appl. Phys. Lett.* **85**, 4127 (2004).
8. M. Zhu, H.-C. Chin, C.-H. Tung, and Y.-C. Yeo, *J. Electrochemical Society* **54**, H879 (2007).
9. M. M. Frank, G. D. Wilk, D. Starodub, T. Gustafsson, E. Garfunkel, Y. J. Chabal, J. Grazul, and D. A. Muller, *Appl. Phys. Lett.* **86**, 152904 (2005).
10. J. R. Arthur, *Surface Science* **43**, 449 (1974).

Silicon and Other Materials on Silicon

Mater. Res. Soc. Symp. Proc. Vol. 1068 © 2008 Materials Research Society 1068-C07-08

Nondestructive defect measurement and surface analysis of 3C-SiC on Si (001) by electron channeling contrast imaging

Yoosuf N. Picard[1], Christopher Locke[2], Christopher L. Frewin[2], Rachael L. Myers-Ward[1], Joshua D. Caldwell[1], Karl D. Hobart[1], Mark E. Twigg[1], and Stephen E. Saddow[2]
[1]Electronics Science and Technology, Naval Research Lab, Code 6812, 4555 Overlook Ave. SW, Washington, DC, 20375
[2]Electrical Engineering Dept., University of South Florida, 4202 E. Fowler Ave., Tampa, FL, 33620

ABSTRACT

The electron channeling contrast imaging (ECCI) technique was utilized to investigate atomic step morphologies and dislocation densities in 3C-SiC films grown by chemical vapor deposition (CVD) on Si (001) substrates. ECCI in this study was performed inside a commercial scanning electron microscope using an electron backscatter diffraction (EBSD) system equipped with forescatter diode detectors. This approach allowed simultaneous imaging of atomic steps, verified by atomic force microscopy, and dislocations at the film surface. EBSD analysis verified the orientation and monocrystalline quality of the 3C-SiC films. Dislocation densities in 3C-SiC films were measured locally using ECCI, with qualitative verification by x-ray diffraction. Differences in the dislocation density across a 50 mm diameter 3C-SiC film could be attributed to subtle variations during the carbonization process across the substrate surface.

INTRODUCTION

The cubic silicon carbide polytype, 3C-SiC, offers many advantages for high power and high temperature device applications. However, utilizing 3C-SiC films in devices requires careful control of crystallographic defects and surface morphology. Developing high quality 3C-SiC films on commercially viable substrates like Si requires optimizing the many processing parameters for chemical vapor deposition (CVD), such as precursor chemistry, process pressure and temperature, as well as optimizing processing procedures, such as the carbonization and growth steps. Progress in optimizing these multiple parameters would greatly benefit from materials analysis approaches that are fast, nondestructive, and provide useful details regarding both the crystalline and surface quality of variously processed films. Recently, the electron channeling contrast imaging (ECCI) technique was demonstrated in a commercial scanning electron microscope (SEM) using a forescatter geometry [1]. This approach has allowed direct imaging of both dislocations and atomic steps in GaN [1-2] and 4H-SiC [3] without sample preparation or significant alteration of the material surface. ECCI is demonstrated in this study for defect and surface analysis of 3C-SiC films grown by different CVD processes on Si (001) substrates.

EXPERIMENT

In this study, two heteroepitaxial 3C-SiC films roughly 10 μm thick were grown by CVD on Si(001) substrates and then analyzed by ECCI. Details of the reactor and processing steps used are available in the literature [4]. The substrate surfaces were prepared prior to growth using the standard RCA cleaning procedure. Deposition was conducted in a low-pressure horizontal hot-wall CVD reactor utilizing propane (C_3H_8) and silane (SiH_4) as the carbon and silicon precursors, respectively, which were transported by a hydrogen carrier gas. Growth was performed on two samples, namely Sample A and B. Sample A was processed using an atmospheric-pressure carbonization step followed by low pressure 3C-SiC heteroepitaxial growth, while Sample B was processed entirely under low pressure conditions. Sample A was carbonized in a H_2/C_3H_8 atmosphere for 2 minutes at elevated temperature (1170°C) and at a pressure of 760 Torr. The pressure was then lowered to 100 Torr and a Si/C ratio of 0.9 was maintained at 1392°C during 40 min of 3C-SiC growth. Sample B was carbonized in a H_2/C_3H_8 atmosphere for 2 minutes at elevated temperature (1135°C) and at a pressure of 400 Torr followed by growth at 100 Torr and a Si/C ratio of 1.0 was maintained at 1385°C during 240 min of 3C-SiC growth. For this sample two consecutive 240 min. growths were conducted, with the sample being manually rotated 180° to achieve a more uniform film thickness (the reactor used does not have wafer rotation). This growth interrupt, the total growth time, the propane mole fraction and the pressure used during the carbonization step are specific processing parameters that distinguish Sample A from Sample B. Carbonization and 3C-SiC growth details are outlined in Table I.

Table I. Carbonization and 3C-SiC growth parameters for samples analyzed by ECCI.

Sample	Carbonization*			3C-SiC Growth		
	C_3H_8 mole fraction	T [°C]	P [Torr]	Si/C ratio	T [°C]	Growth Rate [µm.h]
A	6.00E-04	1170	760	0.9	1392°C	14
B	1.60E-03	1135	400	1.0	1380°C	1.5

*2 min carbonization time at temperature

The surfaces of the grown heteroepitaxial 3C-SiC films were analyzed by ECCI to both image the morphology and to quantify the dislocation density. ECCI was conducted inside a commercial SEM using an electron backscatter detector (EBSD) system equipped with forescatter diode detectors. Specimens were oriented with the [110] direction parallel to the 20 keV electron beam and sample surfaces tilted ~75° relative to the incident electron beam operating at 2.4 nA. Figure 1 depicts this experimental arrangement for conducting ECCI in a forescatter geometry. Based on the position of the diode detectors relative to the various diffraction bands imaged by the EBSD phosphor screen, we initially estimated that one of the (221) Kikuchi bands crosses both detectors. This observation provides an initial guide for determining the diffraction condition yielding channeling contrast for the ECCI images in this study. EBSD showed that both Sample A and Sample B were monocrystalline 3C-SiC films with the following epitaxial relationship to the Si substrate: 3C-SiC (001) || Si (001); 3C-SiC [110] || Si [110]. Atomic force microscopy (AFM) was performed to confirm atomic step imaging by ECCI. Finally, x-ray diffraction (XRD) was performed to further verify the crystalline quality of the 3C-SiC films.

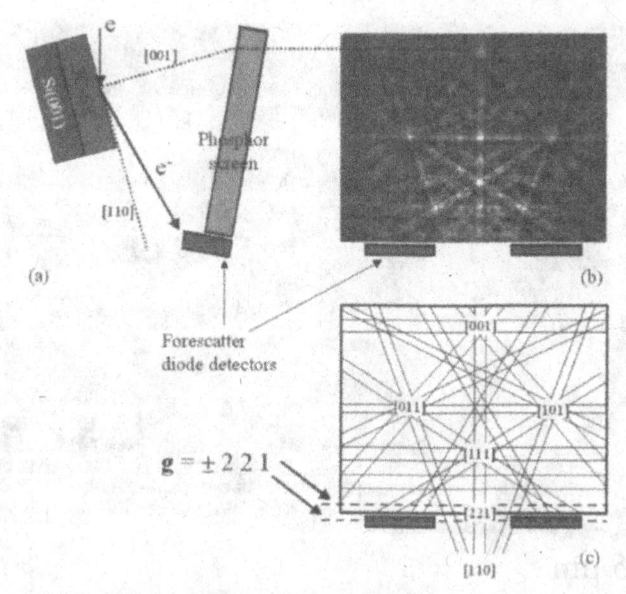

Figure 1. Schematic illustration of (a) the projected side-view of the ECCI experimental configuration, (b) an experimental Kikuchi pattern with the forescatter diode detector positions projected, and (c) a depiction of the primary Kikuchi lines with zone axes labeled.

RESULTS AND DISCUSSION

A characteristic ECCI image of the surface of Sample A is presented in Figure 2. The film morphology exhibits evidence of sporadic clusters of hillocks with very distinct boundaries. Dark/light spots observed across the hillock surfaces are qualitatively identical to threading dislocations previously observed in GaN [2] and 4H-SiC [3]. Thus, each individual dark/light feature observed in Fig. 1 can be correlated to a single dislocation penetrating the (001) surface plane. Atomic steps are also weakly visible as line patterns across the sample surface in Fig. 2. A number of dislocations are found to be centered on concentric atomic steps, with the implication that these dislocations might be acting as atomic step sources during CVD growth. Evidence of axial screw dislocations with Burgers vector c/3 [111], where the *c*-axis follows the <111> direction, were reported by AFM analysis of triangular shaped atomic spirals on (111) 3C-SiC surfaces [5]. Such screw dislocations, penetrating the (001) surface and propagating along <111> directions, might in fact be the dislocations denoted by white arrows in Fig. 2 exhibiting concentric or spiraling atomic steps. Other dislocations penetrating the (001) surface in Fig. 2 are likely to be the previously observed dislocations with Burgers vector a/2 [110] [6-7]. By assuming a **g** = 221 diffraction condition for ECCI in this study (Fig. 1), both types of dislocations mentioned here would be visible based on simple **g** • **b** = 0 invisibility criterion. Total dislocation density estimates were made by measuring the number of characteristic dark/light intensity fluctuations within defined areas between the hillock boundaries. Dislocation densities ranged from $1.8x10^8$ cm^{-2} to $2.5x10^8$ cm^{-2} for Sample A.

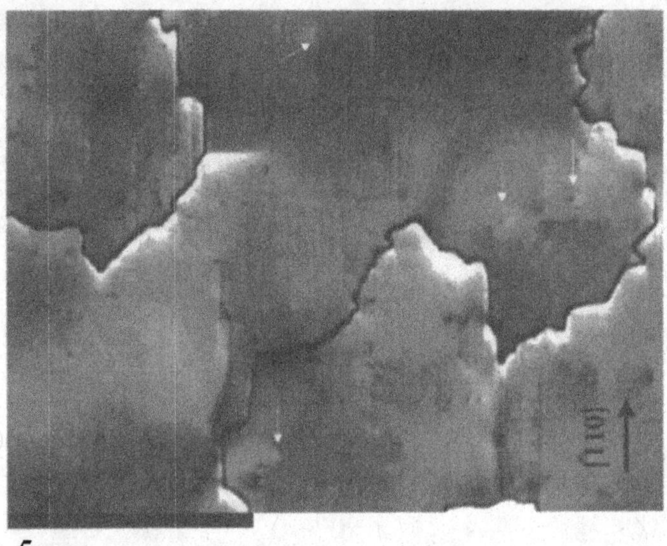

Figure 2. ECCI image of the Sample A surface where dark/light spots indicate dislocation positions and weak line patterns are atomic steps. White arrows denote dislocations centered about concentric atomic steps or atomic step spirals.

Atomic step imaging by ECCI in Fig. 2 also indicated some degree of faceting. Many atomic steps appeared as straight lines parallel to the <110> direction. Such faceting has been commonly observed by 3C-SiC films, especially after H_2 etching [8]. This faceting occurs during growth or etching due to the preferential formation of low surface energy {111} surfaces. Another possible reason for the observed <110> atomic steps is the penetration of the (001) surface by <111> stacking faults that could create corrugated step-bunched surfaces. An example of a possible step-bunched edge on the surface of Sample A is denoted by a white arrow in Fig. 3. The step-bunched edge is clearly visible at the center of the ECCI image as a dark vertical line (Fig. 3(a)). This feature is weakly visible by conventional SEM imaging using secondary electron detection (Fig. 3(b)). AFM analysis of this same area (Fig. 3(c)) shows that the dark vertical line feature visible by ECCI is indeed a step-bunched ledge ~20 nm high. The dark/light spot features visible in ECCI could not be correlated to any surface features discernible by AFM, indicating that these features are indeed generated by channeling/diffraction contrast and not by topological artifacts.

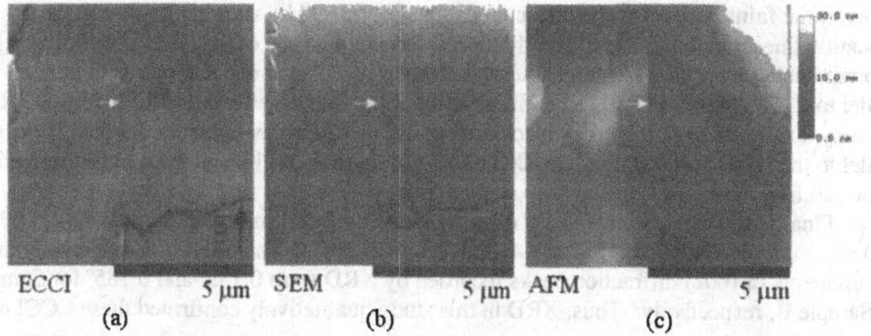

Figure 3. Images of the surface of Sample A recorded by (a) ECCI, (b) conventional SEM, and (c) AFM.

High magnification ECCI images of two separate areas of Sample B are presented in Figure 4. These ECCI images were recorded from two separate regions of the sample surface spaced ~25 mm apart. The area imaged in Fig. 4(a) is upstream relative to the area in Fig. 4(b) when considering the gas flow direction during the carbonization step. Consumption of gaseous reactants is expected to produce a steadily lower C content at the carbonizing surface downstream from the gas-flow direction. Relative to the upstream areas, the downstream areas exhibited more dark/light spots indicating a higher defectivity. Dislocation density estimates for the upstream areas ranged from 4.5×10^8 cm^{-2} to 9.0×10^8 cm^{-2} while the downstream areas ranged from 1.8×10^9 cm^{-2} to 2.0×10^9 cm^{-2}.

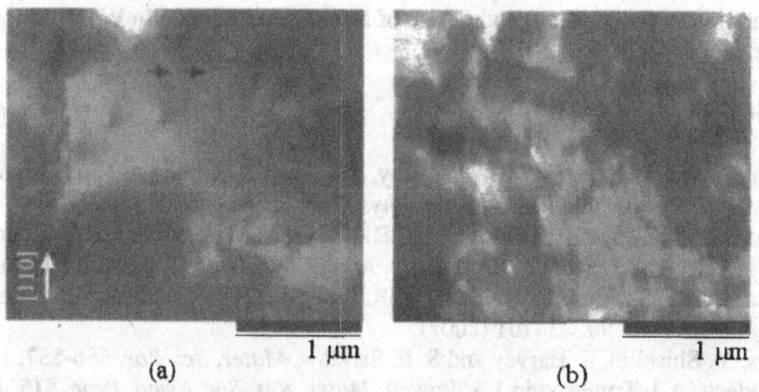

Figure 4. ECCI images of Sample B recorded (a) upstream and (b) downstream from the gas flow direction during the carbonization step. Dislocation densities ranged from (a) 4.5×10^8 cm^{-2} to 9.0×10^8 cm^{-2} and (b) from 1.8×10^9 cm^{-2} to 2.0×10^9 cm^{-2}. Possible line features denoting dislocations are highlighted by black arrows in Fig. 4(a).

Resolving individual dislocations at these higher densities (~10^9 cm^{-2}) in Sample B did prove problematic. Sample B exhibited a rougher surface with more faceting, obscuring channeling contrast features tied to dislocations. Also, the observed horizontal and vertical lines by ECCI were largely attributed to faceted <110> steps, corroborated by AFM analysis.

However, the faint lines might be dislocations that bound <111> stacking faults and have dislocation lines parallel to the <110> direction. Recent analysis of 3C-SiC (111) films by ECCI performed using a backscattering geometry directly imaged such dislocations as faint lines parallel to <110> directions [9]. Therefore, some of the faint line features imaged by ECCI in this study, examples highlighted by black arrows in Fig. 4(a), may in fact be buried dislocations parallel to the (001) surface. Future ECCI and AFM analysis will be pursued to further verify this possibility.

Finally, the dislocation density measurements by ECCI indicated Sample B (4.5-20 x 10^8 cm^{-2}) was more defective than Sample A (1.8-2.5 x 10^8 cm^{-2}). The full-width half-maximum measurements of (002) diffraction peaks recorded by XRD were 0.133° and 0.185° for Sample A and Sample B, respectively. Thus, XRD in this study qualitatively confirmed these ECCI results.

CONCLUSIONS

ECCI analysis demonstrated direct imaging of atomic steps and dislocations in 3C-SiC films grown by CVD on Si (001) substrates. Dislocations imaged by ECCI exhibited characteristic dark/light spots that allowed dislocation densities to be measured locally and nondestructively. Comparative analysis by XRD qualitatively corroborated this result. Step morphologies imaged by ECCI were confirmed by AFM and exhibited preferentially faceting and step-bunching along <110> directions.

ACKNOWLEDGMENTS

We would like to thank Suzie Harvey, formerly of the USF SiC Group, for performing the growth for Sample A. The CVD growth conducted at USF was supported by the Army Research Laboratory (Bruce Geil) and Office of Naval Research (Colin Wood) whose support is gratefully acknowledged.

REFERENCES

1. C. Trager-Cowan, E. Sweeney, P.W. Trimby, A.P. Day, A. Gholinia, N.-H.Schmidt, P.J. Parbrook, A.J. Wilkinson, and I.M. Watson, *Phys. Rev. B.* **75**, 085301 (2007).
2. Y.N. Picard, J.D. Caldwell, M.E. Twigg, C.R. Eddy, Jr., M.A. Mastro, R.L. Henry, R.T. Holm, P.G. Neudeck, A.J. Trunek, J.A. Powell, *Appl. Phys. Lett.* **91**, 094106 (2007).
3. Y.N. Picard, M.E. Twigg, J.D. Caldwell, C.R. Eddy, Jr., P.G. Neudeck, A.J. Trunek, and J.A. Powell, *Appl. Phys. Lett.* **90**, 234101 (2007).
4. M. Reyes, Y. Shishkin, S. Harvey and S. E. Saddow, *Mater. Sci. For.* **556-557**, 191 (2007).
5. P.G. Neudeck, A.J. Trunek, and J.A. Powell, *Mater. Res. Soc. Symp. Proc.* **815**, J5.32.1 (2004).
6. R. Stevens, *J. Mater. Sci.* **5**, 474 (1970).
7. X.J. Ning and P. Pirouz, *J. Mater. Res.* **11**, 884 (1996).
8. C. Coletti, C.L. Frewin, S.E. Saddow, M. Hetzel, C. Virojanadara, and U. Starke, *Appl. Phys. Lett.* **91**, 061914 (2007).
9. K.M. Speer, P.G. Neudeck, M.A. Crimp, C. Burda, and P. Pirouz, *Phys. Stat. Sol. A.* **7**, 2216 (2007).

Mater. Res. Soc. Symp. Proc. Vol. 1068 © 2008 Materials Research Society 1068-C07-09

P Implantation Effect on Specific Contact Resistance in 3C-SiC Grown on Si

Anne-Elisabeth Bazin[1,2], Jean-François Michaud[1], Marc Portail[3], Thierry Chassagne[4], Marcin Zielinski[4], Jean-Marc Lecoq[2], Emmanuel Collard[2], and Daniel Alquier[1]

[1]Laboratoire de Microélectronique de Puissance, Université François Rabelais de Tours, 16 Rue Pierre et Marie Curie, BP 7155, Tours Cedex 2, 37071, France

[2]STMicroelectronics, 16 Rue Pierre et Marie Curie, BP 7155, Tours Cedex 2, 37071, France

[3]Centre de Recherche sur l'Hétéro-Epitaxie et ses Applications CNRS–UPR10, Rue Bernard Grégory, Valbonne, 06560, France

[4]NOVASiC, Savoie Technolac, Arche Bât 4, BP 267, Le Bourget du Lac Cedex, 73375, France

ABSTRACT

In this work, non-intentionally doped 3C-SiC epilayers were implanted using phosphorus at different energies and subsequently annealed at temperatures between 1100°C and 1350°C in order to form n^+ implanted layers. Different techniques such as Fourier Transformed InfraRed spectroscopy (FTIR) and Secondary Ion Mass Spectroscopy (SIMS) were used to characterize implanted 3C-SiC epilayers after the different annealing steps. Successively, metal layers were sputtered in order to form the contacts. The specific contact resistance (ρ_C) was determined by using circular Transfer Length Method (c-TLM) patterns. Specific contact resistance values were investigated as a function of doping and contact annealing conditions and compared to those obtained for highly doped 3C-SiC epilayers. As expected, ρ_C value is highly sensitive to post-implantation annealing and metal contact annealing. This work demonstrates that low resistance values can be achieved using phosphorus implantation and, hence, enabling device processing.

INTRODUCTION

Since last decades, silicon carbide (SiC) is the subject of intensive research and development activities. This growing attention is motivated by attractive mechanical and electrical properties which make silicon carbide a promising material for high power and high temperature electronic devices. According to the stacking sequence of the Si-C bilayers, silicon carbide exists in more than 200 different polytypes. Compared to the various existing structures, the 3C-SiC is the only one that can be hetero-epitaxially grown on cheap silicon substrates. The SiC growth capability on low cost and large diameter silicon substrates becomes then a very attractive solution for manufacturing [1]. Indeed, in such conditions, only the required silicon carbide thickness has to be grown according to the targeted application. For such a material, one of the main challenges is the achievement of high quality ohmic contacts in order to create efficient electronic devices. To reach that, ohmic contacts both in highly doped epitaxial or implanted layers may be required depending on process flow. Indeed, doping is a key process in semiconductor manufacturing. Ion implantation is a method of choice to obtain selective doping and the only available for silicon carbide. To obtain n^+ doping in 3C-SiC, both nitrogen and phosphorus implantation were already carried out [2, 3]. Even if nitrogen is often preferred for SiC devices, phosphorus exhibits the advantage of being widely used in all the semiconductor industry. Post-implantation annealing is also a great challenge in 3C-SiC on Si. In fact, this annealing must lead to dopant electrical activation that generally occurs at high temperature and must remain well below the Si melting point. In this work, we investigate the influence of phosphorus implantation and the associated annealing on specific contact resistance in 3C-SiC grown on Si.

EXPERIMENTAL DETAILS

The 3C-SiC films, used for this study, were hetero-epitaxially grown on 2 inches (100) silicon wafers by using a resistively heated hot wall CVD reactor [4]. The growth was performed using silane (SiH_4) and propane (C_3H_8) as precursor gases and purified hydrogen (H_2) as a carrier gas following the classical two stages process defined by Nishino [5] but without initial surface de-oxydation. The growth details are reported elsewhere [6]. Two types of 7 μm thick epilayers with different n doping levels were grown to carry out the ohmic contacts: a non-intentional doping of 6×10^{15} cm^{-3} and a highly doping of 3.5×10^{19} cm^{-3} [7]. The 3C-SiC epilayers were then polished using NOVASiC know-how. Non-intentionally doped samples were then implanted at room temperature with phosphorus using a commercial implanter. Implantations were carried out at multiple energies of 30, 50, 100 and 150 keV for a respective dose of 0.5, 1.2, 2.1 and 4.5×10^{13} cm^{-2} in order to form a phosphorus box-like profile. Parameters were calculated using SRIM simulation. Post-implantation annealing, with an Ar flow of 1.5 slm and at a pressure of 200 mbar, has then been accomplished to activate phosphorus implanted ions. Both temperature and duration of annealing have been investigated as summarized in table I.

Table I: Post-implantation annealing conditions and associated RMS roughness values for 3C-SiC phosphorus implanted samples.

Sample	A	B	C	D	E
Temperature (°C)	1100	1150	1250	1350	1350
Duration (min)	60	60	60	60	120
RMS roughness (nm)	1.36±0.26	1.33±0.37	1.23±0.39	1.23±0.21	1.31±0.21

Different techniques have then been applied to characterize the modifications of the 3C-SiC properties following the implantation. Fourier Transformed InfraRed (FTIR) measurements were performed on a Avatar 370 Thermo Nicolet spectrometer. Spectra were recorded with a 4 cm^{-1} resolution on a 700-1100 cm^{-1} spectral range in order to follow the evolution of the crystal disorder. Secondary Ion Mass Spectroscopy (SIMS) measurements were used to determine both phosphorus dopant concentration and layer homogeneity. The roughness was checked using a Fogale Nanotech "Photomap 3D" optical profiler.

Circular Transfer Length Method (c-TLM) patterns were prepared to determine specific contact resistance. This method presents a serious advantage over linear TLM structure as mesa isolation is not required for c-TLM pattern. Figure 1 shows a representation of the c-TLM of Marlow and Das [8] which consists of nine contact patterns. Each one has a constant contact inner diameter of 160 μm and a space (d) ranging from 12 μm to 48 μm between inner and outer contact.

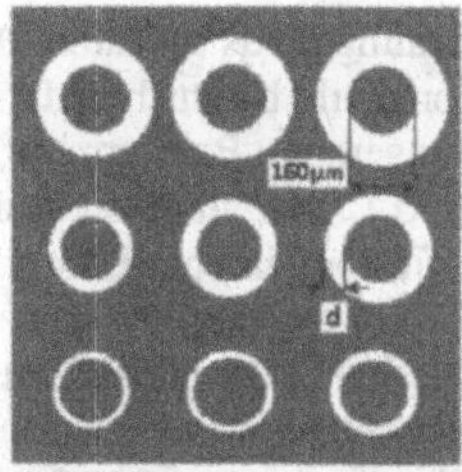

Figure 1: Contact patterns used to make c-TLM measurements.

A 150 nm thick nickel-titanium bilayer has been deposited by sputtering. Afterward, the samples were subjected to a Rapid Thermal Annealing (RTA) step of 1 minute in inert ambient (Ar) at 1000°C or 1050°C. The so-formed c-TLM structures were used to extract the specific contact resistance. A Keithley 2400 Sourcemeter was then used as a current source and voltage measurer. On each sample, more than 6 c-TLM patterns were measured and a regression method was employed to extract the specific contact resistance values as detailed in [9].

RESULTS AND DISCUSSION

Before metal contact step, all samples were first analyzed using different techniques in order to evaluate the implantation step impact of the 3C-SiC layers.

Physical characterization of 3C-SiC implanted layers

Figure 2 presents the obtained FTIR spectra for a non-intentionally doped epilayer, as-implanted sample, 1250°C and 1350°C annealed samples.

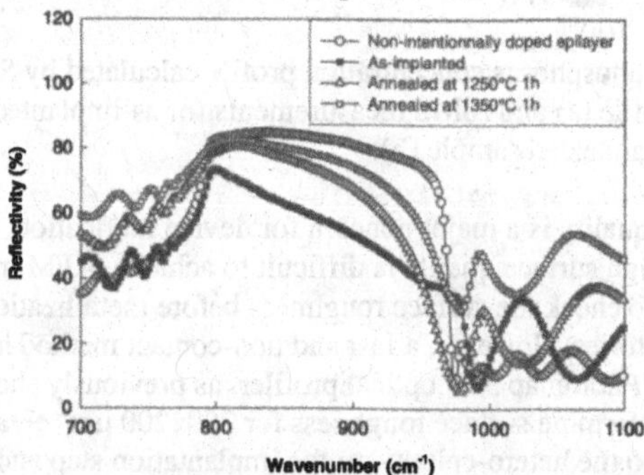

Figure 2: FTIR measurements for a non-intentionally doped epilayer, as-implanted sample, 1250°C annealed sample and 1350°C annealed sample.

We observe in this figure that the reflectivity in this spectral range is soundly modified with respect to the treatment (implantation and/or annealing). A qualitative evolution is noticeable between all the spectra considering the reststrahlen band (spectra between 790 and 970 cm^{-1}). It is straightforward to notice the strong evolution of the reststrahlen band in accordance to the surface treatment [10]. Indeed, if the implantation step drastically modifies the reststrahlen band, the further annealing tends to restore the spectral response towards a highly crystalline material. This evolution could be attributed to an initial lattice damage followed by a thermal healing of the atomic arrangement. Furthermore, the increase of the overall reflectivity on this spectral range after the 1350°C annealing is consistent with a better re-crystallization and, hence, with a higher activation efficiency in comparison to the activation achieved after the 1250°C thermal annealing.

In order to follow the evolution of the phosphorus profile upon annealing as well as the doping homogeneity, SIMS measurements were performed. First of all, it is important to mention that the as-implanted SIMS profile corroborates extremely well the SRIM calculated one, as presented in figure 3 (a). The phosphorus concentration profiles for as-implanted sample and annealed samples (1 hour at 1250°C and 1350°C) are presented in figure 3 (b). All the phosphorus profiles are very close to each other evidencing the extremely low diffusivity of phosphorus in 3C-SiC at considered temperatures. Moreover, the post-implantation annealing preserves the expected box-shape doping profile.

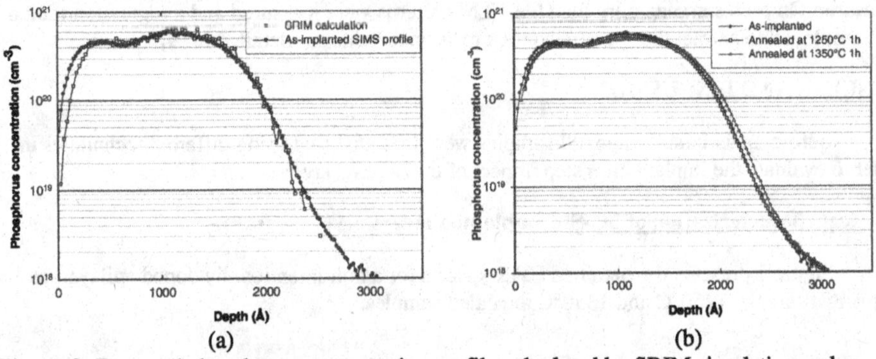

Figure 3: Expected phosphorus concentration profile calculated by SRIM simulation and as-implanted SIMS profile (a) and SIMS measurements for as-implanted sample, 1250°C annealed sample and 1350°C annealed sample (b).

The surface quality is a major concern for device fabrication, even more crucial in the case of SiC where high surface quality is difficult to achieve. AFM measurements were performed in order to check the surface roughness before metallization. This method is fully used in many laboratories. However, a fast and non-contact method has also been exploited using a Fogale Nanotech "Photomap 3D" optical profiler, as previously shown [7]. This 3D tool has been employed to determine surface roughness for 200x200 μm^2 areas. These measures were made subsequently to the hetero-epitaxy, to the implantation step and to the high temperature post-implantation annealing, in order to check potential surface degradation. The results, average of 10 subsequent measures for each sample, are presented in table I. After epitaxy, the RMS roughness value has been evaluated to 1.36 nm. The implantation seems to have no effect on the roughness value which has been determined to 1.35 nm after this step. Subsequently to the post-implantation annealing, the roughness has also been measured, exhibiting no variation. This points out that the high temperature annealing does not damage the silicon carbide surface, whatever is the temperature.

Our results shed light on the implantation modification on the 3C-SiC layers. As expected, implantation enables to create a box-like profile with a limited degradation of surface roughness. The subsequent annealing step, performed as usually to anneal the implantation defects and activate the dopant, presents the best results for the higher temperature without surface degradation. Based on these results, the implementation of this technique for ohmic contacts can be studied.

Electrical characterization of the Ti-Ni contacts

By using the measurement procedure exposed previously, the specific contact resistance has been investigated. Figure 4 presents ρ_C as a function of temperature and post-implantation annealing duration. In the same figure, contact annealing temperature impact is also investigated.

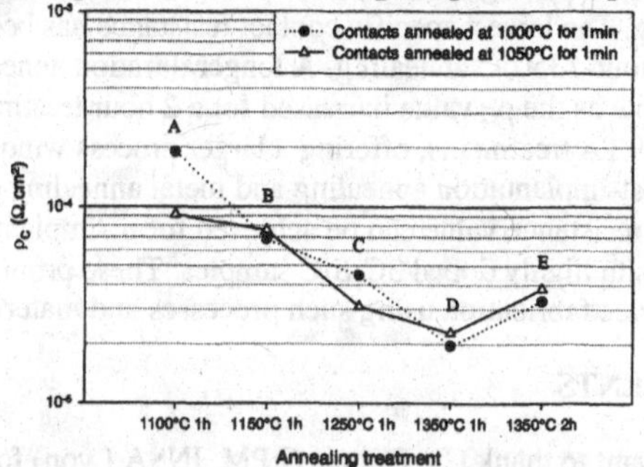

Figure 4: Influence of high temperature post-implantation annealing on the specific contact resistance after a contact RTA treatment of 1 minute in Ar ambient.

The temperature is a key parameter in order to activate phosphorus dopants. As expected, the lowest specific contact resistance value is obtained for the highest implantation annealing temperature for both contact annealing temperatures. Moreover, I-V characteristics (not presented here) underline a perfect ohmic behavior. Due to the silicon substrate melting temperature, 1350°C seems to be the maximal acceptable processing temperature for the annealing. Another parameter that has to be considered is the implantation annealing duration. According to the previous results, a 2 hours annealing at 1350°C has been completed. Subsequently to this treatment, the ρ_C value has drastically increased, around 70% in comparison with the 1 hour annealing at the same temperature. This result is surprising as a lower value was anticipated. A surface morphology modification cannot explain this degradation as roughness measurements do not suggest any surface damage towards the silicon carbide surface after a 2 hours treatment at 1350°C (sample E). Further experiments are necessary to well understand this behavior. Nevertheless, our work demonstrates that a low specific contact resistance value of $2x10^{-5}$ $\Omega.cm^2$ is obtained for samples annealed 1 hour at 1350°C. With the same measurement protocol, the ρ_C value for the highly doped 3C-SiC epilayer was evaluated to $1.7x10^{-5}$ $\Omega.cm^2$. Consequently, these two different doping methods lead to the same results. However, the A-sample points out a different ρ_C value according to the contacts annealing. This tendency is not confirmed for higher post-implantation temperature where the discrepancy in ρ_C values according to the RTA treatment is cut down. These results provide a wider process window to carry out the ohmic contacts in future devices.

CONCLUSION

In this paper, 3C-SiC non-intentionally doped samples were implanted with phosphorus at different energies and subsequently furnace annealed between 1100°C and 1350°C. First, FTIR, SIMS and optical profilometry measurements have been used to determine the influence of the post-implantation annealing on 3C-SiC properties. As expected, a box-like profile is obtained and seems to have the best electrical activation at the highest temperature. Moreover, this annealing step does not damage the silicon carbide surface, whatever is the temperature. Ohmic contacts were then formed using a Ti-Ni bilayer and studied with c-TLM structures. Our measurements demonstrate that good ohmic contacts were obtained successively to the high temperature annealing. The lowest specific contact resistance has been evaluated to $2x10^{-5}\ \Omega.cm^2$ consecutively to a 1 hour-1350°C annealing. A longer duration annealing at this temperature seems not to be suitable as the ρ_C value increased for a 2 hour treatment. Similar behaviors were found for both metal RTA treatments, offering a larger process window. As expected, ρ_C value is highly sensitive to post-implantation annealing and metal annealing conditions. Moreover, we demonstrate that low resistance value can be achieved for n^+-implanted layers, comparable with those obtained on in-situ highly doped 3C-SiC samples. These promising results are of high interest for future device fabrication using such processes and material.

ACKNOWLEDGMENTS

The authors want to thank Ch. Dubois (LPM, INSA Lyon) for her support on SIMS measurements.

REFERENCES

[1] A. Leycuras, Materials Science Forum **338-442,** 241 (2000).
[2] E. Tagushi, Y. Suzuki and M. Satoh, Materials Science Forum **556-557**, 579 (2007).
[3] J. Wan, M.A. Capano and M. R. Melloch, Solid-State Electronics **46**(8), 1227 (2002).
[4] T. Chassagne, A. Leycuras, C. Balloud, P. Arcade, H. Peyre and S. Juillaguet, Materials Science Forum **457-460**, 273 (2004).
[5] S. Nishino, J.P. Powell and H.A. Will, Appl. Phys. Lett. **42**, 460 (1983).
[6] M. Zielinski, M. Portail, H. Peyre, T. Chassagne, S. Ndiaye, B. Boyer, A. Leycuras and J. Camassel, Materials Science Forum **556-557,** 207 (2007).
[7] A.E. Bazin, T. Chassagne, J.F. Michaud, A. Leycuras, M. Portail, M. Zielinski, E. Collard and D. Alquier, Materials Science Forum **556-557**, 721 (2007).
[8] G.S. Marlow, M.B. Das, Solid State Electronics **25**, 91 (1982).
[9] J.H. Klootwijk and C.E. Timmering, Proc IEEE 2004 Int. Conference on Microelectronic Test Structures **17**, March 2004.
[10] R.T. Holm, P.H. Klein and P.E.R Nordquist, Jr., J. Appl. Phys. **60-4**, 1479 (1986).

Mater. Res. Soc. Symp. Proc. Vol. 1068 © 2008 Materials Research Society 1068-C03-20

Simulation of Forescattered Electron Channeling Contrast Imaging of Threading Dislocations Penetrating SiC Surfaces

Mark E. Twigg[1], Yoosuf N. Picard[1], Joshua D. Caldwell[1], Charles R. Eddy[1], Philip G. Neudeck[2], Andrew J. Trunek[3], and J. Anthony Powell[4]

[1]Naval Research Laboratory, Washington, DC, 20375

[2]NASA Glenn Research Center, Cleveland, OH, 44315

[3]OAI, Cleveland, OH, 44315

[4]Sest, Inc., Cleveland, OH, 44315

ABSTRACT

The interpretation of ECCI images in the forescattered geometry presents a more complex diffraction configuration than that encountered in the backscattered geometry. Determining the Kikuchi line that is the primary source of image intensity often requires more than simple inspection of the electron-channeling pattern. This problem can be addressed, however, by comparing recorded ECCI images of threading screw dislocations in 4H-SiC with simulated images. An ECCI image of this dislocation is found to give the orientation of the dominant Kikuchi line, greatly simplifying the determination of the diffraction simulation. In addition, computed images of threading screw dislocations in 4H-SiC were found to exhibit channeling contrast essentially identical to that obtained experimentally by ECCI and allowing determination of the dislocation Burgers vector.

INTRODUCTION

The evolution of Electron Channeling Contrast Imaging (ECCI) and the utilization of detectors positioned to record backscattered electron intensities, especially with respect to dislocation imaging, has been reviewed in detail by Crimp [1]. Recently, ECCI has been demonstrated using diode detectors mounted on commercial electron backscatter diffraction (EBSD) systems housed in conventional scanning electron microscopes [2,3]. This newer configuration used a forescattered geometry that provided orientation contrast and topographic enhancement, as well as channeling contrast from individual threading dislocations. By conducting ECCI in this configuration, the directionality of channeling contrast features can be resolved.

Fully realizing the ability to directly identify extended defects by ECCI based on directionality and/or the invisibility criterion of various channeling contrast features, in a manner analogous to transmission electron microscopy (TEM), requires further experimental and theoretical investigations of contrast generated by specific, well-understood extended defects. Fortunately, many researchers have already addressed the problem of simulating and recording electron channeling patterns and have met with considerable success [4-6]. Building on these previous studies, this paper will address ECCI imaging of threading screw dislocations (TSDs) in 4H-SiC.

EXPERIMENT

The first problem that must be considered is one of geometry, which is addressed, in part, by Fig.1. ECCI is conducted using the same detector configuration as in EBSD analysis. For forescattered ECCI imaging the sample is tilted by 70° so that the electron beam is incident at a 20° angle to the sample surface. The incident electrons undergo inelastic scattering as well as diffraction as they make their way to the detector. It is the inelastic scattering process, particularly impact ionization, which is responsible for redirecting the electrons by 40° or more so that they are able to impinge on the detector.
Researchers recently developed a method to acquire step-free surfaces in 4H- and 6H-SiC using patterned reactive ion etching followed by step-flow homoepitaxial growth [7]. The resulting specimens exhibit spiral step-patterns centered on individual TSDs, providing easily identifiable TSDs for experimental imaging and analysis by ECCI. The reciprocity theorem lets us establish the boundary conditions for diffraction contrast in ECCI. Electrons are imagined as incident from the EBSD detector, then undergoing diffraction before inelastic scattering directs them towards the electron gun. Note that both bright field (BF) and dark field (DF) intensities are combined to form the final image.

DISCUSSION

Image simulations

In order to calculate the modulation of electron intensities, however, the necessary boundary conditions must be imposed at the crystal surface. Applying the constraints, mandated by the presence of this surface and the adjacent vacuum, is more easily realized through the reciprocity principle--essentially the idea that the electron amplitude should be the same

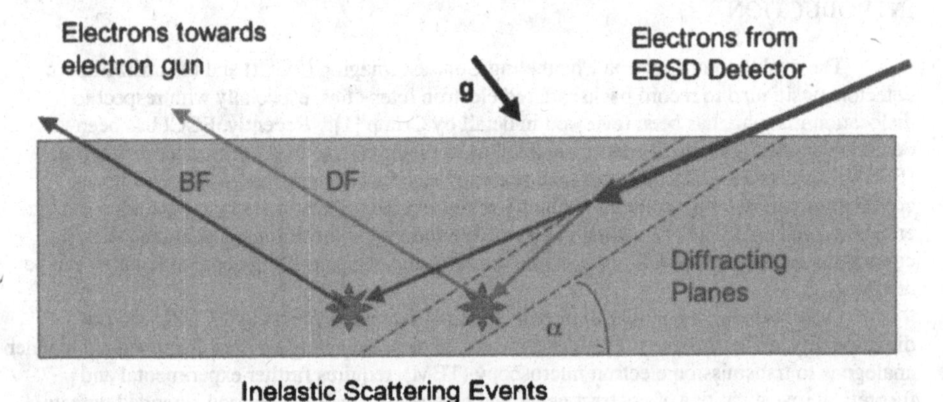

Fig.1 The reciprocity theorem establishes the boundary conditions for diffraction contrast in ECCI. Electrons are imagined as incident from the EBSD detector, then undergoing diffraction before inelastic scattering directs them towards the electron gun.

regardless of whether the electrons follow a given path from source to detector, or from detector to source, as shown in Fig.1 [8,9]. By imagining the electron as originating at the EBSD detector and then scattered toward the electron gun, as allowed by the reciprocity principle, the diffraction geometry is able to assume a form similar to that of TEM imaging, where the electron wave function can be calculated using the Howie-Whelan Equations.

Using the column approximation, the Howie-Whelan equations are integrated over the thickness of the crystal in order to evaluate the wave function at a given depth [9]. In the reciprocal configuration used in evaluating the wave function, diffracted electrons incident from the detector are scattered towards the electron gun by impact ionization. Because impact ionization may occur at any point below the surface, electrons modeled as emanating from the EBSD detector may be diffracted into an impact ionization site after propagating down from the surface to a depth corresponding to a specific impact ionization site. Rossouw et al. established the formalism for integrating over the depth from the surface in order to calculate diffracted channeling intensities [4]. Because of role the strain field of a TSD in these calculations, however, it is necessary to numerically integrate the Howie-Whelan equations [9].

As shown in Fig.2a, the Howie-Whelan equations were integrated over depth in order to trace the propagation of the electron wave function for each specific depth. In these calculations a running total of 1000 depths from 1 nm to 10 mm (in multiples of 1 nm) are tallied. At each of these 1000 depths, wave function propagation is imagined as leading to an inelastic scattering event that directs the electron back to the electron gun. The deviation parameter is assumed to be zero. The acting reflection chosen for this simulation is $g = \bar{2}11\bar{1}\bar{6}$, for reasons that are only fully apparent after the simulated image has been computed and compared with an empirical ECCI image.

After the calculating the BF and DF intensity profiles, these profiles are integrated over the depth from the surface to obtain the BF and DF simulated images shown in Fig.2b. Because inelastic scattering is largely dominated by impact ionization, the differential cross section for impact ionization is used to weight the relative contributions of BF and DF images [4,5]. The weighted contributions of the BF and DF images are then added to produce the total image (BF+DF) shown in Fig.2b.

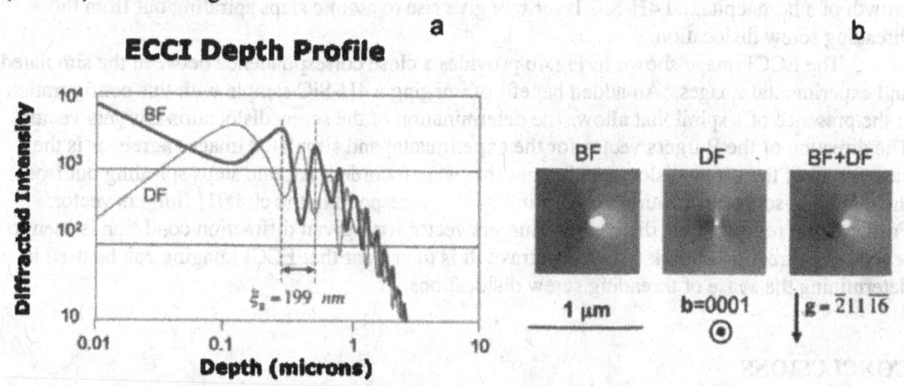

Fig.2 Simulated ECCI images. (a) Profile of BF and DF intensities with depth. The extinction distance is 199 nm. (b) BF and DF simulations of TSD added to form expected ECCI image.

Determination of the diffraction vector

The majority of the experimental parameters are known before recording the first ECCI image of a TSD. The approximate tilt of the sample and detector are known. The atomic structure factors of stoichiometric materials, of a known crystal structure (such as 4H-SiC), can be determined by routine calculations. The energy and direction of the incident electron beam are also known. The principal set of parameters, which are unknown before recording an ECCI image, are those describing the acting Bragg reflection. Selecting the specific Bragg reflection for recording an ECCI image is more straightforward in a backscattered geometry [1]. For forescattered ECCI, the choice of acting reflections can be narrowed down to those associated with the Kikuchi lines crossing the diode detectors. The relative positions of the Kikuchi lines on the phosphor screen (Fig.3) adjacent to the detectors, and the geometry of the sample and screen, allow the determination of which Kikuchi lines cross the diode detectors.

In order to determine which Kikuchi line crossing the detectors corresponds to the dominant reflection, the TSD image must be analyzed and recorded. The key to such an analysis is shown in Fig.4a, where simulated TSD images are shown for four different reflections corresponding to Kikuchi lines that cross over the diode detector positions shown in Fig.3. From the simulations shown in Fig.4a, the TSD contrast is seen to change from dark to light following the direction of the Burgers vector **b** crossed into the diffraction vector **g** (**bxg**). Given the contrast observed in an experimentally recorded image, such as that shown in Fig.4b, the dominant diffraction vector (perpendicular to the corresponding horizontal Kikuchi line) must point in either in either the six o'clock or twelve o'clock direction.

Interpretation of experimental images

The comparison of experimental and simulated images of threading screw dislocations in 4H-SiC has been aided by the approach of Powell *et al.* in growing CVD SiC on mesa substrates [7]. If the mesa is free of screw dislocations, the right growth conditions will result in the formation of a step free substrate. If a screw dislocation penetrates the mesa substrate, however, growth of a homoepitaxial 4H-SiC layer will give rise to atomic steps spiraling out from the threading screw dislocation.

The ECCI image shown in Fig.4b provides a close correspondence between the simulated and experimental images. An added benefit of imaging a 4H-SiC sample with this configuration is the presence of a spiral that allows the determination of the screw dislocation Burgers vector. The direction of the Burgers vector for the experimental and simulated images agree, as is the case for all of the screw dislocation images that were recorded. Atomic steps spiraling out from the threading screw dislocation, as seen in Fig.4b, correspond to the c[0001] Burgers vector. Furthermore, reversing the dislocation Burgers vector for a given diffraction condition is seen to reverse the direction of dark to light contrast. It is then clear that ECCI imaging can be used in determining the sense of threading screw dislocations.

CONCLUSIONS

ECCI imaging is capable of high contrast imaging of threading screw dislocations in 4H-SiC. What is especially significant, however, is that these images are easily interpretable. A

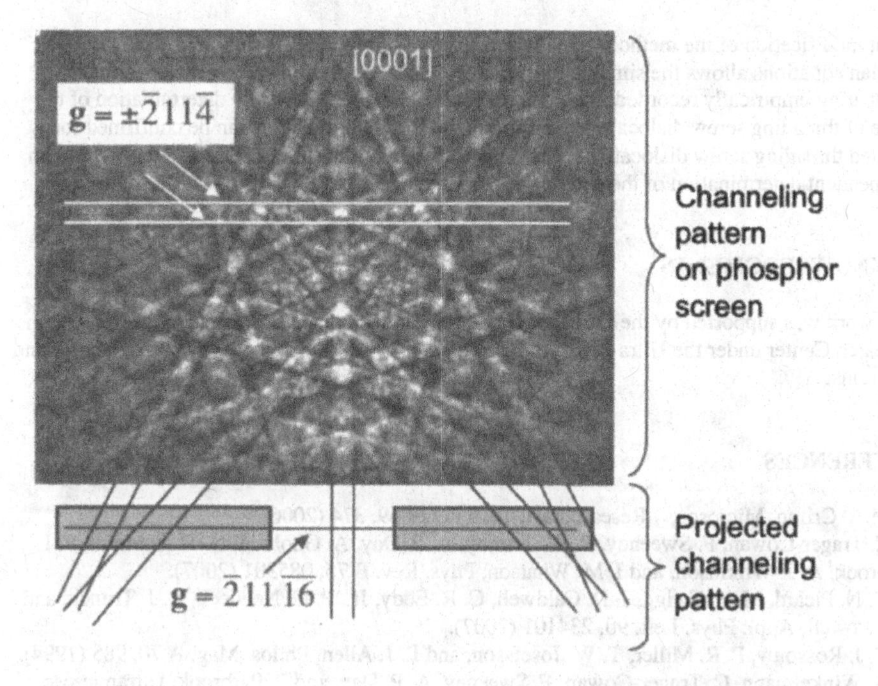

Fig.3 Kikuchi pattern of 4H-SiC with the position of diode detectors. From the experimental geometry it was determined that the $g=\bar{2}11\overline{16}$ reflection crosses the diode detectors.

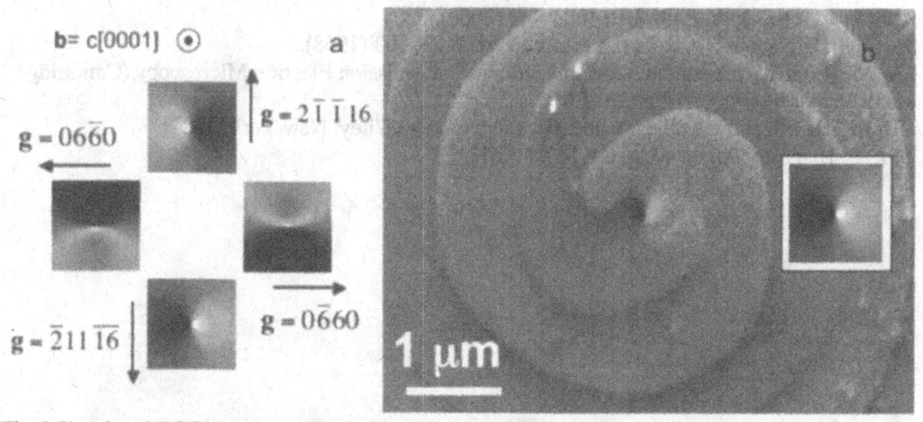

Fig.4 Simulated ECCI images. (a) Simulated images for four different reflections. (b) The experimental and simulated images, assuming the $g=\bar{2}11\overline{16}$ reflection.

slight modification of the methods already established for numerical solution of the Howie-Whelan equations allows the simulation of ECCI images of threading screw dislocations. Comparing empirically recorded images and simulated images allows the determination of the sense of threading screw dislocations. Furthermore, this determination can be confirmed for isolated threading screw dislocations where the direction of the spiral of atomic steps allows an independent determination of the Burgers vector direction.

ACKNOWLEDGMENTS

This work was supported by the Office of Naval Research, and also supported by NASA Glenn Research Center under the Ultra Efficient Engine Technology and Director's Discretionary Fund Programs.

REFERENCES

1. M.A. Crimp, Microscopy Research and Techniques 69, 374 (2006).
2. C. Trager-Cowan, F. Sweeney, P. W. Trimby, A. P. Day, A. Gholinia, N.-H. Schmidt, P. J. Parbrook, A. J. Wilkinson, and I. M. Whatson, Phys. Rev. B 75, 085301 (2007).
3. Y. N. Picard, M. E. Twigg, J. D. Caldwell, C. R. Eddy, Jr., P. G. Neudeck, A. J. Trunek, and J. A. Powell, Appl. Phys. Lett. 90, 234101 (2007).
4. C. J. Rossouw, P. R. Miller, T. W. Josefsson, and L. J. Allen, Philos. Mag. A 70, 985 (1994).
5. A. Winkelmann, C. Trager-Cowan, F. Sweeney, A. P. Day, and P. Parbrook, Ultramicrosc. 107, 414 (2007).
6. S. L. Dudarev, J. Ahmed, P. B. Hirsh, and A. J. Wilkenson, Acta Cryst. A55, 234 (1999).
7. J. A. Powell, P. G. Neudeck, A. J. Trunek, G. M. Beheim, L. G. Matus, R. W. Hoffman, Jr., and L. J. Keys, Appl. Phys. Lett. 77, 1449 (2000).
8. A. P. Pogany and P. S. Turner, Acta Cryst. A 24, 103 (1968).
9. M. De Graef, Introduction to Conventional Transmission Electron Microscopy (Cambridge University Press, Cambridge, 2003).
10. J. P. Hirth and J. Lothe, Dislocations in Crystals (Wiley, New York, 1982).
11. E. H. Yoffe, Philos. Mag. 6, 1147 (1961).

Mater. Res. Soc. Symp. Proc. Vol. 1068 © 2008 Materials Research Society 1068-C07-03

A Comprehensive Study of Growth Techniques and Characterization of Epitaxial $Ge_{1-x}C_x$ (111) Layers Grown Directly on Si (111) for MOS Applications

Mustafa Jamil, Joseph P. Donnelly, Se-Hoon Lee, Davood Shahrjerdi, Tarik Akyol, Emanuel Tutuc, and Sanjay K. Banerjee
Microelectronics Research Center, The University of Texas at Austin, 10100 Burnet Road, Bldg # 160, Austin, TX, 78758

ABSTRACT

We report the growth and characterization of thin germanium-carbon layers grown directly on Si (111) by ultra high-vacuum chemical vapor deposition. The thickness of the films studied is 8-20 nm. The incorporation of small amount (less than 0.5%) of carbon facilitates 2D growth of high quality Ge crystals grown directly on Si (111) without the need of a buffer layer. The $Ge_{1-x}C_x$ layers were grown in ultra high vacuum chemical vapor deposition chamber, at a typical pressure of 50 mTorr and at a growth temperature of 440 °C. CH_3GeH_3 and GeH_4 gases were used as the precursors for the epitaxial growth. The $Ge_{1-x}C_x$ films were characterized by atomic force microscopy (AFM), secondary ion mass spectroscopy, x-ray diffraction, cross-sectional transmission electron microscopy and Raman spectroscopy. The AFM rms roughness of $Ge_{1-x}C_x$ grown directly on Si (111) is only 0.34 nm, which is by far the lowest rms roughness of Ge films grown directly on Si (111). The dependence of growth rate and rms roughness of the films on temperature, C incorporation and deposition pressure was studied. In Ge, (111) surface orientation has the highest electron mobility; however, compressive strain in Ge degrades electron mobility. The technique of C incorporation leads to a low defect density Ge layer on Si (111), well above the critical thickness. Hence high quality crystalline layer of Ge directly on Si (111) can be achieved without compressive strain. The fabricated MOS capacitors exhibit well-behaved electrical characteristics. Thus demonstrate the feasibility of $Ge_{1-x}C_x$ layers on Si (111) for future high-carrier-mobility MOS devices that take advantage of high electron mobility in Ge (111).

INTRODUCTION

The use of high mobility materials in channel has emerged as one of the most aggressively pursued areas to improve CMOS performance. Germanium is one of the most promising materials as both electron and hole mobility is 2X and 4X higher in Ge, respectively, than in Si. Moreover, the mobility of the carriers depends strongly on the crystallographic orientation of the surface [1]. The influence of surface orientation on inversion layer mobility in metal-oxide semiconductor field-effect transistor (MOSFET) channels is being investigated [2,3]. Ge (111) is predicted to demonstrate 2X-3X higher electron mobility than Ge (001) and expected to be suitable for ultra-scaled nMOSFETs [3]. Although some initial work has been done on MOSFETs fabricated on bulk Ge substrates, the use of bulk Ge wafers is not as attractive as the use of Si wafers from mechanical strength, cost, and integration perspectives. Therefore, in order to be used in the mainstream CMOS industry Ge should be incorporated epitaxially on Si. Besides, high quality Ge films grown epitaxially on Si (111) would allow the effects of orientation and strain to be studied in greater detail. However, growth of pure Ge directly on Si is a difficult challenge because of 4.2 % lattice mismatch. Moreover, dislocation formation during heteroepitaxial growth is strongly dependent on surface orientation and this phenomenon makes it more difficult to grow pure Ge on Si (111) [4, 5]. Above a certain critical thickness, hetero-epitaxial growth of Ge on Si proceeds in a Stranski-Krastanov growth mode and results in the formation of three-dimensional islands and misfit dislocations. Techniques like the use of extremely thick, graded, relaxed $Si_{1-x}Ge_x$ virtual substrates [5] and surfactant mediated epitaxy [6] have been proposed to achieve two-dimensional growth of Ge on Si (111) and reduce the number of crystal defects. The disadvantage is that the substrate preparation process is complicated or calls for the use of chemical mechanical polishing. Our group has recently reported high performance pMOSFETs fabricated on diamond-structured germanium-carbon alloy ($Ge_{1-x}C_x$) on Si (100) [7]. In this work we explored the technique of incorporating trace amount of C to develop thin $Ge_{1-x}C_x$ layer on Si (111) substrates in order to take the advantage of Ge (111) orientation for high mobility nMOSFETs. The addition of C to Ge facilitates high-quality, two-dimensional growth directly on Si (111), without the extremely thick, relaxed $Si_{1-x}Ge_x$ buffer layers or cyclic annealing as predicted [5].

EXPERIMENT

In this work we developed $Ge_{1-x}C_x$ layers thin enough (8-20 nm) to prevent formation of threading dislocations. However, the films were thick enough to accommodate high mobility inversion layer for MOS applications. 4 inch lightly doped (1-10 Ω·cm) prime grade p-type Si (111) wafers were cleaned using a 2:1 solution of H_2SO_4 and H_2O_2 followed by a 40:1 H_2O:HF dip for 40 s. The wafers were then immediately loaded into the load-lock of a custom-built cold-wall ultra high vacuum chemical vapor deposition (UHVCVD) reactor. Before transferring the wafers to the main reactor, the load-lock was pumped down to less than 10^{-7} Torr. The base pressure of the process chamber was below $8X10^{-10}$ Torr. The deposition pressure for the $Ge_{1-x}C_x$ layer was 50 mTorr and the growth temperature was 440 °C. The $Ge_{1-x}C_x$ films were characterized by atomic force microscopy (AFM) to study the surface morphology and cross-sectional transmission electron microscopy (X-TEM) to study the crystalline structure. Xray-diffraction (XRD) 2-Theta scans and Raman spectroscopy studies were performed to study the orientation and strain condition of the grown film. MOS capacitors were fabricated on 8 nm thick $Ge_{1-x}C_x$ film grown on Si (111) and had 1 nm Si passivation layer. Immediately after the growth, the wafers were cleaned using a HF dip and then the Si cap was partially oxidized using rapid thermal oxidation (RTO). Then ~11 nm thick HfO_2 was deposited in an atomic layer deposition (ALD) system using tetrakis (dimethyl-amino hafnium Hf $(NMe_2)_4$) as the metallic precursor. A post-deposition anneal (PDA) was performed for 5 min in N_2 ambient. About 200 nm thick TaN metal was then deposited using a DC magnetron sputterer, and capacitors were fabricated using conventional lithography and reactive ion etching using CF_4.

DISCUSSION

Figure 1 shows the AFM measurement of (a) Ge film grown directly on Si (111) and (b) $Ge_{1-x}C_x$ film directly on Si (111). The AFM image of pure Ge films shows 3D islanding with an rms roughness of 6.78 nm. Whereas the films grown with small amount of C incorporation show 2D growth behavior for comparable thickness with an rms roughness of 0.34 nm. This extremely low surface roughness was experimentally repeatable and is much lower than the roughness of our control samples grown without any C incorporation. It appears that the C atoms in the $Ge_{1-x}C_x$ promote two-dimensional growth and improve the crystalline quality.

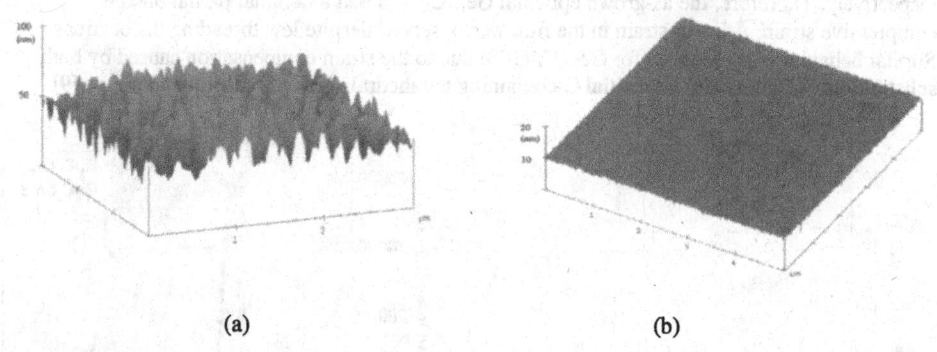

Figure 1. The effect of C incorporation is evident in the AFM images of (a) Ge and (b) $Ge_{1-x}C_x$ layers grown directly on Si (111).
[scale: X-axis: 1 μm/div ; Y-axis : 50 nm/div (Figure 1(a)), 10 nm/div (Figure 1(b))]

Figure 2 compares X-TEM images for (a) $Ge_{1-x}C_x$ film on Si and (b) pure Ge film on Si. Defect formation seems to be located only at the $Ge_{1-x}C_x$ /Si substrate interface for the sample with C. No significant threading dislocation was observed in the $Ge_{1-x}C_x$ sample. Figure 2 demonstrates that the $Ge_{1-x}C_x$ has drastically fewer threading dislocations than the pure Ge film. Although misfit dislocations were formed at the interface they do not seem to thread toward the surface. The $Ge_{1-x}C_x$ films grown in this study is much thinner than other CVD $Ge_{1-x}C_x$ films reported in literature [8]. The lower thickness might have played a role in lower defect densities in the films of this study.

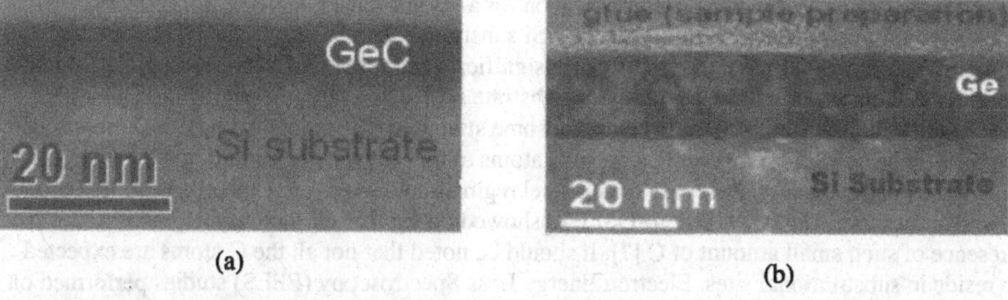

(a) (b)

Figure 2. X-TEM images of (a) 9 nm thick $Ge_{1-x}C_x$ grown on Si (111), (b) pure Ge layer grown directly on Si. Pure Ge film shows higher threading dislocations.

The 2Theta XRD scans of $Ge_{1-x}C_x$ films grown on Si (111) are presented in Figure 3. The scans were performed on (111) plane. The substrate and layer peak positions were at 28.4 and 27.3 degrees respectively. Thus the 2-Theta scan confirms the (111) orientation of the grown $Ge_{1-x}C_x$ layer. The separation between the substrate and the layer is 1.1 degree which suggests that a partially relaxed epitaxial layer was formed. Figure 4 presents Raman spectroscopy study, which further corroborates that the grown Ge film was partially relaxed. Raman spectroscopy spectrum of the bulk Ge substrate is shown for comparison. Wave numbers of bulk Ge and strained-Ge layers with a thickness of 8 nm were found to be 297.38 cm^{-1} and 300.22 cm^{-1}, respectively. Therefore, the as-grown epitaxial $Ge_{1-x}C_x$ film had a nominal partial biaxial compressive strain. A lower strain in the film was observed despite low threading dislocations. Similar behavior was observed for $Ge_{1-x}C_x$ (100) due to the strain compensation caused by both substitutional C atoms and interstitial C-containing tetrahedrals near the substrate interface [9].

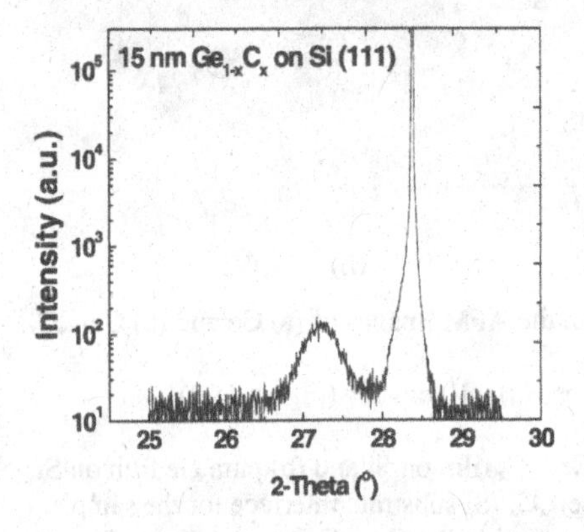

Figure 3. Theta-2Theta XRD scans of a 15 nm thick $Ge_{1-x}C_x$ layer grown on Si (111).

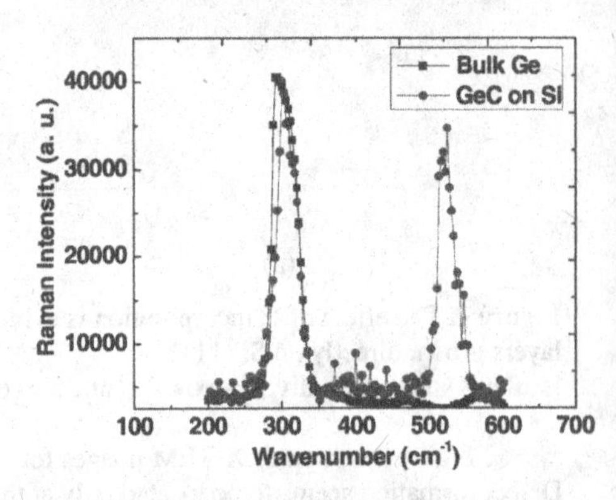

Figure 4. Raman spectroscopy study of an 8 nm $Ge_{1-x}C_x$ layer grown directly on Si (111).

Secondary Ion Mass Spectroscopy (SIMS) depth profile shown in Figure 5 provided an estimate of 0.25 % for the peak C concentration for a 90-nm-thick $Ge_{1-x}C_x$ film. Higher C concentration was observed near the $Ge_{1-x}C_x$/Si substrate interface. It is believed that the high strain fields around individual C atoms cause significant retardation of dislocation glide, which confines defects at the interface with the Si substrate [10]. Therefore, the segregation of C near the Si substrate interface is assumed to cause some strain compensation and help 2-D growth of Ge directly on Si (111). The segregation of C atoms to the substrate has additional benefit. As only trace amount of C is present in the channel region, the devices are expected to be less prone to alloy scattering. $Ge_{1-x}C_x$ (100) pMOSFETs showed no significant mobility degradation for the presence of such small amount of C [7]. It should be noted that not all the C atoms are expected to reside in substitutional sites. Electron Energy Loss Spectroscopy (EELS) studies performed on $Ge_{1-x}C_x$ films grown on Si (100) suggested about 30% C substitutionality [9].

Figure 6 shows that higher amount of C (based on the gas flow ratio) leads to lower surface roughness. The surface roughness of $Ge_{1-x}C_x$ layers increased with the increase of growth temperature at a fixed CH_3GeH_3 flow rate. An increase in the growth temperature from 440 °C to 520 °C resulted in the surface roughness increasing from 0.34 nm to 3.3 nm. A general trend of increasing growth rate with decreasing methyl-germane flow, increasing temperature and pressure was observed. From these AFM results we conclude that the smoothness of $Ge_{1-x}C_x$ has a stronger dependence on the growth temperature than on the C concentration, although the C concentration is also an important variable.

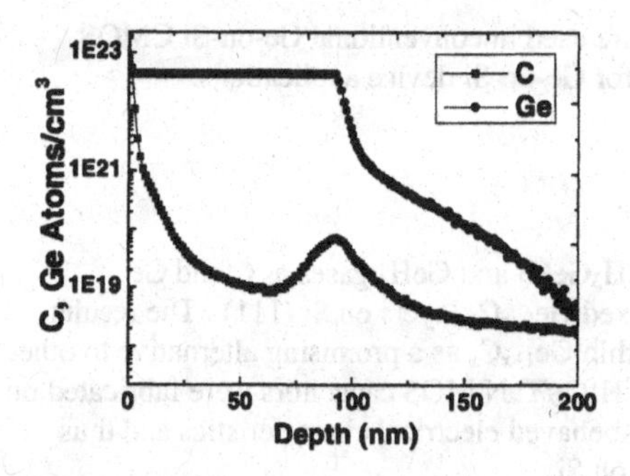

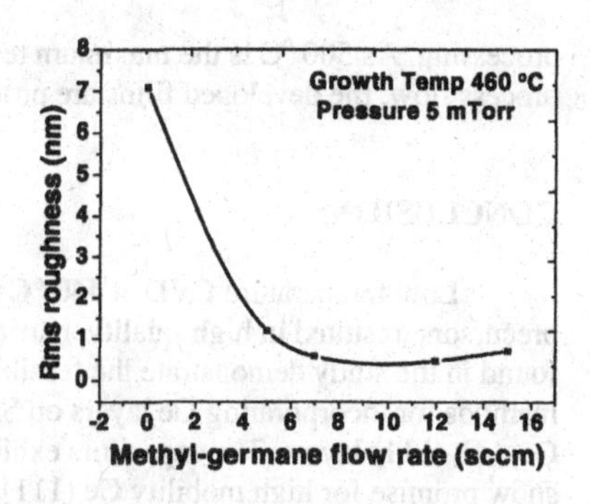

Figure 5. Carbon concentration profile extracted from SIMS data for a 90 nm thick $Ge_{1-x}C_x$ film.

Figure 6. RMS surface roughness measurements acquired on $Ge_{1-x}C_x$ films as a function of methyl-germane flow.

Figure 7 shows the electrical characteristics of the capacitors fabricated on $Ge_{1-x}C_x$ and Ge films on Si (111). The interface state density D_{it} was extracted using the NCSU cvc [11] simulation for the $Ge_{1-x}C_x$ devices. A reasonably low D_{it} of $6.28X10^{11}/cm^2$ indicates an acceptable-quality $Ge_{1-x}C_x/SiO_2/HfO_2$ interface. Bidirectional voltage sweep measurements showed 40-100 mV hysteresis (not shown) for typical $Ge_{1-x}C_x$ devices. Pure Ge devices showed C-V stretch-out, larger hysteresis (over 100 mV) and more frequency dispersion.

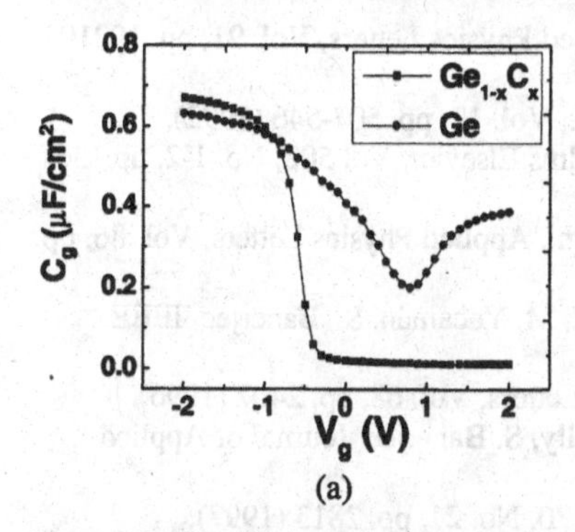

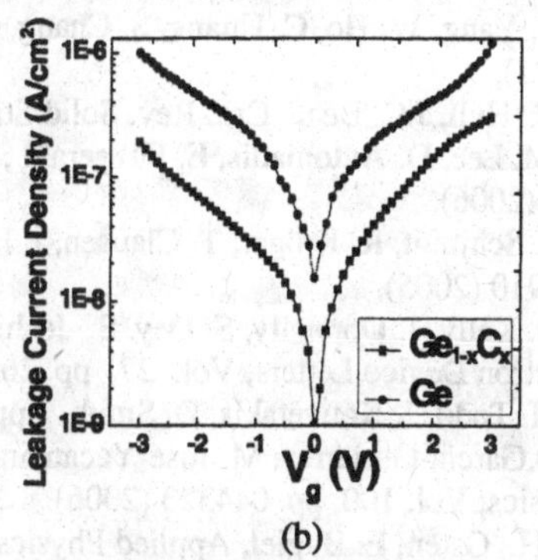

Figure 7. Electrical characteristics of capacitors fabricated on $Ge_{1-x}C_x$ layer: (a) Capacitance-Voltage and (b) Gate leakage current density on the same capacitors.

The leakage current density of $Ge_{1-x}C_x$ capacitor measured in accumulation at 1 V was $2.1X10^{-8}$ A/cm^2 [Figure 7(b)], which is an order of magnitude lower than that of the Ge MOS device with same EOT. The poor C-V behavior and higher leakage in the pure Ge on Si devices can be attributed to the poor surface roughness and higher trap density than the $Ge_{1-x}C_x$ devices.

Carbon segregation during high temperature processing could be an issue for these meta-stable films. The MOS capacitors investigated in this study has undergone 5-30 min 500 °C anneals in N_2 ambient and do not show any electrical degradation due to such thermal processing. As 500 °C is the maximum temperature used in conventional Ge-on-Si CMOS process flow, the developed films are promising for Ge-on-Si device applications.

CONCLUSIONS

Low-temperature CVD at 440 °C using CH_3GeH_3 and GeH_4 gases as C and Ge precursors, resulted in high-quality, partially relaxed $Ge_{1-x}C_x$ layers on Si (111) . The results found in the study demonstrate the feasibility of thin $Ge_{1-x}C_x$ as a promising alternative to other methods for incorporating Ge layers on Si (111). HfO_2/TaN MOS capacitors were fabricated on $Ge_{1-x}C_x$ (111) layers. The capacitors exhibit well-behaved electrical characteristics and thus show promise for high mobility Ge (111) nMOS on Si.

ACKNOWLEDGEMENTS

This work was supported in part by the DARPA.

REFERENCES

1. S. Takagi, A. Toriumi,M. Iwase, H. Tango, IEEE Transactions on Electron Devices, Vol. 41, No.12, pp.2363-2368 (1994).
2. M. Fischetti, S. Laux, P. Solomon, A. Kumar, 10th International workshop on computational electronics, October (2004).
3. Y. Yang, W. Ho, C. Huang, S. Chang, C. Liu , Applied Physics Letters, Vol. 91, pp. 102103 (2007).
4. R. Hull, J.C. Bean, Crit. Rev. Solid State Mater. Sci., Vol. 17, pp. 507-546 (1992).
5. M. Lee, D. Antoniadis, E. Fitzgerald , Thin Solid Films Elsevier, Vol 508, No. 1-2, pp.136-139 (2006).
6. T. Schmidt, R. Kröger, T. Clausen, J. Falta, A. Janzen , Applied Physics Letters, Vol. 86, pp. 111910 (2005).
7. D. Kelly, J. Donnelly, S. Dey, S. Joshi, D. Gutiérrez, M. Yacamán, S. Banerjee, IEEE Electron Device Letters, Vol. 27, pp. 265 (2006).
8. M. Todd, J. Kouvetakis, D. Smith, Applied Physics Letters, Vol. 68, pp. 2407 (1996).
9. D.Garcia-Gutierrez, M. José-Yacamán, S. Lu, D. Kelly, S. Banerjee, Journal of Applied Physics, Vol. 100, pp. 044323 (2006).
10. H. Osten, E. Bugiel, Applied Physics Letters, Vol. 70, No. 21, pp. 2813 (1997).
11. R. Hauser, K. Ahmed, Characterization and Metrology for ULSI Technology, edited by D. G. Seiler et al. AIP, Woodbury, NY, pp. 235–239 (1998).

Mater. Res. Soc. Symp. Proc. Vol. 1068 © 2008 Materials Research Society 1068-C03-18

Condensation Mechanism for the Formation of Relaxed SiGe Layer Grown-on-Insulator

Hun-Joo Lee, Gon-Sub Lee, Young-Soo Han, Seuck-Hoon Hong, Tae-Hun Shim, and Jae-Gun Park
Nano Scale Semiconductor Engineering, Hanyang University, Nano SOI Process Laboratory, Room #101, HIT. Hanyang University 17 Haengdang-dong, Seoungdong-gu, Seoul, 133-791, Korea, Republic of

ABSTRACT

The use of the condensation method to grow a relaxed SiGe layer-on-insulator (ε-SGOI) for making high-speed complementary metal–oxide–semiconductor field-effect transistors (C-MOSFETs) has attracted interest because of its high quality and cost effectiveness. Many reports have presented its superiority in a device performance to bonding and dislocation sink technologies. However, in case of the condensation method, the mechanism by which the method produces ε-SGOI has also not been clearly explained and the surface properties have not been evaluated. Thus, we investigated the condensation mechanism and the effect of temperature in detail by characterizing the surface property and the Ge profile in the SiGe layer. A SiGe layer on silicon-on-insulator layer was epitaxial grown at 550°C, and three oxidation thicknesses at 40, 60, and 90 nm were grown at 950°C. The Ge concentration was increased from 15 to 38.6%, 46.4%, and 63.2%. In the experiment to measure the effect of temperature, the root mean square decreased from 0.175 to 3.412 nm, and the uniformity of Ge improved when the oxidation temperature was increased from 950 to 1100°C. Therefore, our talk will focus on the explanation for the mechanism by which using the condensation method produces ε-SGOI by characterizing the surface property, the thickness of the SiGe, the remaining Si thickness on the insulator, the Ge concentration in the SiGe layer, and the oxidation temperature.

INTRODUCTION

SiGe-on-insulator (SGOI) is a typical template substrate for strained Si-SOI MOSFET structures, which benefits from both enhanced mobility provided by its strained channels and from the low junction capacitance of its SOI structures [1-2]. In order to fabricate a strained-SOI MOSFET that has high performance, we need to form highly strained Si layers that are free from dislocations and defects on SGOI substrates [3]. To have free of dislocations and no defective wafers, it is also important to know about the condensation mechanism. A condensation mechanism was revealed by using Ge diffusion, relaxation, Ge profile uniformity, the amount of the oxide growth rate (by dislocation generation), and the time it took the Ge profile in the SiGe layer to change [4-5]. We studied the Ge profile and the roughness of the SiGe layer after the oxidation by changing the oxidation time on an ultra thin SOI wafer. Moreover, we also observed the effect of the oxidation temperature [17].

EXPERIMENT

A condensation method using Ge diffusion is the best means to significantly reduce the process time compared with the dislocation sink and the bonding methods previously reported [4-10]. Controlling the final concentration of Ge concentration is done relatively easily by optimizing the initial SiGe thickness and the oxidation process time. The schematic procedure of the condensation method of the strained-Si on SGOI substrates is shown in Fig. 1. First, a 50 nm-thick strained $Si_{0.85}Ge_{0.15}$ layer was grown on a 10-nm SOI substrate by ultra-high vacuum chemical vapor deposition (UHV-CVD) at 550°C. For the change from strained to relaxed SiGe, the SiGe layer was oxidized in dry O_2 at 950°C at 40, 60, and 90 nm. The oxidized samples were evaluated by using sputtering Auger electron spectroscopy for depth profiling of Ge, X-ray diffraction for the relaxation evaluation, transmitting electron microscopy for structural analysis, and an atomic force microscope for measuring the surface roughness.

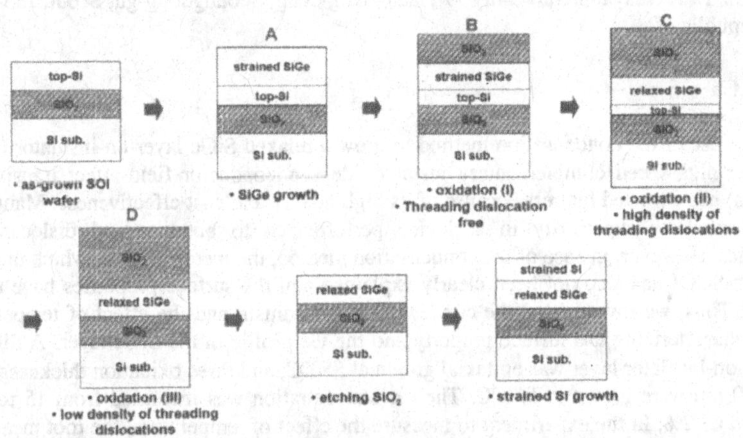

Figure 1. Fabrication process of SGOI substrate using condensation method.

A Cap-Si (10 nm)/SiGe (50 nm, 20 at%)/top-Si (30 nm)/SiO_2 (145 nm)/Si substrate was fabricated by UHV-CVD at 550°C on a SOI wafer, which was fabricated using the SIMOX method. We evaluated the samples after a 110-nm-thick oxidation process with temperatures at 950, 1000, 1050, and 1100°C.

DISCUSSION

Variation of the Ge profile during oxidation

During oxidation, Ge atoms are rejected from the SiGe-oxide layer and condensed in the remaining SGOI layer because Ge diffusion out of the layer is blocked by the top and bottom SiO_2 layers [11-12]. The Ge oxidation profile is shown in Fig 2. The Ge profile of a layer with a 40-nm oxide thickness, widened through diffusion down into the top Si layer, and the Ge concentration increased by 38.6 at%. Although Ge atoms diffused into the top Si layer, because of an insufficient oxidation time, the layer was too thick to relax the SiGe layer. For this reason, the SiGe layer was not transformed into relaxed SiGe through Ge diffusion. For a layer of a 60-nm oxide thickness, the top Si thickness gradually decreased gradually while the Ge concentration increased relatively. The SiGe layer proceeded to the relaxation stage because of the high Ge concentration and reduction of the top Si thickness. Consequently, misfit dislocations were generated at the interface between the SiGe and the top Si layer, and the Ge concentration is shown in a graded profile. For a 90-nm oxide thickness, all of the Si atoms in the top Si layer were consumed, and the Ge profile was sharp (similar to a Gaussian distribution) because of a very high Ge concentration. This means that the SiGe layer becomes a completely relaxed layer. The rejection of Ge atoms from the oxide layer causes no change to the total Ge amounts. Although the free energy of formation of SiO_2 is much lower than that of GeO_2, SiO_2, rather than GeO_2, is formed [13-14]. Therefore, we were able to determine a strain relaxation through the transformation of the Ge oxidation profile in the SiGe layer.

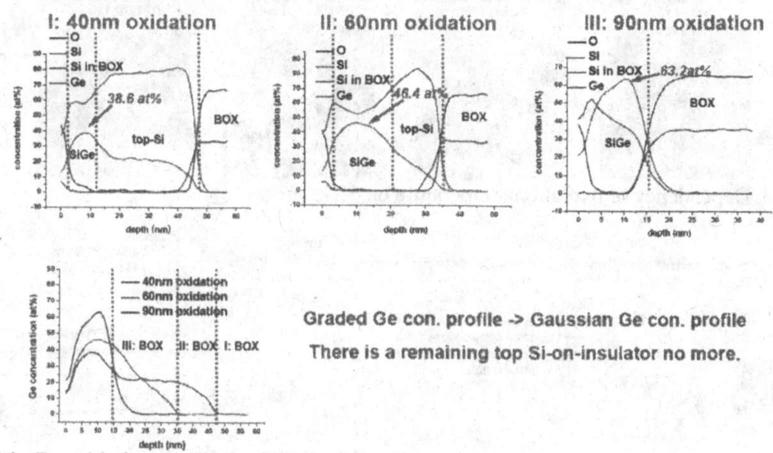

Figure 2. Ge oxidation profile for SiGe and top Si layers.

Effect of oxidation temperature

The change of the oxidation temperature affects the diffusion rate of Ge atoms in a SiGe layer toward the top Si layer and Si atoms in the top Si layer toward the SiGe layer [15]. (These two diffusion rates can be dealt using the same parameters.) The dependency of the oxidation temperature on the SiGe surface is shown in Fig. 3. It demonstrates that the surface roughness was improved as the oxidation temperature increased. This is attributed to the increase in oxidation temperature from 950 to 1100°C, the higher diffusion rates of Ge atoms from the SiGe layer into the top Si layer, and the increase in the number of Si atoms from the top Si layer into the SiGe layer [16]. Thus, as shown in Fig. 4(a), the Ge concentration profile becomes flat. We found that as temperature increases the Ge concentration profile in a depth becomes gradually uniform in shape. Simultaneously, the Ge concentration decreases with increasing temperature. As shown in Fig. 4(b), at the second Y-axis, the surface roughness was improved because there were fewer dislocations generated at lower Ge concentrations than at higher ones. Moreover, the oxidation time for the same thickness was less because of the increase in the oxidation temperature. This could be an important factor for the suppressing dislocation generation. Finally, once the oxidation temperature increases, surface migration occurs and this leads to a smoother surface.

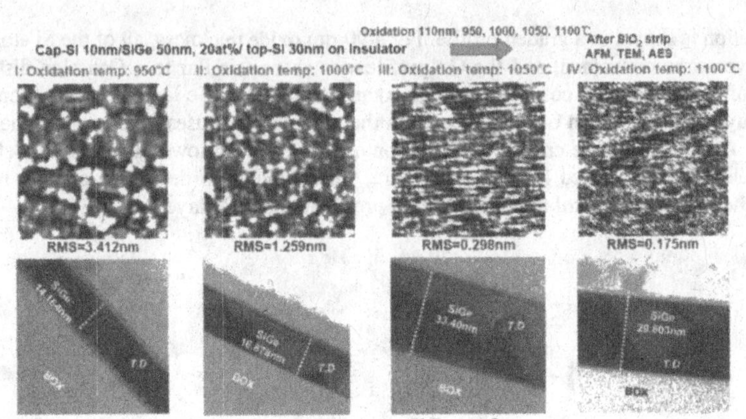

Figure 3. Dependency of oxidation temperature on SiGe surface.

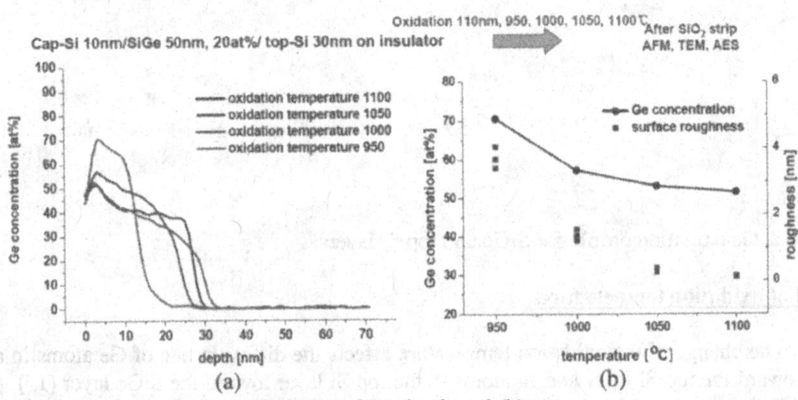

Figure 4. Ge concentrations as a function of (a) depth and (b) temperature.

Consequently, a high oxidation temperature, i.e., 1100°C, is suitable for forming a ε-SGOI structure that has excellent surface properties. In particular, the fact that a high oxidation temperature leads to a uniform Ge concentration profile is very useful for mass production of ε-SGOI and ε-SOI structures.

CONCLUSION

The oxidation temperature and the process time are affective the diffusion length because the Ge diffusion model is used in the condensation method. We found that the Ge profile was changed from a graded profile to a Gaussian one on the basis of the change in the oxidation time. Furthermore, the surface roughness was smoother. The oxidation time increased, the diffusion was longer, and the top Si thickness was reduced. This meant that the Ge atoms diffused into the bottom of the top Si and accumulated on the buried oxide (BOX) surface while the remaining top Si layer caused the misfit dislocations from the lattice mismatch that occurred at the interface between the SiGe and top Si layers. However, the all of the remaining top Si layer was consumed and Ge atoms accumulated on the surface of the BOX layer to change from a strained to a relaxed SiGe layer. Also, maintaining a high oxidation temperature ensures that we obtain a uniform SiGe layer. At the high temperature, the Ge atoms have a longer diffusion length. Therefore, the Ge atoms do not accumulate at the surface of the SiGe layer, and this leads to a fully relaxed SGOI with good uniformity.

ACKNOWLEDGMENTS

This research was supported by the Korea Ministry of Commerce, Industry and Energy, through the National Development Program for 0.1-Terabit Non-volatile Memory.

REFERENCES

1. Th. Volelsang, K. R. Hormann, "Electron transport in strained Si layers on $Si_{1-x}Ge_x$ substrates" Appl. Phys. Rev., Vol. **63**, p. 186, (1993).
2. S. M. Sze, Physics of Semiconductor Devices, John Wiley & Sons, (1976), p. 849.
3. Semiconductor Industry Association, International Technology Roadmap for Semiconductor (2004); available at http://public.itrs.net.
4. S. Takagi, T. Mizuno, T. Tezuka, N. Sugiyama, T. Numata, K. Usuda Y. Moriyama, S. Nakaharai, J. Koga, A. Tanbe, N. Hirashita and T. Maeda, IEDM (2003).
5. J. G. Park, G. S. Lee, T. H. Kim, S. H. Hong, S. J. Kim, J. H. Song, T.H. Shim, Materials Science and Engineering B **134** (2006).
6. L. K. Bera et al., IEEE Electron Device Lett., Vol. **27**, No.5, MAY (2006).
7. Zengfeng Di, Paul K., J. Appl. Phys., **97**, 064504 (2005).
8. Lijuan Huang, Jack O. Chu S. A. Goma, C. P. D' Emic, Steven J. Koester, Donald F. Canaperi, Patricia M. Monney, S. A. Cordes, James L. Speidell, R. M. Anderson, and H. –S. Philip Wong, IEEE TRANSACTIONS ON ELECTRON DEVIECS, Vol. **49**, No. 9, SEPTEMBER (2002).
9. G. Taraschi, Z. –Y. Cheng, M. T. Currie, C. W. Leitz, T. A. Langdo, M. L. Lee, A. Pitera, J. L. Hoyt, D. A. Antoniadis, and E. A. Fitzgerald, Proceeding of the Tenth International Symposium on Silicon-on-Insulator Technology and Devices, (2001), p. 27.
10. Tomohisa Mizuno, Naoharu Sugiyama, Tsutomu Tezuka, Toshinori Numata, and Shin-ichi Takagi, IEEE TRANSACTIONS ON ELECTRON DEVICES, Vol. **50**, No. 4, APRIL (2003).
11. T. Tezuka, N. Sugiyama, S. Takagi, and T. Kawakubo, "Dislocation-free formation of relaxed SiGe-on-insulator layers," Appl. Phys. Lett., Vol. **80**, no. 19, pp. 3560-3562, May (2002).
12. T. Tezuka, N. Sugiyama, and S. Takagi, Appl. Phys. Lett. **79**, 1798 (2001).
13. I. Barin, O. Knacke, Thermochemical Properties of Inorganic Substances, Soinger, Berlin, (1973).
14. S.J. Kilpatrick, R.J. Jaccodine, P. E. Thompson, J. Appl. Phys. **81** 8018 (1997).
15. N. Sugiyama, T. Tezuka, T. Mizuno, and M. Suzuki. J. Appl. Phys., Vol. **95**, No. 8, APRIL (2004).
16. M. Ogino, Y. Onabe, and M. Watanabe Phys. Status Solidi A 72, 535 (1982).
17. N. Sugiyama, T. Tezuka, T. Mizuno, and M. Suzuki, J. Appl. Phys. **95**, 4007 (2004).

AUTHOR INDEX

SUBJECT INDEX

Printed in the United States
By Bookmasters